# Aquatic Pollution

## *Concerns and Abatement*

*Edited by*

**Imtiyaz Qayoom**

**Adnan Abubakr**

**Anu Gopinath**

**Shabir A. Dar**

**Keezia Khurshid**

CRC Press
Taylor & Francis Group
Boca Raton London New York

CRC Press is an imprint of the
Taylor & Francis Group, an **informa** business

-EPH-

**Elite Publishing House**

First published 2025
by CRC Press
4 Park Square, Milton Park, Abingdon, Oxon, OX14 4RN

and by CRC Press
2385 NW Executive Center Drive, Suite 320, Boca Raton FL 33431



*British Library Cataloguing-in-Publication Data*
A catalogue record for this book is available from the British Library

Print edition not for sale in South Asia (India, Sri Lanka, Nepal, Bangladesh, Pakistan or Bhutan)

ISBN13: 9781032822563 (hbk)
ISBN13: 9781032822594 (pbk)
ISBN13: 9781003503705 (ebk)

DOI: 10.4324/9781003503705

Typeset in Adobe Caslon Pro
by Elite Publishing House, Delhi

-EPH-

# Contents

# Preface

Increase in pollution of aquatic ecosystems is concerning with every passing day. It is a sub-set of expansion of global population that invariably imposes thrust on available earth's resources, not available infinitely. With the advent of time, global population is likely to get increased to 9.7 billion in 2050 against 8.0 billion at the present, that will have certain negative impacts on marine and freshwater ecosystems, all over the world. Concerning issues like global warming, severe weather, sea level rise, water contamination, xenobiotic transfer through food chains and deteriorating water quality are the outcome of population expansion that need to be addressed and controlled at an earliest. Further, the secondary issues that have negatively engulfed large human populations due to deteriorating waters are deadly diseases such as cholera, diarrhea, dysentery, hepatitis A, typhoid, and polio contributing to the spread of 80% of all diseases in nations who consume contaminated waters. It is estimated that about 1 million people die every year from diarrhea. Similarly, *Schistosomiasis*, a severe illness acquired by coming in contact with contaminated water, required prophylactic treatment for around 251.4 million individuals in 2021.

The book in your hand is based on invited book chapters on the burning issues related to pollution of aquatic environments. In addition to the above-mentioned types of pollution, various other types include pesticide pollution, metal pollution, radioactive pollution, thermal pollution, etc. are discussed in the book. Moreover, role of vital model organisms that not only act as pollution indicators, but also help in removal of pollution from lakes has been comprehensively addressed. In this book, readers will also get acquainted with the methods of decontamination of lakes by phytoremediation, pesticide removal techniques from lakes, toxic site reclamation and environmental sustainability using microbial aspects associated with clean-up of wastes.

This book proposes elementary insights to the sources, mechanism, impact and abatement of pollution in waterbodies and lays a base foundation for further research. I am thankful to all the authors who have contributed their valuable chapters for this book. I am indebted to my co-editors to assist me in the editing, drafting and compilation of this book. I am also thankful to the publisher for making the publication of this book a success.

**Dr. Imtiyaz Qayoom**

**(Chief Editor)**

# About the Editors

**Dr. Imtiyaz Qayoom** works as Assistant Professor cum Junior Scientist Head in the Division of Aquatic Environmental Management, Faculty of Fisheries, Sher-e-Kashmir University of Agricultural Sciences and Technology, Kashmir, SKUAST-K.

He is working in the field of Aquatic Toxicology, Fish toxicity testing, Environmental Monitoring, Residue Analysis and *in-vitro* toxicity assessment in fish models. He is involved in various teaching, research, extension and administrative activities of the University with having experience of more than 5 years. Dr. Imtiyaz has been instrumental in conducting various workshops/ trainings to the researchers, fisher communities and stake holders as awareness campaigns to imbibe in them the aptitude of lake de-contamination, restoration and management. He has been enthusiastic enough to reach remote villages of far-flung areas of Ladakh to interact with farmers and address their concerns. Besides having several externally and University funded projects, Dr. Imtiyaz has privilege to publish 31 research papers articles in high impact factor journals of repute, 05 books and guide masters and PhD scholars for their degrees related to Aquatic/ Fish Toxicology.

**Dr. Adnan Abubakr** is working as Professor of Limnology & Head in the Division of Aquatic Environmental Management, Faculty of Fisheries, Sher-e-Kashmir University of Agricultural Sciences and Technology, Kashmir, SKUAST-K.

He has earned M.Sc. & Ph.D. degrees in Limnology from Department of Limnology, Barkatullah University Bhopal, and have specialised in Water and Wastewater examination, analytical and bacteriological techniques related to water, freshwater and brackish water. He has specialization in designing artificial wetland systems for treatment of wastewaters. Besides this he is involved in teaching, research and extension experience from more than 15 years in the field of aquatic ecology. He has been instrumental in Instituting Masters in Fisheries Sciences (M.F.Sc.) & Ph.D. Degree programmes in Aquatic Environment Management (AEM) in the Division of AEM. Having more than 60 research publication in peer reviewed journals; completed 02 externally funded and 03 university funded research projects, he has guided more than 20 students for their Masters and Ph.D. degrees in Fisheries Sciences.

**Dr Anu Gopinath**, Associate Professor and Head, Department of Aquatic Environment management, KUFOS has twenty years of teaching and research experience in the field of Chemistry as well as Chemical Oceanography. She started her career as an Assistant Professor in Chemistry, Department of Chemistry, St Teresa's College, and Ernakulam in 2005 and later in 2015 moved to Kerala University of Fisheries and Ocean Studies to join as a faculty in Chemical Oceanography. Dr Anu is a commonwealth fellow and has participated in Indian Arctic Expedition thrice. She has received DST Young Women Scientists fellowship and DST Fast Track Scheme for Young Scientists. She is collaborating with NCPOR -Goa, SAC-ISRO, Ahmedabad and INCOIS Hyderabad in several projects related to the monitoring of coastal oceans. Dr. Anu has successfully guided six Ph.D. students and has several research publications to her credit. She has authored nine books for the students at the undergraduate and postgraduate levels.

**Dr. Shabir Ahmad Dar** is working as Assistant Professor in the Division of Fishery Engineering, Faculty of Fisheries, Sher-e- Kashmir University of Agricultural Sciences and Technology Kashmir (SKUAST-K). He did his Ph. D in Fisheries Resource Management (FRM) with specialization in Fishing Technology from Central Institute Fisheries Education (CIFE) Mumbai. He has more than 10 years' experience in teaching, research and extension in the field of Fishing Technology. He has a total 45 publications to his credit in well reputed National and International Journals, which includes 25 research articles, 13 popular articles, 7 book chapters. He has organized, coordinated and co-ordinated several training programmes in Fisheries for the benefit of fish farmers, rural folks/fishers and entrepreneurs.

**Ms. Keezia Khurshid** is a Ph. D. Research Scholar in Division of Aquatic Animal Health Management, Faculty of Fisheries, SKUAST-K, J&K. She has done her graduation and M.F.Sc. from the SKUAST-K.

She has published number of research papers in national reputated publications, presented number of papers and got positions in various national and international conferences. She is working in the field of cell cultures and established 1st time a Primary cell culture of one of the endemic fish species in Kashmir.

# Chapter - 1

## Freshwater Faunal Diversities as Pollution Indicators: A Special Reference to Microfauna

**Shafia Jan[1*], Rabia Rasool[1], Inain Jaies[2] and Feroz A. Shah[2]**

[1]Department of Zoology, Central University of Kashmir, Ganderbal-191201, J&K, India
[2]Division of Aquatic Animal Health Management, Faculty of Fisheries, SKUAST-K Rangil, Ganderbal, J&K, India
*Corresponding email: tantrayshafia111@gmail.com

## Abstract

Over two thirds of our planet's surface is taken up by aquatic ecosystems, which are crucial in regulating the climate as well as offering a wide range of functions to a rapidly expanding human population. But anthropogenic activities are having a negative impact on aquatic environments more and more. One of the biggest challenges to the global biota is pollution of freshwater environments. The deterioration and pollution of freshwater is further exacerbated by population development and human activity. The hazardous contaminants harm aquatic ecosystems and are solely or jointly responsible for the deadly diseases that affect humans. Water pollution is the result of significant damage that man has done to the aquatic environment through development, civilization, and rapid industrialization. Indicators are crucial in the effort to define and quantify the presence and impacts of contaminants on aquatic systems. The fish, macrophytes, microflora (phytoplankton and algae), birds, man, microfauna (Cladocera and Rotifer) show different levels of sensitivity and can be successfully employed as indicators to access and predict freshwater pollution in a timely manner.

**Keywords:** Pollution, Aquatic Ecosystem, Freshwater, Microflora, Microfauna.

## Introduction

Water pollution is the act of rendering a body of water unfit for biotic communities to live in by contaminating it with sewage, fertilizer, or industrial effluents. Natural water contains some dissolved gases and minerals. The term "polluted water" refers to water characterized by the number of dissolved salts and other hazardous substances artificially increased to the point where it is unsafe for drinking, horticulture, or agricultural use. The contamination of streams, rivers, lakes, and groundwater is referred to as "freshwater pollution." The ecosystem, aquatic life, and human health are all harmed by dirty water. In the majority of poor nations, almost 70% of industrial waste that has not been cleaned is thrown into freshwater bodies. Consequently, the water supply that can be used for drinking, bathing, fishing, and swimming becomes contaminated. Freshwater sources make up less than 3% of the total amount of water on Earth and are essentially the only source of all drinking water. The criteria used to categorize pollution determine how it should be classified. There are two main categories of water pollution: fresh water pollution and marine pollution. These categories are based on the medium in which pollutants occur. There are two types of fresh water contamination: surface water pollution and ground water pollution. Surface water contamination is what happens when a pollutant enters a lake, pond, or river. However, if the pollutant mixes with percolating water and enters an aquifer, it lowers the quality of the groundwater and is referred to as groundwater pollution.

One of the biggest issues that people worldwide, particularly in developing nations, face is pollution. Numerous human activities result in the release of different contaminants into aquatic habitats, endangering the lives of living things and degrading the ecosystem by making the water bodies unfit. Aquatic ecosystems are scavengers for anthropogenic contamination, industrial waste, and leaks, whether they be chemical or solid wastes. These wastes can include "heavy metals, detergents, microfibers, plastic or non-plastic origin," etc., and they contribute to "aquatic contamination problems" (Hampel *et al.*, 2015). Many contaminants and their outrageously harmful consequences are directed at aquatic environments. The radioactive elements strontium, caesium, and radon, metals cadmium, mercury, and lead, industrial solvents and "volatile organic compounds" (benzene, xylenes, and formaldehyde), "agrochemicals" (fertilisers and pesticides), household products (detergents, cleaners, and paints), "fuel combustion" (nitrogen and sulphur oxides, polycyclic aromatic hydrocarbons, carbon monoxide and "carbon dioxide"), "nanoparticles," microplastics, antibiotics, "as well as huge variety of prescription and "non-prescription drugs and pharmaceuticals of human and veterinary medicine". Aquatic ecosystems are water-based habitats where biotic and abiotic elements of the ecosystem interact. The "marine ecosystem" and the "freshwater ecosystem" are the two categories into which aquatic ecosystems are typically subdivided. Over 70% of the Earth's surface is covered by the marine ecosystem, which is the largest water ecosystem. The oceans, estuaries, coral reefs, and coastal ecosystems are subdivided into the marine ecosystem. On the Earth's surface, freshwater ecosystems make up less than 1% of the total area.

## Global Rise in Freshwater Pollution

Freshwater is a key natural resource that we use for a variety of daily activities, including drinking and various developmental goals. Our natural environment becomes unstable as a result of rising pollution levels, which also harms the ecosystem's living creatures and physical processes. When undesirable substances enter water bodies, it causes water pollution and changes the water's quality (Malik *et al.*, 2020). There are about 126,000 species in freshwater environment, which include lakes, ponds, rivers, streams, springs, and marshes. The health, livelihood, and general well-being of billions of people are supported by the providing, sustaining, regulating, and cultural ecosystem services that these aquatic ecosystems provide in addition to being an essential habitat for biodiversity. Freshwater habitats are among the most drastically altered ecosystems on planet, despite their significance. Freshwater ecosystems are essential to the proper operation and resilience of other ecosystems, the local generation of freshwater flows, and the global water cycle. A healthy ecosystem is essential for freshwater security. Currently, pollution, excessive water usage and abstraction, rapid infrastructural development, and changes in land use are the main human risks to freshwater ecosystems (Kumaraswamy *et al.*, 2020). Numerous human activities, including as extensive agriculture and industry as well as routine activities like driving vehicles and fertilizing lawns, damage freshwater ecosystems. Point sources, such industrial or municipal sewage outputs, frequently cause significant amounts of pollutants to infiltrate freshwater systems. Non-point source pollution, however, often has a much greater impact. Airborne contaminants have the ability to enter the atmosphere, travel great distances, and end up in lakes and waterways in otherwise clean areas.

The best sink for pollutants is aquatic ecosystems. Human activities including urbanization, industrialization, and agriculture all contribute to water contamination. Pesticides, fertilizers, and sewage from residential and commercial areas eventually make their way to aquatic environments. Thus, the water quality deteriorates and infectious diseases including dysentery, diarrhea, and jaundice spread. One of the main types of pollution that has a substantial detrimental impact on health and mortality is contamination in aquatic environments. It naturally has the ability to neutralize contamination, but when the contamination gets out of hand, water loses its ability to generate itself. Therefore, regular pollution discharge monitoring and management into the local aquatic environment are required. Our diverse ecosystems are almost contaminated by human activity, which leads to pollution. According to Bashir *et al.*, 2020 "pollution is defined as the production and or introduction of substances or energy, into the environment by man, directly or indirectly, resulting in the adverse effects to living resources, including humans, or interfering with amenities and other uses of the environment."

## Causes of Pollution

Sometimes the natural and human-made causes of water pollution are so thoroughly entwined that it is impossible to distinguish them from each other. For instance, siltation, which denotes the active movement of suspended particles by water or wind through a series of bounces, together with sedimentation, which denotes the passive accumulation of silt in a water body, are two issues often encountered by the waterbodies. During their travel toward plains, rivers carry silt from mountains due to the rumbling of rocks. When water flow abruptly changes, from zero flow to flash floods in a short period of time, silt naturally accumulates in the form of sediments. It's also possible for water bodies undergo sediment deposition artificially. There are times when tonnes of silt are dumped into river beds by sewage, industrial effluents, and agricultural cropland discharge, turning them into marshy, swampy stretches of foul-smelling soil. In this instance, it is impossible to distinguish between the natural and man-made (artificial) factors. Similarly, aluminum companies, phosphate fertilizer facilities, and ceramic industries all produce fluoride, a potent contaminant that causes knock knee disease. Fluoride also occurs naturally in water supplies. Some of the significant pollutants produced by human activity are heated waste waters, radioactive materials, industrial waste, agricultural wastes, and domestic sewage. Although multiple sources of pollution may contribute to the same water body at the same time, the following subheadings can be used to study sources of water pollution for the sake of simplification:

**(a) Domestic Waste Water and Sewage:** It comprises of water-borne wastes from daily chores like bathing, laundry, meal preparation, and utensil washing. Garbage, soaps, detergents, expired food, paper, cloth, used toiletries, and human excrement are all included in household trash. Sewage is the term for this waste water, and it is the main cause of water pollution. Phosphate is a significant detergent component. Phosphate encourages luxuriant growth of algae, sometimes known as "algal blooms", when it is released into water. These give off an unpleasant odor and clog up the water bodies.

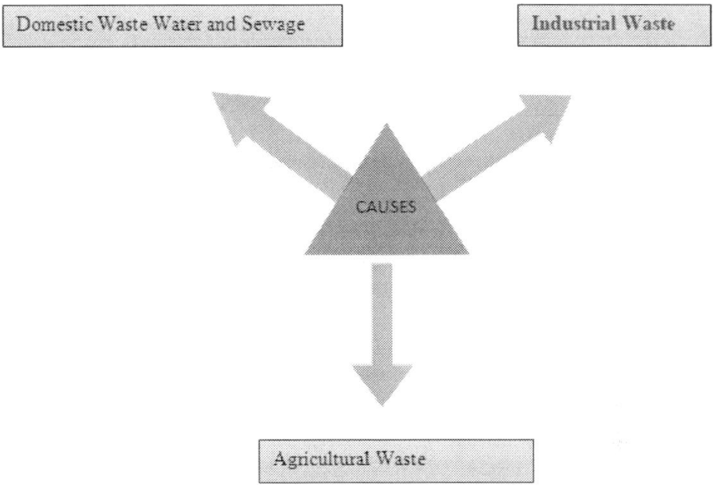


**(b) Industrial Waste:** The majority of rivers and freshwater streams that are found close to big cities, townships, or other populated areas have been contaminated by industrial waste or effluent. Paper manufacturing, textile and sugar mills, distilleries, and thermal power plants are only a few of the common businesses. There are many different types of industrial effluents produced. Aromatic long-chained hydrocarbons are produced by the paint and varnish industries, while the textile industry also produces a variety of dyestuffs and metal salts that are used as mordants. The other industrial effluents are full of pollutants from different industries, including oils, greases, plastics, metallic wastes like copper, zinc, arsenic, cadmium, lead, and mercury, as well as acids, alkalis, cyanides, and chlorides.

**(c) Agricultural Waste:** Manure, along with other wastes from farms and chicken coops, slaughterhouse waste, fertilizer runoff from croplands, harvest wastes, pesticides, and salt and silt drained from irrigated or degraded land are some of the wastes that are included in this category. These wastes are discharged into waterways through agricultural runoff. When fertilisers (phosphates, nitrates, or manures) are added to a water body, the water becomes nutrient-rich, which causes eutrophication and oxygen depletion. Methemoglobinemia, is a dangerous condition caused by high nitrates seeping into groundwater and then being consumed by youngsters. The most serious water contaminants nowadays are pesticides, particularly DDT (Dichloro-diphenyl-trichloro-ethane), which is used to manage agricultural pests and mosquitoes. The insecticide, which has a lengthy half-life under natural circumstances, keeps accumulating in soil and water with subsequent applications. Following the rainy season's pesticide runoff from agricultural fields into neighbouring rivers, there have been severe incidences of fish mortality. Due to their fat solubility, the majority of pesticides enter animals' and humans' adipose tissues. Pesticides are released into the bloodstream during fat breakdown and cause harmful effects.

## Indicators of pollution

Any unfavorable alterations to the physical, chemical, and/or biological components of a certain environment are referred to as pollution. It is currently seen as a significant difficulty that the modern world faces. Physical, chemical, and biological indicators of pollution are codified as illustrative, performance, productivity, policy effectiveness, and total welfare ones. However, logging every variable in a certain ecosystem requires the fusion of biological, signs of physical and chemical contamination. A good physical pollution indicator should be connected with the physical processes of the environment (Zagloul *et al.*, 2019). Organism(s) or community characteristics that can be used to offer information about the state of the environment and the deviations from the norm. Organisms (microbes, plants, insects, bacteria etc.) acting as a benchmark for environmental contamination by providing warnings about dangerous substances or an indication of how polluted the environment is, by having the capacity to accumulate, water, air, dirt, etc. causing pollution are referred to as pollution indicators (Maqbool *et al.*, 2012). Numerous organisms, including bacteria, biotic communities, plants, and animals, exhibit varying degrees of sensitivity and can be successfully used as indicators to acquire environmental pollutant data quickly and forecast it in a timely manner (Ali *et al.*, 2011).

The aquatic environment, especially the water quality, is thought to be the primary influence on human and animal health and disease. From the perspective water pollution, the most significant heavy metals are Zn, Cu, Pb, Cd, Hg, Ni, and Cr. Some of these (e.g., Ni, Cr, and Zn) are necessary for life, but at greater amounts they become poisonous. Others, like Pb and Cd, are harmful substances while having no significance in biology (Maqbool *et al.*, 2012). Biomarkers are crucial in the effort to define and quantify the impacts and presence of contaminants on the aquatic systems. It is frequently challenging to choose a bio-monitor that accurately traces pollution to the right source in the present world of different chemical emissions from numerous industrial operations. Optimizing the selection of a biological indicator in the face of such challenges is a challenging but crucial undertaking.

## Importance and role of Pollution Indicators in Fresh water lakes:

Throughout the history, freshwater ecosystems have been important to the survival of life and the establishment of civilization. In addition to supplying humans with potable water, freshwater systems are essential for farming, transportation, energy production, industrial activities, garbage grinder, and the extraction of fish and other items. This dependence has caused human settlements to be clustered in freshwater environment globally, with more than half of the world's population residing within 20 kilometers of permanent river. Freshwater ecosystems are essential to the existence of not only people but also a vast array of plants, animals, and microorganisms. Rivers, lakes, and wetlands support to disproportionately large share of the earth's biodiversity, despite the fact that freshwater ecosystems account for only 0.01% of the earth's water and a small portion of its surface. More than one-quarter of all living species of vertebrates are fresh water fish alone (McAllister *et al.*, 1997).

Naturally occurring tool for identifying environmental changes, both good and bad, and their effects on human society is the use of the pollution indicators, which are used to evaluate the health of the environment. A few variables, such as light transmission, water flow, temperature, and suspended particles, control the presence of bio-indicators in the environment. Using these bio-indicators, one may predict an area's natural state, as well as the amount and degree of pollution of a particular area (Khatri and Tyagi, 2015). Additionally, environment changes brought in by contaminants, which can harm the biodiversity and the species that live there, can be detected by pollution indicators (Walsh, 1978; Parmar *et al.*, 2016; Gerhardt, 2002; Holt and Miller, 2011). The health of the water quality in fresh water is assessed using naturally occurring bio-indicators, which provides a comprehensive picture of the ecosystem. To select a model bio-indicator, a list of requirements is established, including sedentary lifestyle, abundance, high toxicity tolerance, population stability, high accumulation capacity etc. Chemical analysis provides information on the current state of the environment without considering any prospective harmful effects on the ecosystem. Water status changes are thought to be indicated by changes in the distribution, number, and variety of organisms as well as by the bioaccumulation of poisons in them. There are many different indicators, including as phytoplankton, Zooplanktons, macrophytes, that ranges from plankton to benthic invertebrates (Kumari and Paul, 2020). The use of flora and fauna as bio-indicators provides information to the cumulative impact of the chemical pollutants as well as the gradual transformation. When using a chemical assay, it can be challenging to quantify the indirect effects of bioaccumulation. Metal tends to build up in living things, which causes it to amplify in organisms within the food chain. It is challenging to analyze the presence of such pollutants using chemical or physical tests (Manickavasagam *et al.*, 2019; Gurjar *et al.*, 2010).

## Types of Pollution Indicators

### 1. Fish as pollution indicator:

A fish community is a part of the ecosystem that exclusively consists of fish and their interactions with one another. A specific locations biota interacts with its physical and chemical environment, such as its physical habitat, nutrients, oxygen, and temperature, to form an aquatic ecosystem. These alterations in the ecosystem are probably reflected in the fish communities. In the US, fish assemblages and fishing regulations are the main topics of designations of water quality. For the purpose of describing water quality and stream health in New York State, labels like "swimming/fishing," "trout", and "trout spawning" are employed (Barbour, 1999).

Due to their varying sensitivity to pollution, fish have been extensively studied as potential indicators of the environmental water quality (Naigaga *et al.*, 2011). Ecology's core concept still revolves around estimating the number of species in a given area. Understanding the mechanism and impacts of the environmental disturbances like pollution is aided by species variety, which is related to the operation of the ecological systems (McGill *et al.*, 2007; Tokeshi and Arakaki, 2007; Mora *et al.*, 2008). Comparing the highly impacted sites to the less impacted sites, low levels of diversity in the fish community structure were observed. Fish diversity could therefore, be a helpful biological indicator of water quality and map out fish species hotspot areas (Naigaga *et al.*, 2011). Fish serve as a biological indication of the quality of water due to their propensity to store metals in their muscles, which modify their physiology, biochemistry, and genetic characteristics (Zhao *et al.*, 2012; Javeed *et al.*, 2017). The benefit of being able to integrate and show cumulative change over time, as proposed to a range of other methodologies, such as flow regimen, energetic resources, and biotic monitoring. (Cairns and Pratt, 1993; Zamora and Alba-Tercedor, 1996). Fish and macro-invertebrate assemblages have been singled out among the flora as excellent bio-indicator for tracking how farming, forestry, urban and industrial effluents, and climate change are causing ecosystems to deteriorate (Fierro *et al.*, 2018; Carvalho *et al.*, 2012). Additionally, according to recent studies, the stomach contents of Salmonids, such as *Oncorhynchus mykiss* and *Salmo trutta*, contain a variety of invertebrate prey that is present in the benthos of non-intervened basins reflecting anthropogenic impact on the basin (Vargas Chacoff *et al.*, 2013).

### 2. Macrophytes as pollution indicators:

Because they are typically the major producers, have high biomass production, and support biodiversity, aquatic macrophytes make up a significant population in limnic ecosystems (Neckles *et al.*, 1993; D. Esteves, 1998). In shallow lakes, macrophytes participate in a number of feedback systems that work to keep the water clean (Scheffer, 1998). In order to survive in aquatic systems, macrophytes must contend with flow, nutrient availability, and other physical and chemical factors that are beyond their control. Therefore, a collection of these creatures in a lake or river can work as a helpful indicator of the integrated mixture of pressure and stress problems that affect the environment of those organisms (Kennedy, 1999). Recent years have seen a lot of work put into creating predictive models and a biotic index based on macrophytes that may be used as management tools to assess the quality of water bodies (Camargo *et al.*, 2011; Ciecierska and Kolada, 2014; Clayton and Edwards, 2006; Beck *et al.*, 2010; Kocic *et al.*, 2008; Penning *et al.*, 2008).

Depending on their stages of development, the availability of light, and the concentration of nutrients, macrophytes morphological and physiological properties might vary in particular ways (Ali *et al.*, 2021). According to studies, (Pajević *et al.*, 2003; Bytyqi *et al.*, 2020) changes in nutrients and the arrival of organic pollutants are to blame for macrophytes structural changes in bodies of running water. On the other hand, the distribution of aquatic macrophytes is known to be impacted by the presence of pollution and an increase in some nutrients. On the other side, it is known that the distribution of aquatic macrophytes can be impacted by an increase in some nutrients and the presence of contaminants (Onaindia *et al.*, 2005). Given that, changes in the composition of aquatic vegetation are accurate indicators of water quality, aquatic macrophytes may therefore, be highly helpful in the biological monitoring of aquatic systems. They also serve as indicators of the environmental circumstances to which they are exposed, making them suitable for use as long term indicators with great geographical resolution (Rameshkumar *et al.*, 2019; Sarkar *et al.*, 2020).

Macrophytes improve nutrient cycle and sediment stability by accumulating more chemical components, which has an impact on the degree of eutrophication. Given that, the concentration of heavy metals in macrophytic tissues may be 106 times higher than those in an aqueous environment; their involvement in the accumulation of heavy metals is crucial (Rai, 2009).

### 3. Birds as pollution indicators:

The bio-indicators that birds provide are crucial. The public is interested in and concerned about birds because they are noticeable, relatively simple to see, one of the most researched groups of species. Top predators like raptors and seabirds store toxic compounds, which have in the past frequently served as early indicators of environmental change by affecting factors like physiology, reproduction, and even death. These factors all lead to population decrease and endangerment. As a result, birds are an appealing alternative for bio-monitors, signaling specific environmental change such as chemical contamination, marine pollution, fish populations, and other environmental changes (Becker, 2003). Pigeons (e.g., wild pigeons) as an indicator of water pollution monitoring, Pigeon blood, kidney, lung, liver, and other organs contain the amounts of heavy metals, including Zn and Pb, which damage DNA (Artacho *et al.*, 2007). In bio-monitoring, wild birds have served as passive bio-monitors, acting as either accumulative or sensitive indicators. The most eloquent evidence of the value of birds as bio-monitors comes from their use as qualitative and quantitative accumulative indicators of pesticides and heavy metals, based on logistically practical and non-destructive avian matrices like eggs, feathers, or blood, and on high bio-magnification rates in dose-dependent responses. It is now possible to examine contaminants in extremely little amounts of material, such as in small blood samples that may be acquired from caught individuals without injuring them, because to advancements in chemical analytical techniques over the previous few years (Elliott and Shutt, 1993; Kahle and Becker, 1999).

### 4. Man as pollution indicators:

Without a question, man himself holds the central position and is target in the problem of biosphere degradation. The contamination of micro-organisms, plants, and animals, as well as the deterioration of the air, soil, and water pose a serious threat to public health and result in significant financial loses. However, if man him-self got contaminated to levels giving rise to large-scale adverse somatic or genetic effects that would be a real catastrophe. Arsenic, mercury, lead, cadmium, and other trace elements with significant levels of toxicity should receive priority attention because they are likely to have no physiological function (Ryabukhin, 1976).

### 5. Phytoplankton as pollution indicators:

The total biomass of phytoplankton and their several species are used as indicators of the suitability of aquatic habitat because phytoplankton communities are sensitive to changes in their habitat (Chellappa *et al.*, 2009). Phytoplankton monitoring is required since systems that rely only on physical and chemical analyses are occasionally insufficient. The composition of the phytoplankton not only reveals a particular water situation but also prior aquatic ecosystem conditions. Changes in the distribution, community structure, and percentage of sensitive species in phytoplankton serve as indicators of water quality (Gharib *et al.*, 2011). Physical parameters like temperature and light intensity have the greatest impact on the seasonal abundance and diversity of plankton biotypes, whereas chemical factors like dissolved oxygen, pH, salinity, hardness, electric conductivity, and nutrition content have a less significant impact (Reynolds, 1984). As phytoplankton are quick to adopt to environmental changes and have a shot lifespan (Zebek, 2004), they may be a helpful indicator for determining the quality of water and for giving early warning indications of deteriorating circumstances (Jaffari and Gunale, 2006; Bhunia *et al.*, 2000; Kaparapu *et al.*, 2013).

The primary focus of research is phytoplankton, which can be used as an indicator to track not only water quality but also eutrophication and pollution. Phytoplankton should be a good indicator of the conditions present in the aquatic environment (El-Kassas *et al.*, 2016). The progressive decline in the variety and abundance of phytoplankton could be attributed to the stress-inducing effects of rising pollution, the eradication of sensitive species, and the spread of more tolerant species that will eventually take over (Patrick, 1949).

### 6. Algae as pollution indicator:

Algae comprise an essential class of bacteria and plants in aquatic environment, making them a significant component of biological monitoring techniques for evaluating water quality. Algae are excellent for determining the quality of water since they have a brief life cycle, quick reproduction rate, and limited nutritional demand. Due to their fast response to a variety of water conditions brought on by changes in water chemistry, algae are important markers of the health of ecosystems. For instance, the makeup of genera that can withstand these conditions is impacted by changes in water acidity brought on by acid forming compounds that modify lake pH levels as well as heavy metals released from industrial sites (Brunn, 2012). In nutrient affected water bodies, the dominating present, or absent algal taxa are frequently determined by the N: P ratio. Algac are most strongly impacted by physical and chemical conditions since they are main producers. Algae assemblages can rapidly acquire some contaminants and are sensitive to them, and algal metabolism can be affected by a variety of environmental and natural perturbations. Water quality and odor are impacted by algae, so operational managers in charge of distributing and supply potable water must eliminate how the algae will migrate and thrive in river system (Neal *et al.*, 1986). One of the most significant algae found in aquatic ecosystem that are connected to substrate is periphytons which are frequently utilized in biological water quality monitoring system (Leland and Carter, 1985; Newman McIntosh, 1989).

## 7. Microfauna as a pollution indicator

Eukaryotic, single-celled protozoans (such as naked amoebae, flagellates, and ciliates) and multicellular organisms (such as rotifers, tardigrades, and worms) that are too small to study and identify without a microscope make up the microfauna. Protists and tiny metazoans, together known as microfauna, have proven to be an effective method for measuring water contamination. Over the past few years, the prevalence of toxicants in the aquatic environment has increased in frequency. One of the most crucial biotic elements affecting the functioning of an aquatic ecosystem's food chains, food webs, energy flow, and matter cycling is zooplankton (Pal *et al.*, 2015). Freshwater zooplankton adapt swiftly to environmental changes, which causes a fluctuation in their species indices (Avinash, 2014). In freshwater lake environment, the two significant zooplankton species that serve as markers of pollution are cladocera and rotifer. Bioindicators and bio-monitors are desirable instruments for pollution investigators. The biota is seen to be the best indicator of ecological changes since it provides data on the actual biological impacts (Zhou *et al.*, 2008).

## 8. Rotifer as pollution indicator

With almost 1850 species, rotifers are undifferentiated, pseudo-coelomate, and bilaterally symmetrical microscopic invertebrates that live in watery or semi-aquatic environments. These creatures commonly make up more than 60% of the zooplankton population in lakes (Armengol *et al.*, 1998). They can be found in a variety of watery habitats as well as on damp soil, where they live in the thin water films that form around soil particles. In addition to moving water habitats like rivers and streams, rotifers can also live in still water habitats like lake bottoms. Since they frequently make up a bigger portion of the zooplankton biomass at specific periods of the year, rotifers provide a significant contribution to secondary production in many aquatic habitats. Rotifers are desirable species for ecotoxicological investigations due to their fast reproduction, small size, and low volume requirements for culture and toxicity testing. This is in addition to the significance of their ecological roles in aquatic communities, rotifers are global citizens. They have been regarded as reliable biological indicators due to their ability to quickly utilize favourable circumstances and their relative toleration of modest environmental disturbances (Arora *et al.*, 2003). They are significant in freshwater environment as a result of having one of the greatest rates of reproduction among metazoans, which allows them to reach high population densities quickly and become dominant in many zooplanktonic groups. They serve as a bridge connecting the microbial population to higher trophic levels. Rotifers rapidly invade ecosystems and transform primary production (algae and cyanobacteria) into a form that is useful for secondary consumers, making energy accessible for the following trophic levels. They can also serve in river drainage systems as biological sensors of nutrient status and as food for higher-order species. They aid in nutrient recycling in the interstitial water from marshy soils (Wallace *et al.*, 2010). Rotifer population dynamics are well understood both in the wild and in the lab, making them an excellent tool for studying ecological and demographic concepts. Additionally, they are widely employed to evaluate aquatic toxicity from a population perspective. The intrinsic growth rate (r), a commonly utilized end point, has shown to be quite sensitive (Snell *et al.*, 1991; Perez-Legaspi *et al.*, 2012).

## Rotifers as sentinels or bio-indicators

Due to their quicker generation times than fish, as well as the fact that they are smaller in size, aquatic invertebrates provide appealing model species in aquatic toxicity (Janssen *et al.*, 1993; Snell *et al.*, 1991). The porosity of the integument, their small size, and the fact that they feed on filters made rotifers very vulnerable to chemical and physical changes (Nogrady *et al.*, 1993). Because of their significance in aquatic trophic webs, rotifers serve as useful sentinel species to reveal toxicant exposure in harmed ecosystems (Wallace, 2006). The following factors make rotifers useful tools for assessing aquatic ecosystems: (a) their basic biology, (b) their sensitivity to contaminants at the physiological and demographic levels, (c) widespread distribution, (d) high ecological relevance, (e) their high growth rate, (f) availability of neonates, (g) high ingestion rates, (h) their ease of culture and handling, (i) their transparency, (j) short life cycles, and (k) their ease of obtaining (Snell *et al.*,1998; Perez-Legaspi *et al.*,2012; Yufera, 2001).

## Rotifer sensitivity

Because rotifers serve as key ecological indicators in aquatic ecosystem, it has been possible to assess the toxicity of a wide range of substances, including organic compounds, metals, and medicines, that are present in water reservoirs as a result of human activities (Snell and Janssen 1995; Zadorozhnaya *et al.*, 2015). The most frequently used test organisms include rotifers of the genus *Brachionus*, and several rotifers are also included for comparison. Although it is always preferable to have a test organism that is highly sensitive to xenobiotic side effects, sensitivity varies between species. These are typically brought on by the biological traits of test organisms, the kind of chemical, and how it works; thus, one species may react strongly to one toxin but not as strongly to another (Vaal *et al.*, 1997).

The sensitivity of various freshwater rotifers, with the exception of marine species *Brachionus plicatilis* and *B. koreanus* while assessing the toxicity of metals, chemical compounds, and pesticides, taking into account their importance from a toxicological standpoint and the literature that is available. Some examples of these variations on sensitivity are as-Lindane, an organochloride pesticide, and methyl parathion (organophosphorus), which demonstrate mild toxicity and have greater toxicity ranges in comparison to malathion, are two instances of these variations on sensitivity. *B. plicatilis* and *B. koreanus* provide a spectrum of organochloride toxicity (14 and 35.89 mg/L, respectively). Methyl parathion, which exhibits greater acute toxicity, ranged in concentration from 0.607 mg/L *for Euchlanis dilatata* to 29.19 mg/L for *B. calyciflorus*. Due to the published $LC_{50}$ values for *B. koreanus* (3.9 mg/L) and *B. calyciflorus* (11.85 mg/L) of an additional organophosphorus of ecotoxicological concern, chlorpyrifos, tends to have a considerable impact on aquatic invertebrates (Casalderrey *et al.*, 1992; Kim *et al.*, 2015).

Endosulfan and pentachlorophenol (PCP), two organochlorides, are extremely harmful agricultural chemicals. *B. koreanus* and *B. plicatilis* serve as the reference organisms for establishing the endosulfan range due to their 24-hour $LC_{50}$ values that have been documented in the literature (PAN Pesticide Database; Kim *et al.*, 2015). *B. calyciflorus* had the lowest and highest toxicity values for PCP. Furthermore, it is crucial to note that when doing a toxicity test, physical conditions might change how sensitive a test organism is. One example is provided by Preston et al. 1999, who found that as UV-B exposure period was extended, the $LC_{50}$ values for PCP and mercury using the bioindicator *B. calyciflorus* decreased.

## 9. Cladocera as pollution indicator

The ecosystem of freshwater depends heavily on microorganisms like zooplankton. A typical order in terms of systematic position - the Cladocera are a small collection of organisms that are associated with crustaceans. They are commonly called as "water fleas" due to the fact that they may be found in a variety of waterways, including marine environment. They serve as a model species in environmental toxicology due to their great sensitivity to water quality and have a crucial place in the aquatic food chain as the link primary and secondary productivity (Siciliano *et al.*, 2015; Rumes *et al.*, 2011). Environmental variables such water depth, temperature, trophic condition, total phosphorus concentration, and others exhibit a substantial reaction in cladocera (Lingampally *et al.*, 2018). According to Zawisza *et al.*, 2016 cladocera species are particularly sensitive to pH alterations because of their susceptibility to acidification. We are aware of around 700 species, many of which are still unknown due to incomplete descriptions. The Cladocera group is regarded as a crucial freshwater bio-indicator of aquatic ecosystems because they are abundant in high nutrient enriched water and responded to diverse physicochemical ecological parameters when there is interaction with changing water environment (Bhandarkar and Paliwal, 2021). While a few species have adapted to life in marine or brackish conditions, most species of cladocera are found in temporary or permanent freshwater pools (Richter *et al.,* 2007). For determining the level of toxicity of pesticides and other environmental pollutants, Cladocera have been utilised as both an indicator and a test organism (Frear and Boyd, 1967`). They contribute significantly to the aquatic food web, lower phytoplankton biomass since they are filter feeders, and enhance water quality because they eat a significant proportion of phytoplankton (Gonzalez, 2000). Their abundance in the vegetated zone which is higher than the non-vegetated zone (Bozkurt and Guven, 2009) aids in the effective grazing control of algal growth. High levels of environmental heterogeneity, according to Begon *et al.* (2007), support high levels of microhabitat variety which leads to an increase in the number of species. Due to the abundance of macrophytes in the littoral region's microhabitats, there is a great diversity of micro-crustacean species (Lemly and Dimmick, 1982; Nogueira *et al.*, 2003). Cladocera have peculiar biological characteristics that enable their use in ecological research, particularly those addressing stress brought on by the environment from a variety of angles, including biochemical, molecular, toxicological, and physiological. These organisms are able to flourish in a variety of highly variable settings, including ponds and shallow coastal lagoons (Petrusek *et al.*, 2008; Santangelo *et al.*, 2008), making them more desirable as models for studies on environmental stress. Due to its vast occurrence, particularly in temperate zones, the Daphnia has received special attention (Sarma *et al.*, 2005). *D. magna* are crucial model animals in Daphnia, employed in everything from ecology to biomedical sciences (Seda and Petrusek, 2011)

## Cladocera in aquatic environments:

Rotifer and Cladocera are typically offered to the fish seeds first, followed by artificial pelleted meals. Most people agreed that *Moina* (Cladoccra) *Bruchionus plicatilis* (rotifer) is a beginning for many fish larvae. Because of its superior nutritional content and economic viability in mass production, cladocera are employed as a live feed source in fish culture. On Cladocera as live food, the fish seeds of *Clarius batrachus, Ompak bimaculatus*, and *Channa stritatus* grow and survive well (Tawaratmaikul, 1988). Because of their increased protein and fat levels, the Cladocera effectively perform as crucial energy sources, especially for larval nutrition for good growth and metabolic maintenance. In general, cladocera thrive on minute organic particles suspended in water, including bacteria, phytoplankton, fungi, and protozoa. Cow dung, chicken droppings, fish faces, horse manure, rice bran, and mineral fertiliser have all been discovered to support the viability of mass culture in aquaculture (Punia, 1988; Rottmann *et al.*, 2003).

**Relation with environmental variables:**

Cladocera are significant bio-parameters for a variety of environmental indicators (Rumes *et al.*, 2011). Environmental variables including trophic state, total phosphorus concentration, water depth, temperature, etc. have a significant impact on Cladocera. Because of their susceptibility to acidification, Cladocera species are particularly sensitive to pH fluctuations (Zawisza *et al.*, 2016). Dodson (1992) asserted that the size and productivity of the water body limit the Cladocera population. The Cladocera are excellent indicators of alterations in the environment (Eggerment and Martens, 2011). When compared to many other aquatic creatures, they are more vulnerable to water contamination (Brix *et al.*, 2001, Von Derohe and Liess, 2004). The criteria regulating the species composition of cladocera are very well established as a consequence of substantial study on the connections between environmental factors and cladocera community organisation (Leppanen, 2019). Only a few data on the ecology, diversity, and contribution to aquatic productivity of cladocera in India's freshwater environment are available. Cladocerans are typically the focus of zooplankton studies (Sharma and Sharma, 2009). Cladocera are a key component of the trophic cascades in the aquatic system and are highly responsive to contaminants, making them one of the key groups for biomonitoring investigations. Even extremely tiny levels of pollutants can cause it to react (Ferdous and Muktadir, 2009; Sharma and Chandrakiran, 2011). Since the pH has an impact on how aquatic creatures function physiologically, including the exchange of ions with the water and respiration, the pH is significant to aquatic life. In water bodies, there can be a wide range in the sensitivity of organisms to pH change. NaOH spills caused by humans could be one of its causes. Although Daphnia pulex's ability to reproduce was constrained to a relatively small range. Davis and Ozburn's (1969) research showed that it would not thrive below a pH of 7. Fish and crustaceans have experienced growth retardation and skeletal malformation in acidic water (Haines, 1981). Many researchers have tried to explain how aquatic species react to pH variations. Although Daphnia species are more sensitive to low pH than Cyclops species. Bulkowski *et al.* (1985) found that Daphnia species disappear at pH 3 and 4. One of the most common species, *Chydorus* cf. *sphaericus,* is renowned for its broad ecological tolerance, notably to pH (Walseng *et al.*, 2003; Belyaeva and Deneke, 2007). Sharif *et al.*, (2017) found that Cladocera had a weakly positive relationship with pH, a moderately negative link with salinity and TDS, and a weakly negative linear association with DO. The connection between Cladocera and water and air temperature were likewise fairly positive. The largest (3mm) and most widely distributed species that often-live-in hard water is *Daphnia magna*. Life expectancy is heavily influenced by water temperature, which can range from a few weeks to months (MacArthur and Baillie,

1929). It can endure brief periods of time in water that is either 0°C or 39°C (Frear and Boyd, 1967). It can endure pH levels of between 6 and 9.5 (Anderson, 1946). Frear and Boyd (1967) noted that a pH of 7.5 resulted in the lowest mortality in culture. One of the most intriguing ecological characteristics of dystrophic water bodies is the pH. The most acidic-condition tolerant species, such as *Alonella exigua* and *Graptoleberis testudinaria*, were noted in the study by Zawisza *et al.* (2016), whereas *Acroperus harpae* and *Alona nana* displayed noteworthy abundances at high pH. These species' high relative abundance at pH values below 5.5 suggested that they had a high level of acidic tolerance. In contrast, neutral waters are preferred by *Bosmina longirostris,* which likewise become more numerous as pH rose. *Alona affinis* concentrations seemed unrelated to pH levels in terms of their abundance. In the research of a contaminated lake in Darjeeling, Jha and Barat (2003) discovered that the lake's pH had become acidic due to pollutants introduced from outside sources. The abundance of *Moina, Daphnia*, and *Bosmina* was supported by the examination of various physiochemical parameters and planktons. According to Mustaq (1990), several Cladoceran species can thrive in polluted waters and are thus useful as biological markers of water pollution. There is evidence that the dominant cladocera species, *Bosmina longirostris*, can survive a wider range of salinities than many other cladocera species (Aladin, 1991; Paturej and Gutkowska, 2015). When compared to Daphnia, a greater salinity has been observed (Boronat *et al.*, 2001; Amsinck *et al.*, 2003).

## Conclusion

Water pollution is one of the key issues facing modern society. It is possible to determine the effects of anthropogenic stress on aquatic ecosystem by evaluating the water quality and phytoplankton structure. A useful technique used in conservation research to safeguard and maintain the biological integrity of the natural ecosystem, which includes preventative measures, is biological monitoring. Water must be protected since it is crucial to the preservation of biodiversity. An aquatic ecosystem's integrity can be evaluated by looking at the phytoplankton's physico-chemistry and composition. The addition of foreign substances that are deadly to aquatic life and human society cause pollution, which lowers the quality of freshwater environment. Fresh water ecosystem is subjected to significant change as a result of persistent pollution and possibly dangerous toxic metals that are transported to remote regions by climatic extremes and mix with native and induced biota. Cladocera and Rotifer, two types of microfauna, serve as the best indicators of contamination in freshwater ecosystem. Programs for monitoring water quality should incorporate indicators obtained from a variety of phyla of organisms to give a complete picture of changes in freshwater ecosystem. Improved comprehension of the mechanisms governing the time and spatial distribution of pollution arrival in freshwater should be the main focus. Changes in fresh water biodiversity and human land regimes are required to supplement the recent distribution of contaminants with adequate seasonal resolution.

## Bibliography


Aladin, N. V. (1991). Salinity tolerance and morphology of the osmoregulation organs in Cladocera with special reference to Cladocera from the Aral Sea. *Hydrobiologia*, *225*(1): 291-299.

Ali, A., Obi-Iyeke, G.E., Keke, U.N. and Arimoro, F.O. (2021). Exploring the use of Macrophytes as Biological Indicators for Organic Pollution of Chanchaga River in North Central Nigeria. *Pollution*, *7*(4): 859-870.

Ali, A.H.A., Hamed, M.A., and El-Azim, A. (2011). Heavy metals distribution in the coral reef ecosystems of the Northern Red Sea. *Helgoland marine research*, *65*(1): 67-80.

Amsinck, S.L., Jeppesen, E. and Ryves, D. (2003). Cladoceran stratigraphy in two shallow brackish lakes with special reference to changes in salinity, macrophyte abundance and fish predation. *Journal of Paleolimnology*, *29*(4): 495-507.

Anderson, B.G. (1946). The toxicity thresholds of various sodium salts determined by the use of Daphnia magna. *Sewage Works Journal*, *18*(1): 82-87.

Armengol, X., Esparcia, A. and Miracle, M.R. (1998). Rotifer vertical distribution in a strongly stratified lake: a multivariate analysis. In *Rotifera VIII: A Comparative Approach* 161-170.

Arora, J. and Mehra, N.K. (2003). Species diversity of planktonic and epiphytic rotifers in the backwaters of the Delhi segment of the Yamuna River, with remarks on new records from India. *Zoological Studies-Taipei*, *42*(2): 239-247.

Artacho, P., Soto-Gamboa, M., Verdugo, C. and Nespolo, R.F. (2007). Blood biochemistry reveals malnutrition in black-necked swans (Cygnus melanocoryphus) living in a conservation priority area. *Comparative Biochemistry and Physiology Part A: Molecular & Integrative Physiology*, ***146***(2): 283-290.

Avinash B. G. (2014). Species diversity indices of zooplankton from Sadatpur reservoir, Ahmednagar, Maharashtra. *Annals of Biological Research*, *5*(4): 58-61.

Barbour, M.T. (1999). Rapid bioassessment protocols for use in wadeable streams and rivers: periphyton, benthic macroinvertebrates and fish. US Environmental Protection Agency, Office of Water.

Bashir, I., Lone, F.A., Bhat, R.A., Mir, S.A., Dar, Z.A. and Dar, S.A., (2020). Concerns and threats of contamination on aquatic ecosystems. In *Bioremediation and Biotechnology* (1-26). Springer, Cham.

Beck, M.W., Hatch, L.K., Vondracek, B. and Valley, R.D. (2010). Development of a macrophyte-based index of biotic integrity for Minnesota lakes. *Ecological Indicators, 10*(5): 968-979.

Becker, P.H. (2003). Biomonitoring with birds. *Trace Metals and other Contaminants in the Environment6*: 677-736.

Begon, M., Townsend, C.R. and Harper, J.L. (2007). Ecologia: de indivíduosaecossistemas. 4 ed. Artmed, Porto Alegre.

Belyaeva, M. and Deneke, R. (2007). Colonization of acidic mining lakes: Chydorussphaericus and other Cladocera within a dynamic horizontal pH gradient (pH 3- 7) in Lake Senftenberger See (Germany). *Hydrobiologia, 594*(1): 97-108.

Bhandarkar, S.V. and Paliwal, G.T. (2021). Shorts notes on cladocera: a sentinel organism. In: Ecology Research (ed.1) *3*: 1-14.

Bhunia, F., Saha, N.C. and Kaviraj, A. (2000). Toxicity of thiocyanate to fish, plankton, worm, and aquatic ecosystem. *Bulletin of Environmental Contamination & Toxicology, 64*(2).

Boronat, L., Miracle, M.R. and Armengol, X. (2001). Cladoceran assemblages in a mineralization gradient. *Hydrobiologia, 442*(1): 75-88.

Bozkurt, A. and Guven, S.E. (2009). Zooplankton composition and distribution in vegetated and unvegetated area in three reservoirs in Hatay, *Turkey Journal of Animal and Veterinary Advances, 8*(5): 984-994.

Brix, K.V., DeForest, D.K. and Adams, W.J. (2001). Assessing acute and chronic copper risks to freshwater aquatic life using species sensitivity distributions for different taxonomic groups. *Environmental Toxicology and Chemistry: An International Journal, 20*(8): 1846-1856.

Bruun, K. (2012). Algae can function as indicators of water pollution. Nostoca Algae Laboratory, Washington State Lake Protection Association. Available online: www. nostoca. com.

Bulkowski, L., Kraus, K.A. and Krise, W.F. (1985). Purification of Cyclops cultures by pH shock (Copepoda). *Crustaceana, 48*(1-3): 179-182.

Bytyqi, P., Czikkely, M., Shala-Abazi, A., Fetoshi, O., Ismaili, M., Hyseni-Spahiu, M., ... and Millaku, F. (2020). Macrophytes as biological indicators of organic pollution in the Lepenci River Basin in Kosovo. *Journal of Freshwater Ecology, 35*(1): 05-121.

Cairns, J. and Pratt, J.R. (1993). A history of biological monitoring using benthic macroinvertebrates. *Freshwater biomonitoring and benthic macroinvertebrates,* 10: 27.

Camargo, J.A., Gonzalo, C. and Alonso, Á. (2011). Assessing trout farm pollution by biological metrics and indices based on aquatic macrophytes and benthic macroinvertebrates: A case study. *Ecological Indicators, 11*(3): 911-917.

Carvalho, C.D.S., Bernusso, V.A., de Araújo, H.S.S., Espíndola, E.L.G. and Fernandes, M.N. (2012). Biomarker responses as indication of contaminant effects in Oreochromis niloticus. *Chemosphere,*89(1): 60-69.

Casalderry, A.F., Ferrando, M.D. and Andreu-Moliner, E. (1992). Acute toxicity of several pesticides to rotifer (Brachionus calyciflorus). *Bulletin of Environmental Contamination and Toxicology*, *48*(1): 14-17.

Chellappa, N.T., Chellappa, T., Fabiana, R.A.C., Rocha, O. and Chellappa, S. 2009. Impact of stress and disturbance factors on the phytoplankton communities in Northeastern Brazil reservoir. *Limnologica, 39*(4): 273-282.

Ciecierska, H. and Kolada, A. (2014). ESMI: a macrophyte index for assessing the ecological status of lakes. *Environmental Monitoring and Assessment, 186*(9): 5501-5517.

Clayton, J. and Edwards, T. (2006). Aquatic plants as environmental indicators of ecological condition in New Zealand lakes. In Macrophytes in Aquatic Ecosystems: From Biology to Management (pp. 147-151). Springer, Dordrecht.

Davis, P. and Ozburn, G.W. (1969). The pH tolerance of Daphnia pulex (Leydig, emend., Richard). *Canadian Journal of Zoology*, *47*(6): 1173-1175.

Dodson, S. (1992). Predicting crustacean zooplankton species richness. *Limnology and Oceanography*, *37*(4): 848-856.

Eggermont, H., and Martens, K. (2011). Preface: Cladocera crustaceans: sentinels of environmental change. *Hydrobiologia*, *676*(1): 1-7.

El-Kassas, H.Y. and Gharib, S.M. (2016). Phytoplankton abundance and structure as indicator of water quality in the drainage system of the Burullus Lagoon, southern Mediterranean coast, Egypt. *Environmental monitoring and assessment,188*(9): 1-14.

Elliott, J.E. and Shutt, L. (1993). Monitoring organochlorines in blood of sharp-shinned hawks (Accipiter striatus) migrating through the great lakes. *Environmental Toxicology and Chemistry, 12*(2): 241-250.

Ferdous, Z. and Muktadir, A.K.M. (2009). A review: potentiality of zooplankton as bioindicator. *American Journal of Applied Sciences, 6*(10): 1815-1819.

Fierro, P., Arismendi, I., Hughes, R.M., Valdovinos, C., and Jara-Flores, A. (2018). A benthic macroinvertebrate multimetric index for Chilean Mediterranean streams. *Ecological Indicators, 91*:13-23.

Frear, D.E. and Boyd, J.E. (1967). Use of Daphnia magna for the microbioassay of pesticides. I. Development of standardized techniques for rearing Daphnia and preparation of dosage-mortality curves for pesticides. *Journal of Economic Entomology*, *60*(5): 1228-1236.

Gerhardt, A. (2002). Bioindicator species and their use in biomonitoring. Environmental monitoring, *1*: 77-123.

Gharib, S.M., El-Sherif, Z.M., Abdel-Halim, A.M. and Radwan, A.A. (2011). Phytoplankton and environmental variables as a water quality indicator for the beaches at Matrouh, south-eastern Mediterranean Sea, Egypt: an assessment. *Oceanologia, 53*(3): 819-836.

González, G., Maze, R.A., Dominguez, J. and Pena, J.C. (2000). Trophic ecology of the tench, Tincatinca, in two different habitats in North-West of Spain. *Cybium*, *24*(2): 123-138.

Gurjar, B.R., Jain, A., Sharma, A., Agarwal, A., Gupta, P., Nagpure, A.S. and Lelieveld, J. (2010). Human health risks in megacities due to air pollution. *Atmospheric Environment*, *44*(36): 4606-4613.

Haines, T.A. 1981. Acidic precipitation and its consequences for aquatic ecosystems: a review. *Transactions of the American Fisheries Society*, 110: 669-707.

Hampel, M., Blasco, J. and Segner, H. (2015). Molecular and cellular effects of contamination in aquatic ecosystems. *Environmental science and pollution research*, *22*(22), pp.17261-17266.

Holt, E.A., and Miller, S.W. (2011). Bioindicators: using organisms to measure. *Nature*, *3*: 8-13.

Jafari, N.G. and Gunale, V.R. (2006). Hydrobiological study of algae of an urban freshwater river. *Journal of Applied Sciences and Environmental Management*, *10*(2).

Janssen, C.R., Rodrigo M.D.F. and Persoone, G. (1993). Ecotoxicological studies with the freshwater rotifer *Brachionus calyciflorus*. In: Gilbert, J.J., Lubzens, E., Miracle, M.R. (eds) Rotifer Symposium VI. *Developments in Hydrobiology*, 83. Springer, Dordrecht.

Javeed, M., Ahmad, M., Usmani, N. and Ahmad, M. (2017). Multiple biomarker responses (serum biochemistry, oxidative stress, genotoxicity and histopathology) in *Channa punctatus* exposed to heavy metal loaded waste water. *Scientific reports*, *7*(1): 1-11.

Jha, P. and Barat, S. (2003). Hydrobiological study of lake Mirik in Darjeeling Himalayas. *Journal of Environmental Biology*, *24*(3): 339-344.

Kahle, S. and Becker, P.H. (1999). Bird blood as bioindicator for mercury in the environment. *Chemosphere*, *39*(14): 2451-2457.

Kaparapu, J. and Geddada, M.N.R. (2013). Seasonal Distribution of Phytoplankton in Riwada Reservoir, Visakhapatnam, Andhra Pradesh, India. *Notulae Scientia Biologicae*, *5*(3): 290-295.

Kennedy, A.D. and Jacoby, C.A. (1999). Biological indicators of marine environmental health: meiofauna – a neglected benthic component? *Environmental Monitoring and Assessment*, *54*(1): 47-68.

Khatri, N. and Tyagi, S. (2015). Influences of natural and anthropogenic factors on surface and groundwater quality in rural and urban areas. *Frontiers in life science*, *8*(1): 23-39.

Kim, B.M., Lee, J.W., Seo, J.S., Shin, K.H., Rhee, J.S. and Lee, J.S. (2015). Modulated expression and enzymatic activity of the monogonont rotifer Brachionus koreanus Cu/Zn-and Mn-superoxide dismutase (SOD) in response to environmental biocides. *Chemosphere*, *120*: 470-478.

Kočić, A., Hengl, T. and Horvatić, J. (2008). Water nutrient concentrations in channels in relation to occurrence of aquatic plants: a case study in eastern Croatia. *Hydrobiologia*, *603*(1): 253-266.

Kumaraswamy, T.R., Javeed, S., Javaid, M. and Naika, K. (2020). Impact of Pollution on Quality of Freshwater Ecosystems. In Qadri, H; Bhat, R.A; Mehmood, M.A; Dar, G.H (Eds.), *Fresh Water Pollution Dynamics and Remediation,* pp.69–81.

Kumari, D., and Paul, D.K. (2020). Assessing the role of Bioindicators in freshwater ecosystem. *Journal of Interdisciplinary Cycle Research, 12*(9): 58-74.

Leland, H.V. and Carter, J.L. (1985). Effects of copper on production of periphyton, nitrogen fixation and processing of leaf litter in a Sierra Nevada, California, stream. *Freshwater biology,15*(2): 155-173.

Lemly, A.D. and Dimmick, J.F. (1982). Growth of young-of-the-year and yearling centrarchids in relation to zooplankton in the littoral zone of lakes. *Copeia*, 305-321.

Leppänen, J.J. (2019). An overview of Cladoceran studies conducted in mine water impacted lakes. *International Aquatic Research, 10*: 207-221

Lingampally, V., Solanki, V.R., Anuradha, D.L. and Raja, S., (2018). Cladoceran abundance in relation to water quality parameters of Chakki Talab, Bodhan, Telangana. *International Journal of Current Engineering and Technology8*(1): 1-4.

MacArthur, J.W. and Baillie, W.H.T. (1929). Metabolic activity and duration of life. I. Influence of temperature on longevity in Daphnia magna. *Journal of Experimental Zoology, 53*(2): 221-242.

Malik, D.S., Sharma, A.K., Sharma, A.K., Thakur, R. and Sharma, M. (2020). A review on impact of water pollution on freshwater fish species and their aquatic environment. *Advances in Environmental Pollution Management: Wastewater Impacts and Treatment Technologies, 1*, pp.10-28.

Manickavasagam, S., Sudhan, C. and Aanand, S. (2019). Bioindicators in aquatic environment and their significance. *Journal of Aquaculture in the Tropics,34*(1/2), 73-79.

Maqbool, G.S., Bhat, S.J.A, Hanifa G.S. and Haq, S. (2012). Pollution Indicators and their Detection. *Journal of Plant Science and Research, 28*(2): 55-59.

McAllister, D.E., Hamilton, A.L. and Harvey, B. (1997). Global freshwater biodiversity: striving for the integrity of freshwater ecosystems. Sea wind: bulletin of Ocean Voice International; *11*(3).

McGill, B.J., Etienne, R.S., Gray, J.S., Alonso, D., Anderson, M.J., Benecha, H.K., Dornelas, M., Enquist, B.J., Green, J.L., He, F., Hurlbert, A.H., Magurran, A.E., Marquet, P.A., Maurer, B.A., Ostling, A., Soykan, C.U., Ugland, L.I. and White, E. P. (2007). Species abundance distributions: moving beyond single prediction theories to integration within an ecological framework. *Ecology letters, 10*(10): 995-1015.

Mora, C., Tittensor, D.P. and Myers, R.A. (2008). The completeness of taxonomic inventories for describing the global diversity and distribution of marine fishes. Proceedings of the Royal Society B: *Biological Sciences,275*(1631): 149-155.

Mustaq (1990). A study on Cladoceran fauna of River Kaveri, Tiruchirappalli M. Phil., Thesis submitted to Bharathidasan University, Tiruchirappalli 1-65

Naigaga, I., Kaiser, H., Muller, W. J., Ojok, L., Mbabazi, D., Magezi, G. and Muhumuza, E. (2011). Fish as bioindicators in aquatic environmental pollution assessment: a case study in Lake Victoria wetlands, Uganda. *Physics and Chemistry of the Earth,36*(14-15): 918-928.

Neal, C., Whitehead, P., Neale, R. and Cosby, J. (1986). Modelling the effects of acidic deposition and conifer afforestation on stream acidity in the British uplands. *Journal of Hydrology,86*(1-2): 15-26.

Neckles, H.A., Wetzel, R.L., and Orth, R.J. (1993). Relative effects of nutrient enrichment and grazing on epiphyte-macrophyte (Zostera marina L.) dynamics. *Oecologia, 93*(2): 285-295.

Newman, M.C. and McIntosh, A.W. (1989). Appropriateness of aufwuchs as a monitor of bioaccumulation. *Environmental Pollution, 60*(1-2): 83-100.

Nogrady, T., Wallace, R.L. and Snell, T.W. (1993). Rotifera. In: Dumont, editor. Biology, Ecology and Systematics. Volume 4: Guides to the identification of the microinvertebrates of the continental waters of the world. The Netherlands: SPB Academic 142.

Nogueira, M.G., George, D.G. and Jorcin, A. (2003). Estudo do zooplânctonem zonas litorâneaslacustres: um enfoquemetodológico. *Ecótonosnas interfaces dos ecossistemas aquáticos. São Carlos: Rima*, 83-126.

Onaindia, M., Amezaga, I., Garbisu, C. and García-Bikuña, B. (2005). Aquatic macrophytes as biological indicators of environmental conditions of rivers in north-eastern Spain. In *Annales de Limnologie - International Journal of Limnology* (Vol. 41, No. 3: pp. 175-182). EDP Sciences.

Pajević, S.P., Vučković, M.S., Kevrešan, Ž.S., Matavulj, M.N., Radulović, S.B. and Radnović, D.V. (2003). Aquatic macrophytes as indicators of heavy metal pollution of water in DTD canal system. *ZbornikMaticesrpske za prirodnenauke, 104*: 51- 53.

Pal, S., Das, D. and Chakraborty, K., (2015). A comparative study on physico-chemical characteristics and Zooplankton diversities between natural and man-made Wetlands at Cooch Behar, West Bengal, India. *European Journal of Experimental Biology, 5*(5): 85-97.

PAN Pesticides Database. Chemical Toxicity Studies on Aquatic Organisms. Aquatic toxicity. Available from: http://www.pesticideinfo.org/ [Accessed: 2015-08-10]

Parmar, T.K., Rawtani, D., and Agrawal, Y.K. (2016). Bioindicators: the natural indicator of environmental pollution. Frontiers in life science, 9(2): 110-118.

Patrick, R. (1949). A proposed biological measure of stream conditions, based on a survey of the Conestoga Basin, Lancaster County, Pennsylvania. *Proceedings of the Academy of Natural Sciences of Philadelphia, 101*: 277-341.

Paturej, E. and Gutkowska, A. (2015). The effect of salinity levels on the structure of zooplankton communities. *Archives of Biological Sciences*, *67*(2): 483-492.

Penning, W.E., Mjelde, M., Dudley, B., Hellsten, S., Hanganu, J., Kolada, A. and Ecke, F. (2008). Classifying aquatic macrophytes as indicators of eutrophication in European lakes. *Aquatic Ecology*, *42*(2): 237-251.

Pérez-Legaspi, I.A., Quintanar, J. L. and Rico-Martínez, R. (2012). Comparing toxicity endpoints on Lecanequadridentata (Rotifera: Monogononta) exposed to two anticholinesterases pesticides. *Environmental Toxicology*, *27*(9): 518-525.

Petrusek, A., Hobæk, A., Nilssen, J.P., Skage, M., ČErný, M., Brede, N. and Schwenk, K. (2008). A taxonomic reappraisal of the European *Daphnia longispina* complex (Crustacea, Cladocera, Anomopoda). *Zoologica Scripta*, *37*(5): 507-519.

Preston, B. L., Snell, T. W. and Kneisel, R. (1999). UV-B exposure increases acute toxicity of pentachlorophenol and mercury to the rotifer *Brachionus calyciflorus*. *Environmental Pollution*, *106*(1): 23-31.

Punia, P. (1988). Culture of Moinamicrura on various organic waste products. *Journal of the Indian Fisheries Association*, 18: 129-134.

Rai, P.K. (2009). Heavy metal phytoremediation from aquatic ecosystems with special reference to macrophytes. *Critical Reviews in Environmental Science and Technology*, *39*(9): 697-753.

Rameshkumar, S., Radhakrishnan, K., Aanand, S. and Rajaram, R. (2019). Influence of physicochemical water quality on aquatic macrophyte diversity in seasonal wetlands. *Applied Water Science*, *9*(1): 1-8.

Reynolds, C.S. (1984). The ecology of freshwater phytoplankton. Cambridge university press.

Richter, S., Olesen, J. and Wheeler, W.C., (2007). Phylogeny of Branchiopoda (Crustacea) based on a combined analysis of morphological data and six molecular loci. *Cladistics*, *23*: 301-336.

Rottman, R.W., Graves, J.S., Watson, C. and Roy, P.E.Y. (2003). Culture techniques of *Monia*: The ideal *Daphnia* for feeding freshwater fry. CIR 1054/FA024 2-9

Rumes, B., Eggermont, H. and Verschuren, D., 2011. Distribution and faunal richness of Cladocera in western Uganda crater lakes. *Hydrobiologia*, *676*(1): 39-56.

Ryabukhin, Y.S. (1976). Activation analysis of hair as an indicator of contamination of man by environmental trace element pollutants. *International Atomic Energy Agency*.

Santangelo, J.M., Bozelli, R.L., Rocha, A.D.M. and Esteves, F.D.A. (2008). Effects of slight salinity increases on Moinamicrura (Cladocera) populations: field and laboratory observations. *Marine and Freshwater Research*, *59*(9): 808-816.

Sarkar, R., Ghosh, A.R. and Mondal, N.K. (2020). Comparative study on physicochemical status and diversity of macrophytes and zooplanktons of two urban ponds of Chandannagar, WB, India. *Applied Water Science*, *10*(2): 1-8.

Sarma, S.S.S., Nandini, S. and Gulati, R.D. (2005). Life history strategies of cladocerans: comparisons of tropical and temperate taxa. In *Aquatic Biodiversity II* (315-333). Springer, Dordrecht.

Scheffer, M. (1998). Ecology of shallow lakes (Vol. **1**). London: Chapman and Hall.

Seda, J. and Petrusek, A. (2011). Daphnia as a model organism in limnology and aquatic biology: introductory remarks. *Journal of Limnology*, *70*(2): 337-344.

Sharif, A.S.M., Islam, M.S. and Bhuyan, M.S. (2017). Spatio-temporal occurrence and distribution of copepod in the Karnaphuli river estuary, Bangladesh. *Journal of Biodiversity and Environmental Sciences*, *10*(1): 271-282.

Sharma, B.K. and Sharma, S. (2009). Faunal diversity of Cladocera (Crustacea: Branchiopoda) of Loktak Lake (a Ramsar site), Manipur (NE India). *Journal of the Bombay Natural History Society*, *106*(2): 156.

Sharma, K.K. and Chandrakiran, S. (2011). Comparative analysis of Cladoceran communities from three subtropical freshwater ponds of Jammu: Patterns, composition and Diversity. *The Bioscan*, *6*(2): 233-237.

Siciliano, A., Gesuele, R., Pagano, G. and Guida, M., 2015. How Daphnia (Cladocera) assays may be used as bioindicators of health effects? *Journal of Biodiversity & Endangered Species*, 2-6.

Snell, T.W. and Janssen, C.R. (1995). Rotifers in ecotoxicology: a review. *Hydrobiologia*, *313*(1): 231-247.

Snell, T.W. and Janssen, C.R. (1998). Microscale toxicity testing with rotifers. In: Wells, P.G., Lee, K., Blaise, C. Microscale Testing in Aquatic Toxicology, Advances, Techniques and Practice. Boca Raton, Florida: CRC Press. 409–422.

Snell, T.W., Moffat, B.D., Janssen, C. and Persoone, G. (1991). Acute toxicity tests using rotifers: IV. Effects of cyst age, temperature, and salinity on the sensitivity of Brachionus calyciflorus. *Ecotoxicology and environmental safety*, *21*(3): 308-317.

Tawaratmanikul, P., Viputanimat, T., Mewan, A. and Pokasap, K. (1988): Study on the Suitable Moina Density in Nursing the Giant Catfish, Pangasianodon gigas. Technical Paper No. 6/1988, Thailand: Pathumthani Freshwater Fisheries Station, Inland Fisheries Division, Department of Fisheries, Ministry of Agriculture and Cooperatives, 6 p. (in Thai with English abstract).

Tokeshi, M. and Arakaki, S. (2007). On estimating diversity in species-rich fish assemblages: an example with coral-reef fishes of eastern Indonesia. *Journal of Fish Biology*, *71*(6): 1879-1887.

Vaal, M., van der Wal, J.T., Hermens, J. and Hoekstra, J. (1997). Pattern analysis of the variation in the sensitivity of aquatic species to toxicants. *Chemosphere*, *35*(6): 1291-1309.

Vargas-Chacoff, L., Quilodrán, L., Bertrán, C., Arismendi, I., Fierro, P., Tapia, J. and Hauenstein,

E. (2013). Food of rainbow trout and changes in soil use: the Chilean Example. Trout: From Physiology to Conservation. Nova Science Publishers, New York, 65-80.

Von der Ohe, P.C. and Liess, M. (2004). Relative sensitivity distribution of aquatic invertebrates to organic and metal compounds. *Environmental Toxicology and Chemistry: An International Journal*, 23(1): 150-156.

Wallace, R.L. and Snell, T.W. 2010. Rotifera. In *Ecology and classification of North American freshwater invertebrates* (pp. 173-235). Academic Press.

Wallace, R.L., Snell, T.W., Ricci, C. and Thomas, N. (2006). *Rotifera. 1, biology, ecology and systematics*. Backhuys Publishers, *5*: 90-178.

Walseng, B., Yan, N.D. and Schartau, A.K. (2003). Littoral microcrustacean (Cladocera and Copepoda) indicators of acidification in Canadian Shield lakes. *AMBIO: A Journal of the Human Environment*, *32*(3): 208-213.

Walsh, R.G. (1978). Option values, preservation values and recreational benefits of improved water quality: A case study of the South Platte River Basin, Colorado. Environmental Resources Center, Colorado State University.

Yúfera, M. (2001). Studies on Brachionus (Rotifera): an example of interaction between fundamental and applied research. *Hydrobiologia*, *446*(1): 383-392.

Zadorozhnaya, O., Kirsanov, D., Buzhinsky, I., Tsarev, F., Abramova, N., Bratov, A., ... and Legin, A. (2015). Water pollution monitoring by an artificial sensory system performing in terms of *Vibrio fischeri* bacteria. *Sensors and Actuators B: Chemical*, *207*, 1069-1075.

Zaghloul, A., Saber, M., and El-Dewany, C. (2019). Chemical indicators for pollution detection in terrestrial and aquatic ecosystems. *Bulletin of the National Research Centre*, *43*(1): 1-7.

Zamora-Munoz, C. and Alba-Tercedor, J. (1996). Bioassessment of organically polluted Spanish rivers, using a biotic index and multivariate methods. *Journal of the North American Benthological Society*, *15*(3): 332-352.

Zawisza, E., Zawiska, I. and Correa-Metrio, A. (2016). Cladocera community composition as a function of physicochemical and morphological parameters of dystrophic lakes in NE Poland. *Wetlands*, *36*(6): 1131-1142.

Zębek, E. (2004). Species biodiversity of net phytoplankton as an indicator of trophic changes in the urban lake JeziorakMały. *TekaKomisji OchronyiKs ztałtowania Środowiska Przyrodniczego*, *1*: 316-321.

Zhao, Y., Xia, X.H., Yang, Z.F. and Wang, F. (2012). Assessment of water quality in Baiyangdian Lake using multivariate statistical techniques. *Procedia Environmental Sciences*, *13*: 1213-1226.

Zhou, Q., Zhang, J., Fu, J., Shi, J. and Jiang, G., (2008). Biomonitoring: an appealing tool for assessment of metal pollution in the aquatic ecosystem. *Analytica chimica acta*, *606*(2): 135-150.

# Chapter - 2

# Heavy Metal Toxic Site Reclamation Using Interventional Macrophytic Diversity in Lakes

**[1*]Asif I. Bhadarka, [2]Rohini Kalyani, [3]Vishva N. Barad and [1]Vivek Shrivastava**

[1]Department of Aquaculture, College of Fisheries Science Himmatnagar, Kamdhenu University Gandhinagar, Gujrat India
[2]Department of Fish Nutrition and Feed Technology, ICAR -Central Institute of Fisheries Education (CIFE) Mumbai, Maharashtra India
[3]College of Fisheries Science, Veraval Kamdhenu University, Gandhinagar
[*]Corresponding email: asifbhadarka7@gmail.com

## Abstract

Heavy metals like arsenic, copper, cadmium, chromium, nickel, zinc, lead, and mercury are important contaminants of fresh water reservoirs because they are toxic, non-biodegradable, and persistent. Heavy metals are major environmental pollutants that enter the air, water, soil, and biosphere through several environmental channels, including industrial expansion. As they are highly soluble in aquatic environment, heavy metals are readily absorbed by fish and plants. As a result, they could enter the human body through food chain. To reduce the levels of heavy metal in water and wastewater, various methods have been implemented. These technologies include electrochemical, advanced oxidation, ion-exchange, adsorption, chemical precipitation, membrane filtration, and nanotechnology treatments. Recent years have seen the development of phytoremediation as a technology for decontaminating soil, surface and groundwater, air, or other polluted media that is both energy-efficient and environmentally beneficial. When compared to conventional remediation approaches, this technology has been considered the most promising because it causes less site disturbance, is less expensive, and enjoys greater public acceptance. It can be used to clean up and/or stabilize both inorganic and organic contaminants.

**Keywords:** Heavy metals, Acidification, Effluents, Aquatic Pollution, Reclamation

## Introduction

### Heavy metal toxicity in water

Heavy metal is any metallic element that has high density and is toxic even at lower amount. Some of the common heavy metals are mercury, lead, arsenic, chromium, cadmium, thallium, etc. These are also known as trace elements. These heavy metals are necessary to maintain body metabolism in low concentration but are toxic to the body when they are present in higher concentration. Heavy metals enter in the body through food, air, and drinking water. There are different ways for these metals to get deposited into the aquatic environment like various natural and anthropogenic sources, through industrial and domestic sewage, leaching from landfills, storm runoff, shipping, and harbor activities and through atmosphere (Yilmaz, 2003; Marcovecchio, 2004). Metals have a bio accumulative nature. Bioaccumulation means the accumulation of a substance over time and especially of a contaminant (such as a pesticide or heavy metal) in a living organism. Such compounds in living things are stored at a faster rate and are not easy to metabolize or excrete (Baby *et al.*,2011) The excess quantities of heavy metals are detrimental as these destabilize the ecosystem because of their bioaccumulation in organisms and they elicit toxic effects on biota and even death in most living organisms (He *et al.*, 2005). The contaminants of metals remain either insoluble or in suspension form in aquatic systems. These contaminants descend at the bottom or are carried by aquatic organisms (Olaifa *et al.*, 2010). Heavy metal toxicity in aquatic organisms is greatly affected by several factors, including alkalinity, hardness, pH, dissolved oxygen, temperature, and turbidity (Goyal *et al.*, 2020). The pH of an aquatic environment influences heavy metal speciation. It also affects the physiology of aquatic biota, which in turn can modify the susceptibility of these organisms to heavy metals (Gadd *et al.*, 1977). Fishes are broadly used for the evaluation of the well-being of aquatic ecosystems. Pollutants in the food chain are accountable for unfavorable effects and death in aquaculture (Sankhla *et al.*, 2020). Fishes cannot remain free from the harmful effects of these pollutants among other aquatic organisms. Thus growth, health, survival, and development of fishes is affected (Farkas *et al.*, 2001). Heavy metals are potent neurotoxins in fishes. The synergy of heavy metals with biochemical inducements in fish may prevent the communication of fish with their surroundings (Baatrup, 1991). These heavy metals in the aquaculture can cause diseases like Minamata (Organic Mercury Poisoning) and many more. The bioaccumulation of heavy metals may pose great hazards to the health of humans and animals that rely on water bodies (Baby *et al.*, 2011; Sankhla *et al.*, 2016a).

## Sources of heavy metal contamination in aquatic environment

Heavy metals in surface water systems can come from natural or anthropogenic sources. Geological/natural sources include volcanic eruptions, weathering of metal-containing rocks, sea-salt sprays, forest fires, natural weathering processes that can commence to the discharge of metals from their endemic skies to different environmental sections. Hydroxides, oxides, sulfides, sulfates, phosphates, silicates, and organic compounds are some of the forms in which heavy metals can be found out (Sonone *et al.*, 2020).

**a) Volcanic eruptions:** Volcanic ash is the most widely distributed product of explosive volcanic eruptions, and hundreds of kilometers distant areas from an erupting volcano can receive ash falls. Even minor quantities of ash can disrupt the water system. Volcanic ash falls on the water system causing water contamination with turbidity, acidification, and low pH. Surface coating on fresh volcanic ash is highly acidic due to the influence in the plume of aerosols composed of the strong mineral acids $H_2SO_4$ (Sulphuric acid), HCl (Hydrochloric acid) and HF (Hydrogen fluoride). Therefore, when freshly erupted ash comes into contact with water, it has the potential to lower the pH beyond acceptable limits (Stewart *et al.*, 2006).

**b) Industrial effluents:** Municipal waste, domestic sewage, industrial waste directly discharged into the natural water system are some of the major causes of pollution. Untreated waste discharge leads to water contamination. The major source of contamination of surface and groundwater water is the discharge of manufacturing effluents without treatment inside the water bodies (Afzal *et al*, 2018). Wastewater, which comprises of several microorganisms, heavy metals, nutrients, radionuclide, pharmaceutical, and personal care products, all find their way to surface water resources causing irreversible damage to the aquatic ecosystem and to humans as the aesthetic value of such water is compromised. These pollutants decrease the supply of usable water, increase the cost of purifying it, contaminate aquatic resources, and affect food supplies (Edokpayi *et al.*,2017). Pollutants like acid, toxic metal, agrochemicals, dyes, and other untreated waste discharged from industries cause water pollution. Discharged products which often result in pollution, loss of biodiversity in the aquatic ecosystem, and also cause possible health risks like cholera, diarrhea, etc. in humans.

**c) Agriculture activities:** Agricultural systems have expanded and intensified in response to the ever-increasing demand for food. The overuse and misuse of agrochemicals, water, animal feeds, and drugs designed to increase productivity have resulted in higher pollution loads in the environment, including rivers, lakes, aquifers, and coastal waters. Aquatic ecosystems are also affected by agricultural pollution; for example, eutrophication caused by the accumulation of nutrients in lakes and coastal waters have negative impacts on aquatic biodiversity. Farms discharge large quantities of agrochemicals, organic matter, drug residues, sediments, and saline drainage into water bodies. The resultant water pollution poses demonstrated risks to aquatic ecosystems, human health, and productive activities. Agricultural waste released into aquatic habitats has led to several deleterious effects on aquatic organisms, including fish, concentrates pollutants right from polluted water, and lengthily through the food chain. Insecticides, herbicides, and fungicides are applied intensively in agriculture in many countries. When improperly selected and managed, they can pollute water resources with carcinogens and other toxic substances that can affect humans (Schreinemachers *et al.*, 2012)

**d) Mining:** Environmental pollution by heavy metals is very prominent in areas of mining and old mine sites, and this pollution reduces with increasing distance from mining sites. Mining is the extraction of minerals and other geological material from deposits of earth. Mining industry works to extract the metals and minerals that our society needs for agriculture, housing, music, telecommunication, the environmental industry, construction, space exploration, and medicines. Water pollution is a major concern for mining activity. Large amounts of water produced from mine drainage, mine cooling, aqueous extraction, and other mining processes increases the potential for these chemicals to contaminate ground and surface water. Surface water is principally degraded by various factors like an accidental spill of toxic chemicals, erosion of waste material, or discharging of contaminant water from mines.

**e) Electroplating:** Electroplating is a plating method that utilizes electrical flow to decrease cations of substance from a solution and coat a conductive gadget with a thin layer of the material, such as a metal. It is principally employed for dropping a layer of metal below the desired part (e.g., abrasion and wear resistance, corrosion protection, lubricity, aesthetic qualities, etc.) to surface that otherwise lacks those properties. The effluent produced from electroplating industries contaminates air, water, and soil. Electroplating is a major polluting industry because it discharges toxic materials and heavy metals through water, air emissions, and solid waste in an environment.

**f) Electronic waste/E-waste:** The uncontrolled dumping and inappropriate recycling of e-waste pose serious threat to human health and the environment at large. The toxic chemicals that exist in e-waste include a wide range of heavy metals, such as lead (Pb), cadmium (Cd), mercury (Hg), arsenic (As), and nickel (Ni), and also persistent organic compounds, such as brominated flame-retardants (BFRs) and phthalates. Other chemicals that appear in e-waste include polychlorinated biphenyls (PCBs), nonylphenol (NP), and triphenyl phosphate (TPPs), among others. E-wastes are the major known source of heavy metals, hazardous chemicals, and carcinogens. Proper management and disposal of E-waste diseases related to skin, respiratory, intestinal, immune, and endocrine and nervous systems, including cancers, can prevent certain. With a view to bridging the digital divide, there is exponential growth in the use of Electrical and electronic equipment (EEE), and so there is an alarming effect on the environment and human health when the ICT wastes are not disposed of scientifically. There is an emergent need to implement the existing policies and guidelines in line with the international standards and practices for a healthy E-waste management system (Sankhla *et al.,* 2016b).

**g) Biomedical waste:** Medical waste contains body parts, organs, tissues, blood, and body fluids along with soiled linen, cotton, bandage, and plaster casts from infected and contaminated areas along with used needles, syringes, and other sharps. It contains pathogens in mass, in their invisible form. Water pollution can also be caused due to biological, chemical, and radioactive substances. Leaching out of pathogens can pollute the water and may cause diseases. Heavy metals present in chemical waste can also cause water pollution. They can enter into biological systems through the process of biological magnification. Pollution of water bodies can alter parameters such as pH, BOD, DO, COD, etc. and may further become toxic for flora and fauna of the waterbody. The improper disposal of biomedical waste may have a negative impact on the water quality as different pollutants may leach out from the waste dumping sites into the groundwater. Biomedical waste contains a high concentration of heavy metals and polycyclic aromatic hydrocarbons (PAHs), and results in unfavorable amounts of hazardous materials and may pollute surface and groundwater (Manzoor *et al.,* 2019).

**h) Power plants:** Nuclear and fossil fuel plants produce significant thermal pollutants to water bodies. Water pollution is the degradation of water quality due to a change in ambient water temperature. According to the Environmental Protection Agency (EPA), thermos electric power plants alone contribute 50-60% of all toxic pollutants discharged to surface waters by all industrial categories under the Clean Water Act (CWA) (Osann and Hayat, 2014). Among the various types of thermoelectric generating units, coal-fired plants are the biggest source of that toxic pollution. Out of1,100 steam-electric facilities currently operating in the United States, about half are coal-fired power plants (Feeley *et al.,* 2008).Every year, these plants dump millions of tons of toxic heavy metals such as arsenic, selenium, lead, mercury, boron, and cadmium into waterways across the nation. Multiple issues occur concurrently when heated water is released into an aquatic ecosystem. The immediate change is the decrease in dissolved oxygen levels and rise in pH. Warm water cannot hold as much dissolved oxygen as cold-water and organic matter decomposes faster in warmer temperatures. The increase in decomposed aqueous nutrient concentrations causes eutrophication, most commonly realized as algae blooms, which blocks sunlight for underlying aquatic plants. The abundance of algae is an easy food source for aerobic microbes that soar in the population and further deplete the dissolved oxygen. Low oxygen levels create hypoxic dead zones that cannot support most aquatic organisms (Jebakumar *et al.*, 2018; Eaton *et al.*, 1996). In addition to this, rapidly heated water accelerates the metabolism of cold-blooded aquatic animals like fish, causing malnutrition due to insufficient food sources. Since the environment usually becomes more inhospitable to the aquatic fauna, many species leave that area while more vulnerable species may die, changing the biodiversity of both the original and invaded locations. These effects are especially dramatic near coral reefs, the home of over 2 million aquatic species and roughly 25% of all marine life (Moberg and Folke, 1999).

## Impact of heavy metal toxicity on aquatic organisms

Contrary to organic molecules, the majority of metals cannot easily decompose into less harmful forms; this is known as the death of biodegradability. Once introduced into an aquatic environment, metals are either absorbed by aggregation or dispersed throughout the water column (Fremion *et al.*, 2016). The activity and remobilization of metals in the sediments operate as a semi-permanent source of pollution for the natural phenomenon. Aquatic ecosystems, including aquatic flora and fauna, have the capacity to bioaccumulate metal traces from contaminated habitats, which could eventually enter into common human occurrences and result in health problems (Hasan *et al.*, 2016; Varol *et al.*, 2012). Depending on the physical and chemical conditions present between the sediment and concurrently associated water column, metal accumulation in sediments occurs through processes such as precipitation of positive compounds, binding of fine solid particles, association with organic molecules, co-precipitation with metal or manganese oxides, or species delimited as carbonates (Equeenuddin *et al.*, 2013). The percentage of a metal's total concentration that has the capacity to recurrently accumulate in the body gives insight into the metal's bioavailability. The biology of the organism (metals assimilation efficiency, feeding strategies, size or age, generative stage); the natural science of the metal (distribution in water—sediment, suspended matters, and metal speciation) (Griscom and Fisher, 2004); physical and chemical factors are the factors that control the bioavailability of metals (temperature, salinity, pH, ionic strength, the concentration of dissolved organic carbon, total suspended solids) (Fu *et al.*,2014; Bonnail *et al.*, 2016).The bioavailability of metals regulates how much it accumulates in aquatic organisms. For dissolved metals, absorption happens through the receptive stratum, while for particulate metals, uptake happens through ingesting food. The presence of organic or inorganic complexes, pH, temperature, salinity, and reaction conditions are the most significant factors that affect metal toxicity. Intake and uptake is influenced by a number of factors that are related, including feeding rate, transit time, and digesting efficiency. Although there are a few known exceptions, it has been demonstrated in numerous studies that the majority of bioavailable forms of metal such as Cd, Zn, and other elements are the free, hydrated metallic particle (Wojtkowska *et al.*, 2016; Wang and Fisher, 1997; Chapman *et al.*, 1998).As a result, the significance of various chemical types of dissolved metals and complexes made with appropriate organic ligands having low relative molecular masses shouldn't be neglected. It has been discovered that the presence of organic binders can enhance the bioavailability of Cd in fish and mussels by promoting the diffusion of the hydrophobic substance at various locations along the lipid membrane. Additionally, organic molecules of metals are more bioavailable than their ionic forms. The toxicity of mercurial organic compounds is higher than that of corrosive sublimate because they are lipid-soluble and readily penetrate lipid membranes. The concentration of metals in water is generally impacted by the action on suspended particles. Metal uptake into organisms through dietary consumption depends on the association between solid particles and metals. The insoluble metal compounds are accumulated in the suspended solids, but under favorable circumstances, the metal dissolves in the water and passes through the gap. Significant metal concentrations from sediments or suspended solids are abundant in water, thus a low fraction of them is also an absolutely essential for bioaccumulation in organisms and benthic animals (Rosado *et al.*, 2016). Further studies are needed to investigate the accumulation/bioaccumulation pathways supported by dissolved or suspended metal forms since the dynamics of various types of metals at distinct intervals in the aquatic environment are not completely understood. Due to their filter-feeding activity, various studies have shown that the bioavailability of metals in bivalve mollusks relies on soil particle size. The bioavailability of Cd, Zn, and Ag dramatically increases if the particles are coated with live organisms, polymers, or fulvic acids. Overall, the bioavailability of metals from the sediment decreases by metal binding (Chapman *et al.*, 1998; Rahman *et al.*, 2012). Heavy metal breakdown of water supplies is a significant environmental issue that adversely impacts the health of humans, animals, and plants. Freshwater fish exhibit numerous harmful heavy metals that are released into water bodies from diverse sources. Because heavy metal pollution poses harm to fish and health risks to fish consumers, it has become a major global issue.

## Effect of heavy metal on aquaculture

Heavy metals are the main cause of water pollution because they indirectly destroy economically valuable species by tearing apart the biological food chain (Srivastava and Srivastava, 1994). Fish can be exposed to heavy metals in three different ways: through the gills, the body surface, and the digestive system (Romeo *et al.*, 1999). Gills immediately absorb metals from the water, whereas the body surface is typically thought to play a lesser role in this process. By altering the activities of various enzymes and metabolites, concentrations of metal in tissues produce numerous biochemical, physiological, and histological modifications in fish and other freshwater fauna. Fish is one of the most widely spread organism in the aquatic ecosystem, and their susceptibility to metal poisoning may indicate how much metal pollution is having a biological impact on the water. Depending on a variety of factors, such as the fish life-span, developmental stage, and psychical agents, the process of absorption in fish varies significantly among species. Fish living in polluted waterways tend to accumulate heavy metals in their tissues. In general, collection is influenced by factors like metal concentration, length of expression, mode of uptake, ambient conditions (temperature, pH, hardness, salinity), and inherent factors like fish age and feeding habits. Different metals exhibit different magnetizations in fish tissues. The liver, kidney, and gills are the organs where majority of certain metals tend to accumulate. Fish muscles are matched to the other tissues and often have the highest metal concentrations. Fish with significant metal accumulation in several organs may have skeletal lesions and functional issues (Jezierska and Witeska, 2006).

### A. Effect on growth rate and development:

Every organ or part of a fish is known to be affected by heavy metals. Lead, arsenic, cadmium, and chromium are some of these individual metals that have negative effects on aquaculture. Both natural and manmade sources contribute to the production of lead (Pb). Depending on the amounts and duration of Pb exposure, the expression of Pb can have a wide range of effects. Large quantities of Pb prevent fish from being mature when they are fry (Huseen and Mohammed, 2019). Lead and cadmium bind to calmodulin, a calcium sensor protein that affects a variety of physiological processes. It is known that cadmium reduces the activity of several oxidative metabolism catalysts; including lactate dehydrogenase, succinate dehydrogenase, glucose-6-phosphate dehydrogenase, and citrate synthase (Behra1993; Couture and Kumar, 2003; Hilmy *et al.*, 1985; Osman *et al.*, 2007). The toxic effects of heavy metals can affect the individual growth rates, physiological functions, mortality and reproduction in fish. Some fishes have poisonous effect of chromium as echoed in the blood changes such as anemia, eosinophilia and lymphocytosis, bronchial and renal lesions (Amundsen *et al.*, 1997).

### B. Effect on the reproductive system:

Higher concentrations of zinc, lead, cadmium, copper and mercury were found in fish gonads, according to numerous studies on fish like *Noemacheilus barbatulus, Perca fluviatilis*, *Catostomus commersoni, Oreochromis mossambicus,* and *Oreochromis aureus*. This resulted in sperm and egg contamination, which probably affected fish fecundity and embryonic development (Ellenberger *et al.*, 1994; Miller *et al.*, 1992; Pelgrom*et al.*, 1995; Allen, 1995). Sperm motility was noticeably reduced by methylmercury. Both fish metabolism and the morphology of embryos have been disrupted by heavy metals (Perry *et al.*, 1988). Reduced larvae hatched as a result of anatomical and functional abnormalities during embryonic development. Due to lead and copper absorptions, several freshly hatched larvae died shortly after hatching (Jezierska *et al.*, 2008). Mercury altered embryos display (1) cardiovascular issues (underdeveloped heart, ventricle inadequacies, and incorrect heart distribution; in severe cases, the heart was a defectively beating tube unable to pump blood; the embryos also exhibited blood clots and hemorrhages); (2) skeletal anomalies; and (3) cerebral distortions (Weis *et al.*, 1981). In fish, heavy metal sensitivity alters androgenic and estrogenic production and causes pathological differentiation. The reproductive tissues of fish gonads are affected histopathologically by a variety of contaminants, including pesticides, heavy metals, industrial and agricultural waste (Johnson *et al.*, 1991; Lye *et al.*, 1998). These problems could interfere with the maturation of the germ cell and diminish the ability of fish to reproduce.

### C. Effect on gills and respiratory damage:

The principal site for the intake of metals from the water is known to be the gill (Romeo *et al.*, 1999). Zinc is a crucial toxin to fish that damages gill tissue, upsets acid-base and ion homeostasis. It implies a depressing impact on tissue respiration that results in hypoxia and death. Alterations in the heart's and ventilator's anatomy are also triggered by zinc pollution (Olaifa *et al.*, 2010). A harmful effect of chromium on some fish species can be seen in blood alterations including eosinophilia, anaemia, lymphocytosis, renal and bronchial diseases. The minor accumulation of chromium in fish bodies is well recognized, but higher concentrations of chromium damage the gills of fish that are swimming near to the distribution point of chromium.

### D. Neurotoxic effect:

Among all heavy metals, mercury is an important heavy metal for neurotoxicity. Mercury (Hg) has a substantial impact on both fish and people. While fetal and postnatal manifestations of monomethyl Hg have resulted in an abortion, congenital deformity, and variations in the development of small fry, monomethyl Hg causes harm to the brain. Significant neurotoxicity was seen in the Minamata illness and Hg toxicity by methyl Hg (Singh *et al.*, 2023).

### E. Effect on blood and circulatory system:

Red blood cells (RBCs) require nickel in small amounts for assembly, but in large amounts, it becomes gradually poisonous. Fish myocardial disease and an increase in blood pressure have been associated to cadmium. According to studies, mercury causes damage to fish blood vessels, hemorrhages, and a decrease in the number of blood cells. Chromium affects the blood following bronchial, renal, eosinophilic, and lymphocytic lesions (Weis *et al.*, 1981)

### F. Genotoxic effect:

Arsenic inhibits the production of adenosine triphosphate (ATP) during respiration, coagulates protein, and makes associations with coenzymes (Sankhla, 2019). In aggregation, all of its oxidation states may be cancer-causing, and high levels of expression may be fatal (Sankhla *et al.*, 2019). Several studies have suggested that heavy metals may have genotoxic effects in addition to their unknown carcinogenic effects. Chromosome rearrangements, sister chromatid transfer, and the cellular formation of micronuclei are all induced by arsenic composites, which inhibit DNA repair. Cadmium, arsenic, nickel, and chromium are carcinogenic metals that have been associated to DNA damage by base-pair deletion, mutation, or radical oxygen attack on DNA (Landolph, 1994)

### G. Physiological damage:

In fish and crustaceans, a number of histological changes, such as gill mortification or fatty liver degeneration, frequently take place (Sevcikova *et al.*, 2016; Mishra *et al.*, 2008). Malformations in fish are caused by cadmium, nickel, mercury, chromium, lead, and arsenic. These heavy metals cause numerous physiological changes when they are found in high concentrations. The changes frequently noted include shortening of fins, underdevelopment of gills, liver, and fin functions in fingerlings. Heavy metals including lead, arsenic, cadmium, and zinc also cause stress in fish, in addition to reproduction, hypoxic conditions, inappropriate stocking, and malnutrition. Pollution and other stressors have an adverse effect on growth, development, and reproduction via alternating metabolic, physiological, and biochemical processes (Heath, 2018).

### H. Effect on lifespan and mortality:

Studies on fish have shown that all heavy metals, despite the fact that some of them are necessary for existence, have negative consequences on living things due to metabolic interference and mutagenesis. These negative outcomes include a loss of fitness, interference with reproduction that causes cancer, and eventually death. The harmful effects of heavy metals have the greatest impact on fish's physiological capabilities, mortality, reproduction, individual growth rates, and reproductive success. There have been unfavorable effects found on the physiological processes and biochemical parameters in the blood and tissue of fish living in metal-polluted waters. It has been reported that fish exposed to metals displayed immune system faults, making them more susceptible to infectious diseases and having a higher death rate.

## Methods of cleaning up heavy metal in water

Researchers have been using various techniques to remove heavy metals from various wastewaters for a long time, that can be classified as electrochemical treatments (electrocoagulation, electro-flotation, and electrodeposition), physicochemical processes (chemical precipitation, Ion exchange), adsorption (activated carbon, carbon nanotubes, and wood sawdust adsorbents), or current methods (membrane filtration processes, photo catalysis processes, and nanotechnology). This section describes these techniques, their benefits, as well as their drawbacks when used in industrial settings.

### A. Electrochemical Treatments

Due to the high initial costs and expensive electricity supply, electrochemical wastewater treatment has not received much attention yet (Holt *et al.*, 2005). The electrochemical technologies are in a condition where they are not only more cost-effective than other technologies, but also more effective and compact. In certain situations, treating wastewaters containing refractory contaminants may need electrochemical technology as a crucial step that cannot be avoided (Khandegar and Saroha, 2013).

» *Electrocoagulation*: In the wastewater treatment industry, electrocoagulation (EC) is a simple and effective process (Emamjomeh and Sivakumar, 2009). Due to its inadequate systematic reactor design and issues with electrode dependability, it was never recognized as a reliable approach (Holt *et al.*, 2005). Recently, EC was recognized as a small-scale wastewater treatment technique with more advanced technical approaches (Khandegar and Saroha, 2013). Electrolysis can eliminate a wide range of irritating contaminants (Gomes *et al.*, 2007). The first American patent for EC using electrodes made of aluminum and iron was issued in 1909 (Mollah *et al.*, 2004). A little electric current was used during the entire EC procedure to operate on the waste solution.

In its simplest form, the EC reactor is an electrolytic cell with a single anode and single cathode, also known as sacrificial electrodes (Emamjomeh and Sivakumar, 2009; Mollah *et al.*, 2004). The materials used to create these electrodes could be same or different. The most common materials are iron and aluminum (Gomes *et al.*, 2007). Heavy metal ions and other contaminants are initially held in the wastewater by their surface electrical charges or hydrogen bonds. The electrostatic charges holding the suspended or emulsified contaminants together are neutralized after the electric field is introduced, causing them to coagulate from the aqueous phase. To create a mass called as sludge or floc (Gomes *et al.*, 2007; Mollah *et al.*, 2004), heavy metal particles join together in a way similar to that of tiny magnets. When compared to other approaches, EC produces flocs that are larger, more stable, and contain less bound water. The benefit is that these flocs are easily eliminated by additional processing, such as filtering, which is occasionally expensive but simple to use (Khandegar and Saroha, 2013). The procedure can be summarized as follows:

- Anode dissolution;
- OH– and $H_2$ generation at the cathode;
- Electrolytic reactions at electrode surfaces;
- Coagulant adsorption on colloidal pollutants;
- Colloids removal by sedimentation or flotation.

EC is regarded as an environmentally friendly process, because it doesn't require new chemical additions, uses clean electrons as reagents and the requirement for chemical additions is reduced, which greatly reduces the potential for the production of additional pollutants and flocs. Although EC cannot remove particles that are indefinitely soluble, it does have the advantage of being able to remove very small colloidal particles since the applied electric field accelerates them.

» *Electro-Flotation*: Heavy metal contaminants in wastewater have recently been removed using electro flotation (EF) (Merzouk *et al.*, 2009; Srinivasan and Subbaiyan, 1989; Oussedik and Khelifa, 2001; Alexandrova *et al.*, 1994;Casqueira*et al.*, 2006; Gao *et al.*, 2005), since very diluted solutions typically do not respond well to other methods for treating wastewaters (concentrations below 50 mgm$^{-3}$) (Da Mota *et al.*, 2015). EF quickly gained popularity because of its adaptability, ease of design and operation, compatibility with the environment, cheap operating costs, and tiny and compact units (Zodi *et al.*, 2009).

Due to its many features, EF can be used in a broad range of industries, such as the treatment of oil from oil-water emulsions (Hosny, 1996; Hosny, 1992; Mostefa and Tir, 2004), disinfection of groundwater, treatment of apple juice and food processing effluents (Araya-Farias *et al.*, 2008), removal of colloidal and suspended particles (Fukui and Yuu, 1985; Sandbank *et al.*, 1974), and treatment of septic tank and pit waters (Kolesnikov *et al.*, 2015; Costaz *et al.*, 1983).

Pollutants are separated by EF by floating to the top of the liquid phase. Three fundamental components account for the separation. Pollutants are initially drawn to a cell or reactor with two electrodes and an electrical source (Chen, 2004). Water electrolysis will be the overall process taking place in the cell, releasing oxygen and hydrogen into the solution. Heavy metals will bind to or adsorb on oxygen and hydrogen molecules, destabilizing the emulsified particles and causing flocs to develop (Da Mota *et al.*, 2015; Hosny, 1996). The separation of created foam and settled flocs for settling or flotation is the second phase. The final phase involves filtering out the contaminants that have been gathered (Zodi *et al.*, 2009). Pollutant removal efficiency and power consumption are used to assess the EF cell's performance, with the former being influenced by the size of the bubbles produced during electrolysis and the latter by the cell's design, the materials used for the electrodes, and the operating conditions (Chen, 2004). The main determinants of operating conditions are current density, pH, and electrode type.

Some researchers combined EC and EF to address the drawbacks of the aforementioned methods, which resulted in better removal efficiency than utilizing each one alone. The combined approach for removing pollutants is more secure and efficient (Khandegar and Saroha, 2013). It is known as electrocoagulation- flotation when EC and EF are combined (ECF). This approach yields higher removal percentages.

» *Electrodeposition*: Among electrochemical techniques for recovering heavy metals, electrodeposition is a useful and effective technique (Oztekin and Yazicigil, 2006; Scott and Paton, 1993). Electrodeposition is favorable because it requires no additional chemicals, produces no sludge, is extremely selective, and is inexpensive (Agarwal *et al.*, 1984). In order to prevent corrosion, it converts dissolved metal ions into solid particles by depositing them on ionic conductors (the cathode and the anode). A cell with one anode, one cathode, one electrolyte cell, and a current source is used in electrodeposition (Chen, 2004), a one-step clean process, to reduce and oxidize heavy metal ions. Onto the cathode, heavy metals are reduced and electroplated. The formation and growth of electrodeposits determine their final size and dispersion. In order to prevent interference with the process of recovering heavy metals, it is preferable for the anodes to be insoluble or inert.

The initial waste solution concentration, physical operational parameters like temperature and pH, and the presence of complexing and chelating agents all have an impact on efficiency of the whole process (Scott and Paton, 1993). The main reason electrodeposition is employed is that it can be used with non-aqueous solutions or solutions that also contain chelating agents. Both of these solutions remove pollutants more effectively than aqueous solutions. Traditional aqueous solutions have issues with hydrogen gas molecule liberation, poor heat stability, and a small electrochemical window. Therefore, non-aqueous solutions are occasionally preferred. The commercialization of non-aqueous solutions is hampered by a few technological challenges, including as low current efficiency, heat balance in the cell, and cell component corrosion (Simka *et al.*, 2009) Another interesting case is the electrodeposition of aqueous solutions containing chelating agents like EDTA, NTA, and citrate. These chelating compounds are beneficial because, by forming stronger complexes with metal ions, they minimize the development of insoluble salts and boost recovery efficiency. There will therefore be no need for traditional separating techniques (Oztekin and Yazicigil, 2006).

**B. Physico-Chemical Processes**

» *Chemical Precipitation:* Chemical precipitation is an automated, simple treatment approach. The removal of heavy metals from wastewaters is a common usage of this treatment (Ku and Jung, 2001). Chemical precipitation frequently fails to meet the permissible level for discharge since it needs a lot of chemicals to do so (Jüttner *et al.*, 2000; Özverdi and Erdem, 2006). These chemicals will contribute significantly to further pollution. In chemical precipitation processes, heavy metal ions react with chemical precipitant agents turning them into solid, insoluble particles (Wang *et al.*, 2009). Sedimentation or filtering will be used to separate the solid phase from the solution (Zamboulis *et al.*, 2004; Matis *et al.*, 2004).

It may be preferred to chemical precipitation because EC doesn't require the addition of chemicals, results in higher efficiencies, generates less sludge, and removes species that chemical precipitation is unable to eliminate (Khandegar and Saroha, 2013). In this process, pH modification is essential; basic conditions with a pH of 11 are favored to maximize heavy metal removal (Kurniawan *et al.*, 2006; Fu and Wang, 2011).The soluble metal ions are converted to the dissolved solid through reaction with a precipitant agent after pH is quite well. Some substances must first undergo reduction or oxidation to valence before being precipitated. Calcium chloride can be used to separate fluorine from phosphate, and phosphate can be separated by precipitation into aluminum or iron salts. In order to create dissolved precipitates that can be easily separated by sedimentation, the approach alters the ionic equilibrium (Fu and Wang, 2011).

*i.* *Hydroxide Precipitation*: In this process the addition of coagulants such as iron salts, alum, and certain polymers may enhance the removal of heavy metals from wastewater. By filtering or sedimentation, a soluble metal can become hydroxide and then precipitate. The pH of the water can be raised by adding alkaline substances. Metal ions become less soluble in alkaline solutions and thus precipitate out of the solvent.

*ii.* *Sulfide Precipitation:* Sulfide precipitation works on a similar principle as hydroxide precipitation. Metal ions can be precipitated through the addition of precipitating agents like sodium sulfide, sodium hydrosulfide, ferrous sulfide etc. Sulfide is employed to precipitate the heavy metal ions as metal sulfides, and the resulting sludge is eliminated from the solution by gravity settling or filtration. Due to the toxicity of the sulfide ion and $H_2S$, the sulfide precipitation process requires pre- and post-treatment as well as accurate control of reagent additions. The precipitation procedures are carried out by adjusting compositions and other conditions so that the ionic constituents that are to be separated change from a soluble ionic to a solid phase (Wang *et al.*, 2009).Despite, the method has many benefits, it is impractical due to its high cost of sludge formation, high sludge disposal costs, poor settling, and slow metal precipitation (Wingenfelder *et al.*, 2005; Yang *et al.*, 2001).

» *Ion Exchange*: The principle behind ion exchange treatment is the reversible exchange of ions between the solid and liquid phases (Kurniawan *et al.*, 2006).

> The process is started with ion-exchange reactions, followed by the physical absorption of the heavy metal ions and the formation of a complex between the counter ion and functional group. In the end, hydration takes place at the solution surface or in the pores of the adsorbent (Ferreira *et al.,* 1999). The ion-exchange process is influenced by a number of factors, including pH, anions, temperature, the starting concentration of the sorbate and adsorbent, and contact time (Kang *et al.*, 2004; Gode and Pehlivan, 2006). In this process, a resin removes ions from an electrolytic solution and then releases chemically comparable amounts of other ions with a similar charge (Rengaraj *et al.,* 2001).Ion exchange can also be used to recover valuable heavy metals from inorganic effluents (DąBrowski *et al.*,2004). After separating the loaded resin, the metal is recovered in a more concentrated form by elution with the appropriate reagents. Sulfonic acid is a component of resins with acidic functional groups; hence it may be assumed that physicochemical interactions may occur when metal ions are captured.

Ion exchange serves a very specific role in wastewater treatment and heavy metal removal (Gode and Pehlivan, 2003; Miller, 2000). Heavy metals can be effectively removed by using an ion-exchange cell with strong acidic cation exchangers (Kang *et al.*, 2004). Due to its great effectiveness, low cost, smaller amount of sludge, ability to recover metals, high selectivity, and value, it is typically chosen to alternative methods like chemical precipitation (Rengaraj *et al.,* 2001; Lacour *et al.*, 2001; Lin *et al.*, 2000; Chiarle *et al.,* 2000).

### C. Adsorption

Adsorption, which has been successful in lowering metal content, is another physicochemical technique for removing heavy metals from wastewaters. Mass transfer between the liquid phase and the solid phase, also known as the adsorbent, is the basis of adsorption (Babel and Kurniawan, 2004).There are three main steps in the sorption of a pollutant onto an adsorbent: (i) the penetration of the pollutant from the bulk solution to the adsorbent surface; (ii) the adsorption of the pollutant on the adsorbent surface; and (iii) the penetration in the adsorbent structure (Barakat, 2011).These adsorbents have a high adsorption capacity and large surface areas. The process is reversible, which is advantageous, as desorption will replenish the adsorbents. Agriculture waste, industrial byproducts, or natural resources can all be used to make adsorbents. Carbon nanotubes (CNT), sawdust, and activated carbon (AC) are some of the most popular adsorbents in this process. Below, we go into further detail about each of these adsorbents.

### 1. Activated carbon

Activated carbons (AC) are often made from agricultural by-products, When $K_2CO_3$ is employed greatest surface area lies between 126 to 3256 $m^2g^{-1}$ (Tsai *et al.*,2001; Hayashi *et al.*,2002; Okada*etal.*,2003). The removal of toxic metals using activated carbon adsorption is a common practice that has received substantial study (Abdel-Halim *et al.*, 2003; Kadirvelu *et al.*, 2000).

### 2. Carbon nanotubes

Carbon nanotubes (CNTs) are renowned for their exceptional qualities and applications. The mechanism through which heavy metals sorb to CNT is quite complex (Iijima, 1991). According to several studies, CNTs offer a tremendous potential for heavy metal removal. CNT simmobilized by calcium alginate (CNT/CA) reduce the risk associated with CNT discharge into the aquatic environment (Kabbashi *et al.*, 2009; Kuoand Lin, 2009; Kandah and Meunier, 2007; Pillay *et al.*, 2009).

### 3. Wood Sawdust

Wood sawdust, a solid waste product produced during mechanical wood processing, can be employed as a low-cost adsorbent for heavy metals because of its lingo-cellulosic makeup. It is mostly made up of cellulose (45–50%) and lignin (23–30%), both of which have the ability to bind metal cations since these structural components contain hydroxyl, carboxylic, and phenolic groups (Pillay *et al.*, 2009). Recently, there has been an increase in interest in removing heavy metals from solutions by binding them with agricultural products such as sawdust, waste wool, nut waste, tree bark, and modified cotton (Yu *et al.*, 2000).

## D. Current methods

### i. Membrane Filtration Process:

These procedures were created and employed because they have higher removal efficiencies, no pollutant loads, and usually lower energy requirements than conventional methods (Marjani *et al.*, 2012; Rezakazemi *et al.*, 2012; Rezakazemi *et al.*, 2004; Rezakazemi *et al.*, 2015). Because of the simple separation technique, membrane treatments are used to treat water and wastewater (Padaki *et al.*, 2015; Murthy and Chaudhari, 2009; Razavi *et al.*, 2016). With a few slight variations, various membrane processes use this same method. In all processes, rejection is not accomplished by allowing fluid to pass through the physical pores of the membrane; rather, a higher transmembrane pressure causes the separation to take place through semi-permeable membranes with different pore sizes that make contact between desired phases and regulate the passage of particles by allowing solvent to escape and catching the solute (Sutherland, 2008; van der Bruggen and Vandecasteele, 2003). These separation methods employ membranes, which come in a wide variety of types, fabrications, and structural configurations.

Reverse osmosis (RO), ultrafiltration (UF), microfiltration (MF), nanofiltration (NF), and electrodialysis (ED) are the five main treatment methods used in membrane filtration technology. Although these processes are basically the same, there are some changes in the membrane permeability, operating pressures, and pore structure (pore size, pore size distribution, and porosity) (Murthy and Chaudhari, 2009).

**ii. Nanofiltration**

Nanofiltration is a relatively recent technology. NF was created to address the operational limitations of traditional approaches and enable the separation of large molecules through extremely small pores (Sutherland, 2008). Fabrication and modification of the membranes have an effect on the solvent permeability, making them important factors in NF processes (Mohammad *et al.*, 2015). NF is employed in energy- and environmental-friendly processes for treating wastewater, surface water, and ground water (Jianquanand Yinhua, 2013). The entire procedure, which separates a variety of organic (uncharged) and inorganic particles from solutions, is known as a liquid phase filtration (van der Bruggen and Vandecasteele, 2003).

In order to remove heavy metals using NF methods, there are three basic phases. The first is a pre- treatment. In order to avoid further potential fouling, feed water must be treated before entering the system. The main membrane separation procedure comes in the second part. The widespread rejection of NF membranes is caused by a confluence of Donnan, dielectric, and steric hindrance effects (Mohammad *et al.*, 2015). The steric hindrance mechanism aids in the removal of neutral solutes. Due to the existing Donnan potential and ion size, charged components are rejected (Fane *et al.*, 2011). Post-treatment is the third and last step (Maher *et al.*, 2014; Ng *et al.*, 2013).

When treated with NF aqueous solutions containing lead in a very wide concentration range, toxic heavy metals can be efficiently removed employing AFC-80 membranes made from 1,2-benzisothiazol-3(2H)-one, sodium salt. They are available in the trade names like AFC40, AFC80 and AFC99. The membranes, AFC40 is a nanofiltration membrane while AFC99 and AFC80 used as Reverse Osmosis membranes. Strongly harmful heavy metals like lead and cadmium can be recovered with AFC80 membranes, and wastewater can also be recovered. Even for concentrated feed solutions containing 250mgPbL$^{-1}$, the lead rejection was found to be extremely high (> 98%) (Gherasim and Mikulášek, 2013)

### iii. Reverse Osmosis

Although RO was initially investigated in the 1920s, but it became practical after 30 years. Today, RO is a fundamental separation technology used to treat a variety of wastewaters (Subramani and Jacangelo, 2014; Bódalo-Santoyo *et al.*, 2003; Gagliardo *et al.*, 1998). RO operates under hydraulic working pressure (Malaeb and Ayoub, 2011). Since the semi-permeable membrane used by RO does not have any clear pores through which the particles can pass, the transformation must take place by diffusion (Greenlee *et al.*, 2009). Three fundamental concepts are used to separate solutes in water through a RO membrane. First, concentrated solution of heavy metal ions' water absorbs onto the membrane surface. Diffusion occurs through the membrane in the second stage as a result of the concentration gradient. To reach the permeate side of the membrane, the water molecules down the gradient. A nearly pure solution is created when water molecules desorb from the semi-permeable membrane. A concentrated heavy metal solution will be present on the feed side following separation, while a treated solution will be present on the permeate side (Shenvi *et al.*, 2015). According to environmental standards, it is necessary to remove and purify both the contaminants and the concentrates (Joo and Tansel, 2015). Size, charge exclusion, and the physico-chemical interactions that take place between the solute, solvent, and membrane all have an impact on separation efficiency (Malaeb *et al.*, 2011). Nowadays, integrated membrane bioreactors (MBR) and RO systems are frequently being employed to remove emerging contaminants from wastewater (Joo and Tansel, 2015).

### iv. Microfiltration and Ultrafiltration

Due to their similarities, MF and UF are discussed together. Both procedures are pressure-driven, have a similar usage area, and essentially the same separation process. The primary difference between these two methods is that MF rejects larger solutes than UF (Basile and Nunes, 2011). UF is beneficial in rejecting species like dissolved macromolecules, colloidal materials, and sometimes suspended species (Shenvi *et al.*, 2015), while MF is used to separate suspended particles of effluent (Basile and Nunes, 2011). Both have several applications in a wide range of sectors, including the filtration of water to create drinkable water, the clarity of various solutions, and the elimination of unwanted particles. The treatment of wastewaters to meet discharge regulations is the most important application (Shi *et al.*, 2014). In terms of MF and UF procedures, selecting a suitable membrane is the most challenging step because it might affect the efficiency of removal (Basile and Nunes, 2011). The porosity, structure, and composition of the membranes vary, hydrophilic and hydrophobic membranes, as well as crystalline polymers, have been widely used as materials of MF and UF membranes (Basile and Nunes, 2011).

The various filtration techniques now in use are covered before the separation mechanism is explained. Dead-end and cross-flow filtration techniques are two different categories. When feed flows perpendicular to the membrane surface and the feed travels tangentially over the membrane, the dead-end mode is employed. In the dead-end mode, there is only a feed stream and a permeate stream. Cross-flow mode has higher flux rates and a longer membrane lifetime than dead-end mode, but it requires more complicated utilities (Basile and Nunes, 2011). In MF and UF procedures, the separation technique is typically based on molecular sieving with porous membranes (Basile and Nunes, 2011). Due to pore size restrictions, particles that are larger than the pore size are rejected. The pore size was reduced as a result of the particles being absorbed onto the membrane surface, making the separated particles considerably smaller than the pore size.

### v. Electrodialysis

Studies on desalination led to the development of electrodialysis (ED) in the 1950s (Moon *et al.*, 2014) which is an innovative liquid hybrid membrane process (Sadyrbaeva, 2015). It is used to cleanse wastewater from various industries (Basile *et al.*, 2015a), produce acid and base from salts by water splitting (Moon *et al.*, 2014), and perform fractionation processes like wine stabilization and whey demineralization, etc. Ion-selective exchange membranes exist in ED (IEMs). Cations and anions are not simultaneously transported by IEMs. These membranes come in two varieties: cation exchange membranes (CEM) and anion exchange membranes (AEM) (Basile *et al.*, 2015b; Mulyati *et al.*, 2013). Electrical potential or concentration gradients cause ions to move through these membranes (Moon *et al.*, 2014). Electrodialysis reversal (EDR), which is sometimes more effective than ED, is the technique of reversing the polarity of ED electrodes. Scaling and fouling are reduced, and recovery rates are increased, using EDR (Basile *et al.*, 2015b; Valero and Arbòs, 2010). On the other side, EDR requires more electrical and plumbing controls (Basile *et al.*, 2015b).

### vi. Photocatalysis Process

Advanced oxidation processes (AOPs), are a recent development in the air and water purification industries (Choi, 2006; Litter, 1999; Herrmann *et al.*, 1999; Herrmann *et al.*, 1993). One well-known AOP on which many researchers have focused is photocatalysis (Herrmann *et al.*, 1993; Ohtani, 2010). Instead of chemical substances, this approach makes use of non-toxic semiconductors to capture light of the proper wavelength (Basile *et al.*, 2015b; Robertson, 1996). It is also favored over chemical procedures because harmful chemicals aren't used as catalysts (Basile *et al.*, 2015b; Robertson, 1996; Gazi and Ananthakrishnan, 2011; Jiang *et al.*, 2012). It has gained popularity as an effective method, because of the photocatalysis process, simple design, inexpensive operation, high stability, and high removal efficiency (Jiang *et al.*, 2012; Chong *et al.*, 2010).

In addition, it is effective in removing pollutants like heavy metals to protect the environment (Khalil *et al.*, 2002; Botta *et al.*, 1999; Kim and Choi, 2002). A five-step process is carried out when a solution containing pollutants is added to a photocatalytic system. The contaminants first move from the aqueous phase to the surface. The semiconductor surface then takes them up and absorbs them. Photocatalytic processes taking place in the absorbed phase make up the third stage. The products are then broken down and taken out of the interface region. (Herrmann *et al.*, 1993; Ohtani, 2010; Basile *et al.*, 2015b; Robertson, 1996; Gazi and Ananthakrishnan, 2011). The third step of the treatment system, where the photocatalysis activities take place, is the most important one as light is used throughout the procedure as an activator. Thus, the thermal activation of old processes is neglected (Herrmann *et al.*, 1993).

The majority of the semiconductors employed in photocatalysis are chalcogenides, which include oxides like $TiO_2$, ZnO, $ZnO_2$, $CeO_2$, $WO_2$, and sulphides like CdS, ZnS, $WS_2$, etc.(Herrmann *et al.*, 2005). Among these substances, $TiO_2$ is an experimentally preferred photocatalyst because of its environment friendly features and characteristics, including its ability to potentially oxidize any organic and inorganic particle, stability against photo-corrosion, chemical inertness, and affordability (Fujishima and Honda, 1972; Chin *et al.*, 2011).

Although photocatalysis technologies have developed, they still have certain drawbacks, such as the inability to absorb visible light and the recombination of electrons or holes (Molinari *et al.*, 2009; Ni *et al.*, 2007).

## Heavy metal remediation from aquatic ecosystem using macrophytes:

1. ***Eichhornia crassipes* (Water Hyacinth):** It is the one of the most prevalent macrophyte which has the large capacity to absorb metal ions particularly arsenic (As) than some other submerged macrophyte, because of this characteristic it is also used in the sewage treatment (Mishra and Mohanty,

2008). Though it is problematic weed as it grows year around but it is effective in absorbing Pb, Zn, Mn, Cu, Ni by root system. It could be the finest choice for removing heavy metals.

2. ***Azolla caroliniana* (Mosquito Fern):** It is a potential biological filter for removing Cd from waste water. It also has the capacity to remove contaminants from the sewage water (Rai and Tripathi, 2009)

3. ***Brassica juncea* (Mustard Green):** Due to the obviously increased biomass, it also has a higher Zn extraction capability. It seems to be effective in bio-remediation of soil and absorbs Cd(Szczygłowska*et al.,* 2011)

4. ***Pistia stratiotes* (Water Lettuce):** It grows rapidly with huge biomass and because of its large rhizosphere; it has higher removal of heavy metal ions. It is one of the worthy options for eliminating Pb and Cd from untreated water.

5. **Lemnoideae (Duckweeds):** It is a surface floating macrophyte which is mostly observed in pools and marshes. *Lemna minor* may absorb approximately 90% of Pb from waste water (Caicedo, 2000). High concentration of ammonia and nitrate may inhibit or decrease the growth rate of the duckweed.

6. ***Hydrilla verticillata* (Hydrilla):** Its entire plant body can remove pollutants and helps in absorbing approximately 98% of the lead (Singh *et al.,* 2011).

7. ***Spirodela intermedia* (Intermediate Duckweed):** It has rapid development capacity under variety of environmental conditions and absorbs Cd, Pb, Cr from the effluent and wastewater.

8. ***Schoenoplectus californicus* (California Bulrush):** It is commonly found all over the world. It can absorb metals from the sediment via roots. It can withstand very high heavy metals concentrations in the water bodies.

**Table 1: Bioaccumulation potentiality of various aquatic macrophyte**

| S.No | Aquatic Macrophyte | Common Name | Metals/Metalloids |
|---|---|---|---|
| 1 | *Ceratophyllum demersum* | Rigid hornwort | Cd, As, Cr |
| 2 | *Eichhornia crassipes* | Water hyacinth | Hg, Cu, Ni, Zn |
| 3 | *Lemna minor* | Common duckweed | Cr, Cu, Ni, Pb |
| 4 | *Mentha aquatica* | Water mint | Pb, Fe, Cr |
| 5 | *Myriophyllum spicatum* | Spiked watermilfoil | Cu, Pb, Cd |

| 6 | *Nasturium officinate* | Watercress | Ni, Cr, Zn |
|---|---|---|---|
| 7 | *Phrangmites austrails* | Common reed | Pb, Cr, Cu, Zn |
| 8 | *Pistia Stratiotes* | Water cabbage | Mn, Cr, Pb, Fe |
| 9 | *Potamogeton crispus* | Curled Pondweed | Cu, Pb, Zn |
| 10 | *Potamogeton pectinatus* | Sago pondweed | Ni, Cu, Mn |
| 11 | *Salvania herzogii* | Water moss | As, Cd, Cr |
| 12 | *Salvania minima* | Water spangles | Cr, Cd |
| 13 | *Scirpus sp.* | Bulrush | Al, Cd, Fe |
| 14 | *Spartina alterniflora* | Smooth cordgrass | Zn, As, Pb, Mn |
| 15 | *Spirodela intermedia* | Intermediate duckweed | Cr, Pb, Zn |
| 16 | *Typha latifoiia* | Broadleaf cattail | Ni, Zn,Mn |
| 17 | *Valliseneria spiralits* | Eel grass | Ar |

(Saha *et al.*, 2022)

**Illustration of using different macrophytes for this remediation:**

In an environmentally responsible manner, revegetating heavy metal-polluted soil with plants is possible through phytoremediation. It has been determined that more than 400 plant species have the ability to improve the soil and water. *Thlaspi, Brassica, Sedum alfredii H., and Arabidopsis* species have received the greatest attention among them. Phyto-remediation employs many methods such as Phyto-extraction, Phyto-stabilization, Phyto-volatilization, Rhizo-filtration or Rhizo-degradation, Phyto-transformation and Phyto-filtration during heavy metal absorption or aggregation in the plants.

1. **Phyto-extraction:** Commonly known as phytoaccumulation, is the process by which heavy metals are absorbed in root system and subsequently translocated to an aerial component of a plants such as buds, leaves, etc. After this process, plants may be gathered and burnt to produce energy as well as, if necessary, recycle metals from the ashes. It is an effective phytoremediation technology.

2. **Phyto-stabilization:** This process helps plants to limit the migration of pollutants in soil. It is often referred to as in-situ inactivation of contaminants. It has no negative effect on ecosystem. It is highly beneficial for the area which is highly contaminated. This is not a long-term solution since only metal transportation is controlled, but metal remain inside the ground.

3. **Rhizo-filtration:** In this process roots have a crucial role to play and plants absorb or adsorb pollutants. When plants have absorbed all toxins, they may be readily collected and discarded. Plants must have rhizomes so as to accumulate large amounts of heavy metals. It is utilized effectively to purify wastewater discharges, industrial emissions, nuclear contaminants and minerals.

4. **Phyto-volatization:** Through this process plant transforms contaminants in a more volatile form and subsequently releases them further into its surroundings via the plant's pores. One of the advantages is that this doesn't need any extra maintenance after cultivation is completed.

5. **Phyto-transformation:** The absorption of pollutants from the ground and water is referred to as phyto-transformation. It is also known as phyto-degradation. It is mostly suitable for the industrial areas.

6. **Phyto-stimulation:** It is also known as improved rhizosphere biodegradation. Here, biological decomposition of organic pollutants in the ground by increased soil microorganisms in the root zone of the plants or rhizosphere.

## Recent Trend of Interventional Usage of Macrophytes for Heavy Metal Remediation:

a. **Chemical assisted Phyto-remediation:** Different chemicals have proven to be an effective method for increasing heavy metal bioavailability in plants. Some chelating compounds, such as ethylene glycol tetra-acetic acid (EGTA) and diethylene triamine penta acetic acid (DTPA) have shown to be effective chelating agents in increasing heavy metal bioavailability and phyto-remediation. But the use of various chemicals might have disadvantage like causing toxicity, leach into underground water and disrupt heavy metal absorption in plants.

b. **Microorganism Assisted Phyto-remediation:** Microorganisms play an important role in heavy metal removal from soil and water. Mycorrhizal fungi and endophytic microbes in soil layers enhance vegetation growth and increase phytoremediation.

c. **Genetically modified plants:** These plants have genes which are modified to level up the metabolic activity, intensification, and absorption of chemical contaminants. Transgenic plants are seen to be effective at treating polyphenols, carbon tetrachloride and exploding pollutants. But they accumulate contaminants instead of absorbing them.

d. **Biomass from non-living plants:** Non-living plant biomass is used fruitfully for metal accumulation and restoration. The use of dehydrated and decaying vegetation to eliminate metals from moisture has attracted increasing attention in recent years since it is simple to handle as well as cost-effective general method (Han *et al.,* 2018; Kaewsarn, 2002).

**Heavy Metals Assimilation by Native and Non-Native Aquatic Macrophyte Species: A Case Study of a River in the Eastern Cape Province of South Africa.**

Tshithukhe *et al.,* 2021 conducted study on the water and sediment quality at 10 sites along the severely modified Swartkops River system in the Eastern Cape Province of South Africa and then quantified the phytoremediation potential by native and non-native macrophyte species over a period of 6 months. They hypothesized that the presence of semi and permanent native and non-native macrophytes mats would reduce water and sediment contamination through assimilation downriver. Results were variable and, thus, inconsistent with hypothesis; there were no clear trends in water and sediment quality improvement along the Swartkops River. Although variable, the free-floating non-native macrophyte, *Pontederiacrassipes* (Eichhornia) recorded the highest assimilation potential of heavy metals in water (e.g., Fe and Cu) and sediments (e.g., Fe and Zn), followed by a submerged native macrophyte, S*tuckenia pectinatus*, and three native emergent species, *Typha capensis, Cyperus sexangularis,* and *Phragmites australis.* Pollution indices clearly showed the promising assimilation by native and non-native macrophytes species; however, the Swartkops River was heavily influenced by multiplenon-point sources along the system, compromising the assimilation effect. Furthermore, they emphasis that excessive anthropogenic inputs compromise the system's ability to assimilate heavy metal inputs leading to water quality deterioration.

**A case study on heavy metal removal from lake water and agricultural return wastewater with the use of duckweed (Lemnacea)**

The removal efficiency of heavy metal pollution in the discharge channel (IP-1) wastewater where Gala Lake waters and agricultural activity return waters in this area are collected was investigated in laboratory conditions with the common duckweed (*L. minor*) plant in the flora of Gala Lake. This study was carried out in two stages. In the first stage, in accordance with the standard methods, at a depth of 30 cm from the IP-1 discharge channel and from two different points (channel entry and exit) at three different times (before paddy planting, planting sequence, and after harvest), the heavy metal concentrations in the water samples taken from five different points from Gala Lake had values close to each other, so the water sample taken from the first point was used in the experiments carried out at this stage. In the discharge channel (IP-1), the water sample taken from the discharge channel outlet and estimated to have high heavy metal pollution was studied. In the second stage, the removal efficiency of characterized heavy metal pollution in the water samples was determined by the *minor* (duckweed) plant collected.

According to all the results obtained from this study, a natural treatment process using the *L. minor* (duckweed) plant, which is commonly found in the flora of Gala Lake, is an effective process in the removal of major heavy metals (i.e., cadmium (Cd), copper (Cu), lead (Pb), nickel (Ni)) from water. According to Water Pollution Control Regulation (WPCR in Turkey), IV and III class quality Gala Lake water and discharge channel (IP-1) wastewater can be transformed into the first quality water after 7 days of natural treatment with *L. minor* (duckweed).

## Conclusion

The presence of excessive levels of heavy metals in aquatic life or aquaculture is a significant problem. Prioritizing the removal of heavy metals from wastewater is necessary. Prior to being discharged into freshwater or water sources, wastewater must be treated to decrease toxins, pollutants, and unwanted components. Before dumping garbage into water bodies, industries must use heavy metal detecting techniques. The number of heavy metals in wastewaters must be determined using a variety of chemical techniques and tools. The effectiveness of each approach for removing heavy metals and the particular factors that have an impact on the removal process varies. The results differ depending on the method and the experimental setting. Removal efficiencies range from 12% to 100%, depending on the variable factors. Phytoremediation comes out more suitable as alternative to conventional physicochemical remediation technologies, when important aspects were considered such as cost-effectiveness, technically suited, secondary pollution prevention, combining remediation, encourages in-situ remediation, restore soil fertility, eco-friendly and thus prevent disruption of agro-ecosystem and compared. All these mentioned attributes of phytoremediation were established and documented by many resulted-oriented studies carried out by researchers. Compilation of the successful evidence amounted to distinguishing phytoremediation from all other techniques. The review elicits the result on phytoremediation when applied solely, enhanced with other techniques using living organisms or agricultural wastes still comes up with huge successes makes it emerge as an outstanding cleaning technique.

## Bibliography

Abdel-Halim, S., Shehata, A. and El-Shahat, M. (2003). Removal of lead ions from industrial waste water by different types of natural materials. *Water Research*, *37*(7), 1678–1683. https://doi.org/10.1016/s0043-1354(02)00554-7

Afzal, S. M. (2018). Characterization of industrial effluents and groundwater of Hattar industrial estate, Haripur. *Advances in Agriculture and Environmental Science: Open Access (AAEOA)*, *1*(2). https://doi.org/10.30881/aaeoa.00013

Agarwal, I., Rochon, A., Gesser, H. and Sparling, A. (1984). Electrodeposition of six heavy metals on reticulated vitreous carbon electrode. *Water Research*, *18*(2), 227–232. https://doi.org/10.1016/0043-1354 (84)90073-3

Alexandrova, L., Nedialkova, T. and Nishkov, I. (1994). Electroflotation of metal ions in waste water. *International Journal of Mineral Processing*, *41*(3–4), 285–294. https://doi.org/10.1016/0301-7516 (94)90034-5

Allen, P. (1995). Accumulation profiles of lead and cadmium in the edible tissues of Oreochromis aureus during acute exposure. *Journal of Fish Biology*, *47*(4), 559–568. https://doi.org/10.1111/j.1095-8649.1995.tb01922.x

Amundsen, P. A., Staldvik, F. J., Lukin, A. A., Kashulin, N. A., Popova, O. A. and Reshetnikov, Y. S. (1997). Heavy metal contamination in freshwater fish from the border region between Norway and Russia. *Science of the Total Environment*, *201*(3), 211–224. https://doi.org/10.1016/s0048-9697(97)84058-2

Araya-Farias, M., Mondor, M., Lamarche, F., Tajchakavit, S. and Makhlouf, J. (2008). Clarification of apple juice by electroflotation. *Innovative Food Science and Emerging Technologies*, *9*(3), 320–327. https://doi.org/10.1016/j.ifset.2007.08.002

Baatrup, E. (1991). Structural and functional effects of heavy metals on the nervous system, including sense organs, of fish. *Comparative Biochemistry and Physiology Part C: Comparative Pharmacology*, *100*(1–2), 253–257. https://doi.org/10.1016/0742-8413(91)90163-n

Babel, S. and Kurniawan, T. A. (2004). Cr (VI) removal from synthetic wastewater using coconut shell charcoal and commercial activated carbon modified with oxidizing agents and/or chitosan. *Chemosphere*, *54*(7), 951–967. https://doi.org/10.1016/j.chemosphere.2003.10.001

Baby, J., Raj, J., Biby, E., Sankarganesh, P., Jeevitha, M., Ajisha, S. and Rajan, S. (2011). Toxic effect of heavy metals on aquatic environment. *International Journal of Biological and Chemical Sciences*, *4*(4). https://doi.org/10.4314/ijbcs.v4i4.62976

Barakat, M. (2011). New trends in removing heavy metals from industrial wastewater. *Arabian Journal of Chemistry*, *4*(4), 361–377. https://doi.org/10.1016/j.arabjc.2010.07.019

Basile, A. and Nunes, S. P. (2011). Advanced membrane science and technology for sustainable energy and environmental applications. *King Abdullah University of Science and Technology (KAUST).* https://doi.org/10.1533/9780857093790

Basile, A., Cassano, A. and Rastogi, N. K. (2015a). *Advances in Membrane Technologies for Water Treatment.* Elsevier Gezondheidszorg.

Basile, A.; di Paola, L.; Hai, F.; Piemonte, V. and di Paola, L. (2015b). *Membrane Reactors for Energy Applications and Basic Chemical Production.* Elsevier Gezondheidszorg.

Behra, R. (1993). In vitro effects of cadmium zinc and lead on calmodulin-dependent actions in Oncorhynchus mykiss, Mytilus sp., and Chlamydomonas reinhardtii. *Archives of Environmental Contamination and Toxicology, 24*(1), 21–27. https://doi.org/10.1007/bf01061085

Bódalo-Santoyo, A., Gómez-Carrasco, J., Gómez-Gómez, E., Máximo-Martín, F. and Hidalgo-Montesinos, A. (2003). Application of reverse osmosis to reduce pollutants present in industrial wastewater. *Desalination, 155*(2), 101–108. https://doi.org/10.1016/s0011-9164(03)00287-x

Bonnail, E., Sarmiento, A., DelValls, T., Nieto, J. and Riba, I. (2016). Assessment of metal contamination, bioavailability, toxicity and bioaccumulation in extreme metallic environments (Iberian Pyrite Belt) using Corbicula fluminea. *Science of the Total Environment, 544*, 1031–1044.

Botta, S. G., Navío, J. A., Hidalgo, M. C., Restrepo, G. M. and Litter, M. I. (1999). Photocatalytic properties of $ZrO_2$ and $Fe/ZrO_2$ semiconductors prepared by a sol–gel technique. *Journal of Photochemistry and Photobiology A: Chemistry, 129*(1–2), 89–99. https://doi.org/10.1016/s1010-6030(99)00150-1

Caicedo, J. (2000). Effect of total ammonia nitrogen concentration and pH on growth rates of duckweed (*Spirodela polyrrhiza*). *Water Research, 34*(15): 3829-3835. https://doi.org/10.1016/s0043-1354(00)00128-7

Casqueira, R., Torem, M. and Kohler, H. (2006). The removal of zinc from liquid streams by electroflotation. *Minerals Engineering, 19*(13), 1388–1392. https://doi.org/10.1016/j.mineng.2006.02.001

Chapman, P. M., Wang, F., Janssen, C., Persoone, G. and Allen, H. E. (1998). Ecotoxicology of metals in aquatic sediments: binding and release, bioavailability, risk assessment, and remediation. *Canadian Journal of Fisheries and Aquatic Sciences, 55*(10), 2221–2243.

Chen, G. (2004). Electrochemical technologies in wastewater treatment. *Separation and Purification Technology, 38*(1), 11–41.

Chiarle, S. (2000). Mercury removal from water by ion exchange resins adsorption. *Water Research, 34*(11), 2971–2978. https://doi.org/10.1016/s0043-1354(00)00044-0

Chin, S. F., Pang, S. C. and Dom, F. E. I. (2011). Sol-gel synthesis of silver/titanium dioxide ($Ag/TiO_2$) core-shell nanowires for photocatalytic applications. *Materials Letters, 65*(17–18), 2673–2675.

Choi, W. (2006). Pure and modified $TiO_2$ photocatalysts and their environmental applications. *Catalysis Surveys from Asia*, *10*(1), 16–28. https://doi.org/10.1007/s10563-006-9000-2

Chong, M. N., Jin, B., Chow, C. W. and Saint, C. (2010). Recent developments in photocatalytic water treatment technology: A review. *Water Research*, *44*(10), 2997–3027. https://doi.org/10.1016/j.watres.2010.02.039

Costaz, P., Miquel, J. and Reinbold, M. (1983). Electroflotation et disinfection simultanee d'eaux residuaires urbaines. *Water Research*, *17*(3), 255–262. https://doi.org/10.1016/0043-1354(83)90178-1

Couture, P. and Kumar, R. P. (2003). Impairment of metabolic capacities in copper and cadmium contaminated wild yellow perch (*Perca flavescens*). *Aquatic Toxicology*, *64*(1), 107–120. https://doi.org/10.1016/s0166-445x(03)00028-6

Dą Browski, A., Hubicki, Z., Podkościelny, P. and Robens, E. (2004). Selective removal of the heavy metal ions from waters and industrial wastewaters by ion-exchange method. *Chemosphere*, *56*(2), 91–106.

Da Mota, I. O., de Castro, J. A., de GóesCasqueira, R. and de Oliveira Junior, A. G. (2015). Study of electroflotation method for treatment of wastewater from washing soil contaminated by heavy metals. *Journal of Materials Research and Technology*, *4*(2), 109–113. https://doi.org/10.1016/j.jmrt.2014.11.004

Eaton, J. G. and Scheller, R. M. (1996). Effects of climate warming on fish thermal habitat in streams of the United States. *Limnology and Oceanography*, *41*(5), 1109–1115.

Edokpayi, J. N., Odiyo, J. O. and Durowoju, O. S. (2017). Impact of Wastewater on Surface Water Quality in Developing Countries: A Case Study of South Africa. *Water Quality*. https://doi.org/10.5772/66561

Ellenberger, S. A., Baumann, P. C. and May, T. W. (1994). Evaluation of Effects Caused by High Copper Concentrations in Torch Lake, Michigan, on Reproduction of Yellow Perch. *Journal of Great Lakes Research*, *20*(3), 531–536. https://doi.org/10.1016/s0380-1330(94)71169-2

Emamjomeh, M. M. and Sivakumar, M. (2009). Review of pollutants removed by electrocoagulation and electrocoagulation/flotation processes. *Journal of Environmental Management*, *90*(5), 1663–1679.

Equeenuddin, S. M., Tripathy, S., Sahoo, P. and Panigrahi, M. (2013). Metal behavior in sediment associated with acid mine drainage stream: Role of pH. *Journal of Geochemical Exploration*, *124*, 230–237.

Fane, A., Tang, C. and Wang, R. (2011). Membrane Technology for Water: Microfiltration, Ultrafiltration, Nan filtration, and Reverse Osmosis. *Treatise on Water Science*, 301–335. https://doi.org/10.1016/b978-0-444-53199-5.00091-9

Farkas, A., Salánki, J. and Specziár, A. (2001). Relation Between Growth and the Heavy

Metal Concentration in Organs of Bream Abramisbrama L. Populating Lake Balaton. *Archives of Environmental Contamination and Toxicology*, *43*(2), 236–243. https://doi.org/10.1007/s00244-002-1123-5

Feeley, T. J., Skone, T. J., Stiegel, G. J., McNemar, A., Nemeth, M., Schimmoller, B., Murphy, J.T. and Manfredo, L., 2008. Water: a critical resource in the thermoelectric power industry. Energy 33, 1–11. https://doi.org/10.1016/j.energy.2007.08.007

Ferreira, S. L., de Brito, C. F., Dantas, A. F., Lopo De Araújo, N. M. and Costa, A. (1999). Nickel determination in saline matrices by ICP-AES after sorption on Amberlite XAD-2 loaded with PAN. *Talanta*, *48*(5), 1173–1177. https://doi.org/10.1016/s0039-9140(98)00339-7

Frémion, F., Bordas, F., Mourier, B., Lenain, J. F., Kestens, T. and Courtin-Nomade, A. (2016). Influence of dams on sediment continuity: A study case of a natural metallic contamination. *Science of the Total Environment*, *547*, 282–294. https://doi.org/10.1016/j.scitotenv.2016.01.023

Fu, F. and Wang, Q. (2011). Removal of heavy metal ions from wastewaters: A review. *Journal of Environmental Management*, *92*(3), 407–418. https://doi.org/10.1016/j.jenvman.2010.11.011

Fu, J., Zhao, C., Luo, Y., Liu, C., Kyzas, G. Z., Luo, Y., Zhao, D., An, S. and Zhu, H. (2014). Heavy metals in surface sediments of the Jialu River, China: Their relations to environmental factors. *Journal of Hazardous Materials*, *270*, 102–109. https://doi.org/10.1016/j.jhazmat.2014.01.044

Fujishima, A. and Honda, K. (1972). Electrochemical Photolysis of Water at a Semiconductor Electrode. *Nature*, *238*(5358), 37–38. https://doi.org/10.1038/238037a0

Fukui, Y. and Yuu, S. (1985). Removal of colloidal particles in electroflotation. *AIChE Journal*, *31*(2), 201–208. https://doi.org/10.1002/aic.690310205

Gadd, G. M. and Griffiths, A. J. (1977). Microorganisms and heavy metal toxicity. *Microbial Ecology*, *4*(4), 303–317. https://doi.org/10.1007/bf02013274

Gagliardo, P., Adham, S., Trussell, R. and Olivieri, A. (1998). Water repurification via reverse osmosis. *Desalination*, *117*(1–3), 73–78. https://doi.org/10.1016/s0011-9164(98)00069-1

Gao, P., Chen, X., Shen, F. and Chen, G. (2005). Removal of chromium (VI) from wastewater by combined electrocoagulation? Electroflotation without a filter. *Separation and Purification Technology*, *43*(2), 117–123. https://doi.org/10.1016/j.seppur.2004.10.008

Gazi, S. and Ananthakrishnan, R. (2011). Metal-free-photocatalytic reduction of 4-nitrophenol by resin-supported dye under the visible irradiation. *Applied Catalysis B: Environmental*, *105*(3–4), 317–325. https://doi.org/10.1016/j.apcatb.2011.04.025

Gherasim, C. V. and Mikulášek, P. (2013). Influence of operating variables on the removal of heavy metal ions from aqueous solutions by nanofiltration. *Desalination*, *343*(22),

67–74. https://doi.org/10.1016/j.desal.2013.11.012

Gode, F. and Pehlivan, E. (2003). A comparative study of two chelating ion-exchange resins for the removal of chromium (III) from aqueous solution. *Journal of Hazardous Materials, 100*(1–3), 231–243. https://doi.org/10.1016/s0304-3894(03)00110-9

Gode, F. and Pehlivan, E. (2006). Removal of chromium (III) from aqueous solutions using Lewatit S 100: The effect of pH, time, metal concentration and temperature. *Journal of Hazardous Materials, 136*(2), 330–337. https://doi.org/10.1016/j.jhazmat.2005.12.021

Gomes, J. A., Daida, P., Kesmez, M., Weir, M., Moreno, H., Parga, J. R., Irwin, G., McWhinney, H., Grady, T., Peterson, E. and Cocke, D. L. (2007). Arsenic removal by electrocoagulation using combined Al–Fe electrode system and characterization of products. *Journal of Hazardous Materials, 139*(2), 220–231. https://doi.org/10.1016/j.jhazmat.2005.11.108

Goyal, D., Yadav, A., Prasad, M., Singh, T. B., Shrivastav, P., Ali, A., Dantu, P. K. and Mishra, S. 2020. Effect of Heavy Metals on Plant Growth: An Overview. In: Naeem, M., Ansari, A., Gill, S. (eds) Contaminants in Agriculture. Springer, Cham. https://doi.org/10.1007/978-3-030-41552-5_4.

Greenlee, L. F., Lawler, D. F., Freeman, B. D., Marrot, B. and Moulin, P. (2009). Reverse osmosis desalination: Water sources, technology, and today's challenges. *Water Research, 43*(9), 2317–2348. https://doi.org/10.1016/j.watres.2009.03.010

Griscom, S. B. and Fisher, N. S. (2004). Bioavailability of sediment-bound metals to marine bivalve molluscs: An overview. *Estuaries, 27*(5), 826–838. https://doi.org/10.1007/bf02912044

Han, Z., Guo, Z., Zhang, Y., Xiao, X., Xu, Z. and Sun, Y. (2018). Adsorption-pyrolysis technology for recovering heavy metals in solution using contaminated biomass phytoremediation. *Resources, Conservation and Recycling, 129*: 20-26. https://doi.org/10.1016/j.resconrec.2017.10.003.

Hasan, M., Khan, M., Khan, M., Aktar, S., Rahman, M., Hossain, F. and Hasan, A. (2016). Heavy metals distribution and contamination in surface water of the Bay of Bengal coast. *Cogent Environmental Science, 2*(1), 1140001. https://doi.org/10.1080/23311843.2016.1140001

Hayashi, J., Horikawa, T., Muroyama, K. and Gomes, V. G. (2002). Activated carbon from chickpea husk by chemical activation with $K_2CO_3$: preparation and characterization. *Microporous and Mesoporous Materials, 55*(1), 63–68. https://doi.org/10.1016/s1387-1811(02)00406-7

He, Z. L., Yang, X. E. and Stoffella, P. J. (2005). Trace elements in agroecosystems and impacts on the environment. *Journal of Trace Elements in Medicine and Biology, 19*(2–3), 125–140. https://doi.org/10.1016/j.jtemb.2005.02.010

Heath, A. G. (2018). Uptake, Accumulation, Biotransformation, and Excretion of Xenobiotics. *Water Pollution and Fish Physiology*, 79–123. https://doi.org/10.1201/9780203718896-5

Herrmann, J. M. (1999). Heterogeneous photocatalysis: fundamentals and applications to the removal of various types of aqueous pollutants. *Catalysis Today*, *53*(1), 115–129. https://doi.org/10.1016/s0920-5861(99)00107-8

Herrmann, J. M. (2005). Heterogeneous photocatalysis: state of the art and present applications in honor of Pr. R.L. Burwell Jr. (1912–2003), Former Head of Ipatieff Laboratories, Northwestern University, Evanston (Ill). *Topics in Catalysis*, *34*(1–4), 49–65. https://doi.org/10.1007/s11244-005-3788-2

Herrmann, J. M.; Guillard, C. and Pichat, P. (1993). Heterogeneous photocatalysis: an emerging technology for water treatment. *Catalysis Today*, *17*(1–2), 7–20. https://doi.org/10.1016/0920-5861(93)80003-j

Hilmy, A.; Shabana, M. and Daabees, A. (1985). Bioaccumulation of cadmium: Toxicity in *Mugil cephalus*. *Comparative Biochemistry and Physiology Part C: Comparative Pharmacology*, *81*(1), 139–144. https://doi.org/10.1016/0742-8413(85)90105-7

Holt, P. K., Barton, G. W. and Mitchell, C. A. (2005). The future for electrocoagulation as a localised water treatment technology. *Chemosphere*, *59*(3), 355–367. https://doi.org/10.1016/j.chemosphere.2004.10.023

Hosny, A. (1992). Separation of oil from oil/water emulsions using an electroflotation cell with insoluble electrodes. *Filtration and Separation*, *29*(5), 419–423. https://doi.org/10.1016/0015-1882(92)80204-v

Hosny, A. Y. (1996). Separating oil from oil-water emulsions by electroflotation technique. *Separations Technology*, *6*(1), 9–17. https://doi.org/10.1016/0956-9618(95)00136-0

Huseen, H. M. and Mohammed, A. J. (2019). Heavy Metals Causing Toxicity in Fishes. *Journal of Physics: Conference Series*, *1294*(6), 062028. https://doi.org/10.1088/1742-6596/1294/6/062028

Iijima, S. (1991). Helical microtubules of graphitic carbon. *Nature*, *354*(6348), 56–58. https://doi.org/10.1038/354056a0

Jezierska, B. and Witeska, M. (2006). The Metal Uptake and Accumulation in Fish Living in Polluted Waters. *NATO Science Series*, 107–114. https://doi.org/10.1007/978-1-4020-4728-2_6

Jezierska, B., Ługowska, K. and Witeska, M. (2008). The effects of heavy metals on embryonic development of fish (a review). *Fish Physiology and Biochemistry*, *35*(4), 625–640. https://doi.org/10.1007/s10695-008-9284-4

Jiang, L., Wang, Y. and Feng, C. (2012). Application of Photocatalytic Technology in Environmental Safety. *Procedia Engineering*, *45*, 993–997. https://doi.org/10.1016/j.proeng.2012.08.271

Jianquan L. and Yinhua W. (2013) Effects of pH and salt on nanofiltration – a critical review. *Journal of Membrane Science, 438*, 18-28

Johnson, L. L., Casillas, E., Myers, M. S., Rhodes, L. D. and Paul Olson, O. (1991). Patterns of oocyte development and related changes in plasma 17-β estradiol, vitellogenin, and plasma chemistry in English sole Parophrysvetulus Girard. *Journal of Experimental Marine Biology and Ecology*, *152*(2), 161–185. https://doi.org/10.1016/0022-0981(91)90213-g

Joo, S. H. and Tansel, B. (2015). Novel technologies for reverse osmosis concentrate treatment: A review. *Journal of Environmental Management*, *150*, 322–335. https://doi.org/10.1016/j.jenvman.2014.10.027

Jüttner, K., Galla, U. and Schmieder, H. (2000). Electrochemical approaches to environmental problems in the process industry. *Electrochimica Acta*, *45*(15–16), 2575–2594. https://doi.org/10.1016/s0013-4686(00)00339-x

Kabbashi, N. A., Atieh, M. A., Al-Mamun, A., Mirghami, M. E., Alam, M. and Yahya, N. (2009). Kinetic adsorption of application of carbon nanotubes for Pb(II) removal from aqueous solution. *Journal of Environmental Sciences*, *21*(4), 539–544. https://doi.org/10.1016/s1001-0742(08)62305-0

Kadirvelu, K. and Namasivayam, C. (2000). Agricultural By-Product as Metal Adsorbent: Sorption of Lead (II) from Aqueous Solution onto Coir pith Carbon. *Environmental Technology*, *21*(10), 1091–1097. https://doi.org/10.1080/09593330.2000.9618995

Kaewsarn, P. (2002). Biosorption of copper (II) from aqueous solutions by pre-treated biomass of marine algae *Padina sp.Chemosphere*, *47*(10): 1081-1085. https://doi.org/10.1016/s0045-6535(01)00324-1.

Kandah, M. I. and Meunier, J. L. (2007). Removal of nickel ions from water by multi-walled carbon nanotubes. *Journal of Hazardous Materials*, *146*(1–2), 283–288. https://doi.org/10.1016/j.jhazmat.2006.12.019

Kang, S. Y., Lee, J. U., Moon, S. H. and Kim, K. W. (2004). Competitive adsorption characteristics of Co2+, Ni2+, and Cr3+ by IRN-77 cation exchange resin in synthesized wastewater. *Chemosphere*, *56*(2), 141–147. https://doi.org/10.1016/j.chemosphere.2004.02.004

Khalil, L., Rophael, M. and Mourad, W. (2002). The removal of the toxic Hg(II) salts from water by photocatalysis. *Applied Catalysis B: Environmental*, *36*(2), 125–130. https://doi.org/10.1016/s0926-3373(01)00285-5

Khandegar, V. and Saroha, A. K. (2013). Electrocoagulation for the treatment of textile industry effluent– A review. *Journal of Environmental Management*, *128*, 949–963. https://doi.org/10.1016/j.jenvman.2013.06.043

Kim, S. and Choi, W. (2002). Kinetics and Mechanisms of Photocatalytic Degradation of $(CH_3)_{nNH4-n}^{+}$ $(0 \leq n \leq 4)$ in $TiO_2$ Suspension: The Role of OH Radicals. *Environmental Science and Technology*, *36*(9), 2019–2025. https://doi.org/10.1021/es015560s

Kolesnikov, V. A., Il'in, V. I., Brodskii, V. A., Guseva, T. V. and Vartanyan, M. A. (2015). Improvement of Electroflotation Treatment of Waste Waters from Ceramic Enterprises. *Glass and Ceramics*, *71*(11-12), 421+. https://link.gale.com/apps/doc/

A425765256/AONE?u=anon~96dc66c3andsid=googleScholarandxid=39bb7788

Ku, Y. and Jung, I. L. (2001). Photocatalytic reduction of Cr(VI) in aqueous solutions by UV irradiation with the presence of titanium dioxide. *Water Research*, *35*(1), 135–142. https://doi.org/10.1016/s0043-1354(00)00098-1

Kuo, C. Y. and Lin, H. Y. (2009). Adsorption of aqueous cadmium (II) onto modified multi-walled carbon nanotubes following microwave/chemical treatment. *Desalination*, *249*(2), 792–796. https://doi.org/10.1016/j.desal.2008.11.023

Kurniawan, T. A., Chan, G. Y., Lo, W. H. and Babel, S. (2006). Physico–chemical treatment techniques for wastewater laden with heavy metals. *Chemical Engineering Journal*, *118*(1–2), 83–98. https://doi.org/10.1016/j.cej.2006.01.015

Lacour, S., Bollinger, J. C., Serpaud, B., Chantron, P. and Arcos, R. (2001). Removal of heavy metals in industrial wastewaters by ion-exchanger grafted textiles. *Analytica Chimica Acta*, *428*(1), 121–132. https://doi.org/10.1016/s0003-2670(00)01215-0

Landolph, J. R. (1994). Molecular mechanisms of transformation of C3H/10T1/2 C1 8 mouse embryo cells and diploid human fibroblasts by carcinogenic metal compounds. *Environmental Health Perspectives*, *102*(suppl 3), 119–125. https://doi.org/10.1289/ehp.94102s3119

Lin, S. H., Lai, S. L. and Leu, H. G. (2000). Removal of heavy metals from aqueous solution by chelating resin in a multistage adsorption process. *Journal of Hazardous Materials*, *76*(1), 139–153. https://doi.org/10.1016/s0304-3894(00)00207-7

Litter, M. (1999). Heterogeneous photocatalysis Transition metal ions in photocatalytic systems. *Applied Catalysis B: Environmental*, *23*(2–3), 89–114. https://doi.org/10.1016/s0926-3373(99)00069-7

Lye, C., Frid, C. and Gill, M. (1998). Seasonal reproductive health of flounder Platichthysflesus exposed to sewage effluent. *Marine Ecology Progress Series*, *170*, 249–260. https://doi.org/10.3354/meps170249

Maher, A., Sadeghi, M. and Moheb, A. (2014). Heavy metal elimination from drinking water using nanofiltration membrane technology and process optimization using response surface methodology. *Desalination*, *352*, 166–173. https://doi.org/10.1016/j.desal.2014.08.023

Malaeb, L. and Ayoub, G. M. (2011). Reverse osmosis technology for water treatment: State of the art review. *Desalination*, *267*(1), 1–8. https://doi.org/10.1016/j.desal.2010.09.001

Manzoor, J. and Sharma, M. (2019). Impact of Biomedical Waste on Environment and Human Health. *Environmental Claims Journal*, *31*(4), 311–334. https://doi.org/10.1080/10406026.2019.1619265

Marcovecchio, J. (2004). The use of Micropogonias furnieri and Mugil liza as bioindicators of heavy metals pollution in La Plata River estuary, Argentina. *Science of The Total*

*Environment*, *323*(1–3), 219–226. https://doi.org/10.1016/j.scitotenv.2003.09.029

Marjani, A., Rezakazemi, M. and Shirazian, S. (2012). Simulation of Methanol Production Process and Determination of Optimum Conditions. *Oriental Journal of Chemistry*, *28*(1), 145–151. https://doi.org/10.13005/ojc/280121

Matis, K. A., Zouboulis, A. I., Gallios, G. P., Erwe, T. and Blöcher, C. (2004). Application of flotation for the separation of metal-loaded zeolites. *Chemosphere*, *55*(1), 65–72.

Merzouk, B., Gourich, B., Sekki, A., Madani, K. and Chibane, M. (2009). Removal turbidity and separation of heavy metals using electrocoagulation–electroflotation technique. *Journal of Hazardous Materials*, *164*(1), 215–222. https://doi.org/10.1016/j.jhazmat.2008.07.144

Miller, G. (2000). A comment on arsenic species separation using ion exchange. *Water Research*, *34*(4), 1397–1400. https://doi.org/10.1016/s0043-1354(99)00257-2

Miller, P. A., Munkittrick, K. R. and Dixon, D. G. (1992). Relationship between Concentrations of Copper and Zinc in Water, Sediment, Benthic Invertebrates, and Tissues of White Sucker (*Catostomus commersoni*) at Metal-Contaminated Sites. *Canadian Journal of Fisheries and Aquatic Sciences*, *49*(5), 978–984. https://doi.org/10.1139/f92-109

Mishra, S., Sharma, S. and Vasudevan, P. (2008). Comparative effect of biofertilizers on fodder production and quality in guinea grass (Panicum maximum Jacq.). *Journal of the Science of Food and Agriculture*. *88*(9): 1667-1673.

Moberg, F. and Folke, C. (1999). Ecological goods and services of coral reef ecosystems. *Ecological Economics*, *29*(2), 215–233. https://doi.org/10.1016/s0921-8009(99)00009-9

Mohammad, A., Teow, Y., Ang, W., Chung, Y., Oatley-Radcliffe, D. and Hilal, N. (2015). Nanofiltration membranes review: Recent advances and future prospects. *Desalination*, *356*, 226–254.

Molinari, R., Caruso, A. and Poerio, T. (2009). Direct benzene conversion to phenol in a hybrid photocatalytic membrane reactor. *Catalysis Today*, *144*(1–2), 81–86.

Mollah, M. A., Schennach, R., Parga, J. R. and Cocke, D. L. (2001). Electrocoagulation (EC) — science and applications. *Journal of Hazardous Materials*, *84*(1), 29–41. https://doi.org/10.1016/s0304-3894(01)00176-5

Mollah, M.; Morkovsky, P.; Gomes, J.; Kesmez, M.; Parga, J. and Cocke, D. (2004). Fundamentals, present and future perspectives of electrocoagulation. *Journal of Hazardous Materials*, *114*(1–3), 199–210.

Moon, S. H., Park, S. J., Hwang, Y. J., Lee, D. K., Cho, Y., Kim, D. W. and Min, B. K. (2014). Printable, wide band-gap chalcopyrite thin films for power generating window applications. *Scientific Reports*, *4*(1).

Mostefa, N. and Tir, M. (2004). Coupling flocculation with electroflotation for waste oil/water emulsion treatment. Optimization of the operating conditions. *Desalination*, *161*(2), 115–121. https://doi.org/10.1016/s0011-9164(04)90047-1

Mulyati, S., Takagi, R., Fujii, A., Ohmukai, Y. and Matsuyama, H. (2013). Simultaneous improvement of the monovalent anion selectivity and antifouling properties of an anion exchange membrane in an electrodialysis process, using polyelectrolyte multilayer deposition. *Journal of Membrane Science*, *431*, 113–120. https://doi.org/10.1016/j.memsci.2012.12.022

Murthy, Z. and Chaudhari, L. B. (2009). Separation of binary heavy metals from aqueous solutions by nanofiltration and characterization of the membrane using Spiegler–Kedem model. *Chemical Engineering Journal*, *150*(1), 181–187. https://doi.org/10.1016/j.cej.2008.12.023

Ng, L. Y., Mohammad, A. W. and Ng, C. Y. (2013). A review on nanofiltration membrane fabrication and modification using polyelectrolytes: Effective ways to develop membrane selective barriers and rejection capability. *Advances in Colloid and Interface Science*, *197–198*, 85–107. https://doi.org/10.1016/j.cis.2013.04.004

Ni, M., Leung, M. K., Leung, D. Y. and Sumathy, K. (2007). A review and recent developments in photocatalytic water-splitting using $TiO_2$ for hydrogen production. *Renewable and Sustainable Energy Reviews*, *11*(3), 401–425. https://doi.org/10.1016/j.rser.2005.01.009

Ohtani, B. (2010). Photocatalysis A to Z—What we know and what we do not know in a scientific sense. *Journal of Photochemistry and Photobiology C: Photochemistry Reviews*, *11*(4), 157–178. https://doi.org/10.1016/j.jphotochemrev.2011.02.001

Okada, K., Yamamoto, N., Kameshima, Y. and Yasumori, A. (2003). Porous properties of activated carbons from waste newspaper prepared by chemical and physical activation. *Journal of Colloid and Interface Science*, *262*(1), 179–193. https://doi.org/10.1016/s0021-9797(03)00107-3

Olaifa, F., Olaifa, A. and Onwude, T. (2010). Lethal and sub-lethal effects of copper to the african catfish (*clarias gariepinus*) juveniles. *African Journal of Biomedical Research*, *7*(2). https://doi.org/10.4314/ajbr.v7i2.54071

Osann, E. and Hayat, B. (2014) Protecting Our Waters from Toxic Power Plant Discharges and Reducing Water Use in the Process. NRDC Fact Sheet 2014.

Osman, A. G. M., Mekkawy, I. A., Verreth, J. and Kirschbaum, F. (2007). Effects of lead nitrate on the activity of metabolic enzymes during early developmental stages of the African catfish, *Clarias gariepinus (*Burchell, 1822). *Fish Physiology and Biochemistry*, *33*(1), 1–13. https://doi.org/10.1007/s10695-006-9111-8

Oussedik, S. M. and Khelifa, A. (2001). Reduction of copper ions concentration in wastewaters of galvanoplastic industry by electroflotation. *Desalination*, *139*(1–3), 383. https://doi.org/10.1016/s0011-9164(01)00337-x

Oztekin, Y. and Yazicigil, Z. (2006). Recovery of metals from complexed solutions by electrodeposition. *Desalination*, *190*(1–3), 79–88. https://doi.org/10.1016/j.desal.2005.07.017

Özverdi, A. and Erdem, M. (2006). $Cu^{2+}$, $Cd^{2+}$ and $Pb^{2+}$ adsorption from aqueous solutions by pyrite and synthetic iron sulphide. *Journal of Hazardous Materials*, *137*(1), 626–632. https://doi.org/10.1016/j.jhazmat.2006.02.051

Padaki, M., Surya Murali, R., Abdullah, M., Misdan, N., Moslehyani, A., Kassim, M., Hilal, N. and Ismail, A. (2015). Membrane technology enhancement in oil–water separation. A review. *Desalination*, *357*, 197–207. https://doi.org/10.1016/j.desal.2014.11.023

Pelgrom, S., Lamers, L., Lock, R., Balm, P. and Bonga, S. (1995). Interactions between copper and cadmium modify metal organ distribution in mature tilapia, *Oreochromis mossambicus*. *Environmental Pollution*, *90*(3), 415–423. https://doi.org/10.1016/0269-7491(95)00022-j

Perry, D. M., Weis, J. S. and Weis, P. (1988). Cytogenetic effects of methylmercury in embryos of the killifish, *Fundulusheteroclitus*. *Archives of Environmental Contamination and Toxicology*, *17*(5), 569–574. https://doi.org/10.1007/bf01055824

Rahman, M. S., Molla, A. H., Saha, N. and Rahman, A. (2012). Study on heavy metals levels and its risk assessment in some edible fishes from Bangshi River, Savar, Dhaka, Bangladesh. *Food Chemistry*, *134*(4), 1847–1854. https://doi.org/10.1016/j.foodchem.2012.03.099

Rai, P. K. and Tripathi, B. D. (2009). Comparative assessment of *Azolla pinnata* and *Vallisneria spiralis* in Hg removal from G.B. Pant Sagar of Singrauli Industrial region, India. *Environmental Monitoring and Assessment,148*(1-4): 75-84. https://doi.org/10.1007/s10661-007-0140-2.

Razavi, S. M. R., Rezakazemi, M., Albadarin, A. B. and Shirazian, S. (2016). Simulation of $CO_2$ absorption by solution of ammonium ionic liquid in hollow-fiber contactors. *Chemical Engineering and Processing: Process Intensification*, *108*, 27–34. https://doi.org/10.1016/j.cep.2016.07.001

Rengaraj, S., Yeon, K. H. and Moon, S. H. (2001). Removal of chromium from water and wastewater by ion exchange resins. *Journal of Hazardous Materials*, *87*(1–3), 273–287. https://doi.org/10.1016/s0304-3894(01)00291-6

Rezakazemi, M., EbadiAmooghin, A., Montazer-Rahmati, M. M., Ismail, A. F. and Matsuura, T. (2014). State-of-the-art membrane-based $CO_2$ separation using mixed matrix membranes (MMMs): An overview on current status and future directions. *Progress in Polymer Science*, *39*(5), 817–861. https://doi.org/10.1016/j.progpolymsci.2014.01.003

Rezakazemi, M., Ghafarinazari, A., Shirazian, S. and Khoshsima, A. (2012). Numerical modeling and optimization of wastewater treatment using porous polymeric membranes. *Polymer Engineering and Science*, *53*(6), 1272–1278. https://doi.org/10.1002/pen.23375

Rezakazemi, M., Vatani, A. and Mohammadi, T. (2015). Synergistic interactions between POSS and fumed silica and their effect on the properties of crosslinked PDMS nanocomposite membranes. *RSC Advances*, *5*(100), 82460–82470. https://doi.org/10.1039/c5ra13609a

Robertson, P. K. (1996). Semiconductor photocatalysis: an environmentally acceptable alternative production technique and effluent treatment process. *Journal of Cleaner Production*, *4*(3–4), 203–212.

Roméo, M. Siau, Y., Sidoumou, Z. and Gnassia-Barelli, M. (1999). Heavy metal distribution in different fish species from the Mauritania coast. *Science of The Total Environment*, *232*(3), 169–175. https://doi.org/10.1016/s0048-9697(99)00099-6

Rosado, D., Usero, J. and Morillo, J. (2016). Assessment of heavy metals bioavailability and toxicity toward *Vibrio fischeri* in sediment of the Huelva estuary. *Chemosphere*, *153*, 10–17.

Sadyrbaeva, T. Z. (2015). Removal of chromium (VI) from aqueous solutions using a novel hybrid liquid membrane—electrodialysis process. *Chemical Engineering and Processing: Process Intensification*, *99*, 183–191. https://doi.org/10.1016/j.cep.2015.07.011

Saha, A., Mukherjee, P., Roy, K., Sen, K. and Sanyal, T. (2022). A review on phyto-remediation by aquatic macrophytes: A natural promising tool for sustainable management of ecosystem. *International Journal of Experimental Research and Review*, *27*, 9–31. https://doi.org/10.52756/ijerr.2022.v27.002

Sandbank, E.; Shelef, G. and Wachs, A. (1974). Improved electroflotation for the removal of suspended solids from algal pond effluents. *Water Research*, *8*(9), 587–592. https://doi.org/10.1016/0043-1354(74)90116-x

Sankhla, M. S. (2019). Contaminant of Heavy Metals in Groundwater and its Toxic Effects on Human Health and Environment. *International Journal of Environmental Sciences and Natural Resources*, *18*(5). https://doi.org/10.19080/ijesnr.2019.18.555996

Sankhla, M. S. Kumar, R. and Biswas, A. (2019). Dynamic nature of heavy metal toxicity in water and sediments of Ayad River with climatic change. *International Journal of Hydrology*, *3*(5), 345–349. https://doi.org/10.15406/ijh.2019.03.00197

Sankhla, M. S., Kumar, R. and Shefali. (2020). New and Advanced Technologies in Aquaculture to Support Environmentally Sustainable Development. *Environmental and Microbial Biotechnology*, 249–263.

Sankhla, M. S., Kumari, M., Nandan, M., Kumar, R. and Agrawal, P. (2016a). Heavy Metals Contamination in Water and Their Hazardous Effect on Human Health-A Review. *SSRN Electronic Journal*.

Sankhla, M. S., Kumari, M., Nandan, M., Mohril, S., Singh, G. P., Chaturvedi, B. and Kumar, D. R. (2016b). Effect of Electronic waste on Environmental and Human health- A Review. *IOSR Journal of Environmental Science, Toxicology and Food Technology*, *10*(09), 98–104.

Schreinemachers, P. and Tipraqsa, P. (2012). Agricultural pesticides and land use intensification in high, middle and low-income countries. *Food Policy*, *37*(6), 616–626. https://doi.org/10.1016/j.foodpol.2012.06.003

Scott, K. and Paton, E. (1993). An analysis of metal recovery by electrodeposition from mixed metal ion solutions—part I. Theoretical behaviour of batch recycle operation. *Electrochimica Acta*, *38*(15), 2181–2189. https://doi.org/10.1016/0013-4686(93)80096-i

Sevcikova, M., Modra, H., Blahova, J., Dobsikova, R., Plhalova, L., Zitka, O., Hynek, D., Kizek, R., Skoric, M. and Svobodova, Z. (2016). Biochemical, haematological and oxidative stress responses of common carp (*Cyprinus carpio)* after sub-chronic exposure to copper. *VeterinárníMedicína*, *61*(No. 1), 35–50. https://doi.org/10.17221/8681-vetmed

Shenvi, S. S., Isloor, A. M. and Ismail, A. (2015). A review on RO membrane technology: Developments and challenges. *Desalination*, *368*, 10–26. https://doi.org/10.1016/j.desal.2014.12.042

Shi, X., Tal, G., Hankins, N. P. and Gitis, V. (2014). Fouling and cleaning of ultrafiltration membranes: A review. *Journal of Water Process Engineering*, *1*, 121–138. https://doi.org/10.1016/j.jwpe.2014.04.003

Simka, W., Puszczyk, D. and Nawrat, G. (2009). Electrodeposition of metals from non-aqueous solutions. *Electrochimica Acta*, *54*(23), 5307–5319. https://doi.org/10.1016/j.electacta.2009.04.028

Singh, D., Gupta, R. and Tiwari, A. (2011). Phytoremediation of Lead from Wastewater Using Aquatic Plants. *International Journal of Biomedical Research*, *2*(7). https://doi.org/10.7439/ijbr.v2i7.124.

Singh, V., Singh, N., Rai, S. N., Kumar, A., Singh, A. K., Singh, M P., Sahoo, A., Shekhar, S., Vamanu, E. and Mishra, V. 2023. Heavy Metal Contamination in the Aquatic Ecosystem: Toxicity and its Remediation Using Eco-Friendly Approaches. *Toxics*, *11*(2), 147. https://doi.org/10.3390/toxics11020147.

Sonone, S. S., Jadhav, S. V., Sankhla, M. S. and Kumar, R. 2020. Water Contamination by Heavy Metals and their Toxic Effect on Aquaculture and Human Health through Food Chain. *Letters in Applied Nano Bioscience*, *10*(2), 2148-2166.

Srinivasan, V. and Subbaiyan, M. (1989). Electroflotation Studies on Cu, Ni, Zn, and Cd with Ammonium Dodecyl Dithiocarbamate. *Separation Science and Technology*, *24*(1–2), 145–150. https://doi.org/10.1080/01496398908049757

Srivastava, A. and Srivastava, A. (1994). Effects of chlordecone on the gonads of freshwater catfish, *Heteropneustes fossilis*. *Bulletin of Environmental Contamination and Toxicology*, *53*(2). https://doi.org/10.1007/bf00192031

Stewart, C., Johnston, D., Leonard, G., Horwell, C., Thordarson, T. and Cronin, S. (2006). Contamination of water supplies by volcanic ashfall: A literature review and simple impact modelling. *Journal of Volcanology and Geothermal Research*, *158*(3–4), 296–306.

Subramani, A. and Jacangelo, J. G. (2014). Treatment technologies for reverse osmosis concentrate volume minimization: A review. *Separation and Purification Technology*, *122*, 472–489.

Sutherland, K. (2008). Developments in filtration: What is nanofiltration? *Filtration and Separation*, *45*(8), 32–35. https://doi.org/10.1016/s0015-1882(08)70298-2

Szczygłowska, M., Piekarska, A., Konieczka, P. and Namieśnik, J. (2011). Use of Brassica plants in the phytoremediation and biofumigation processes. *International Journal of Molecular Sciences, 12*(11): 7760-7771. https://doi.org/10.3390/ijms12117760.

Tsai, W., Chang, C., Wang, S., Chang, C., Chien, S. and Sun, H. (2001). Preparation of activated carbons from corn cob catalyzed by potassium salts and subsequent gasification with $CO_2$. *Bioresource Technology*, *78*(2), 203–208. https://doi.org/10.1016/s0960-8524(00)00111-5

Tshithukhe, G., Motitsoe, S. N. and Hill, M. P. (2021). Heavy Metals Assimilation by Native and Non- Native Aquatic Macrophyte Species: A Case Study of a River in the Eastern Cape Province of South Africa. *Plants, 10*(12), 2676. https://doi.org/10.3390/plants10122676

Valero, F. and Arbós, R. (2010). Desalination of brackish river water using Electrodialysis Reversal (EDR). *Desalination*, *253*(1–3), 170–174. https://doi.org/10.1016/j.desal.2009.11.011

van der Bruggen, B. and Vandecasteele, C. (2003). Removal of pollutants from surface water and groundwater by nanofiltration: overview of possible applications in the drinking water industry. *Environmental Pollution*, *122*(3), 435–445. https://doi.org/10.1016/s0269-7491(02)00308-1

Varol, M. and Şen, B. (2012). Assessment of nutrient and heavy metal contamination in surface water and sediments of the upper Tigris River, Turkey. *CATENA*, *92*, 1–10. https://doi.org/10.1016/j.catena.2011.11.011

Wang, L. K., Hung, Y. T. and Shammas, N. K. (Eds.). (2009). *Handbook of Advanced Industrial and Hazardous Wastes Treatment*, 141–197. https://doi.org/10.1201/9781420072228

Wang, W. X. and Fisher, N. S. (1997). Modeling Metal Bioavailability for Marine Mussels. *Reviews of Environmental Contamination and Toxicology*, 39–65. https://doi.org/10.1007/978-1-4612-1958-3_2

Weis, J. S., Weis, P., Heber, M. and Vaidya, S. (1981). Methylmercury tolerance of killifish (*Fundulusheteroclitus*) embryos from a polluted vs non-polluted environment. *Marine Biology*, *65*(3), 283–287.

Wingenfelder, U., Hansen, C., Furrer, G. and Schulin, R. (2005). Removal of Heavy Metals from Mine Waters by Natural Zeolites. *Environmental Science and Technology*, *39*(12), 4606–4613. https://doi.org/10.1021/es048482s

Wojtkowska, M., Bogacki, J. and Witeska, A. (2016). Assessment of the hazard posed by metal forms in water and sediments. *Science of The Total Environment*, *551–552*, 387–392.

Yang, X. J. Fane, A. G. and MacNaughton, S. (2001). Removal and recovery of heavy metals from wastewaters by supported liquid membranes. *Water Science and Technology*, *43*(2), 341–348.

Yilmaz, A. B. (2003). Levels of heavy metals (Fe, Cu, Ni, Cr, Pb, and Zn) in tissue of *Mugil cephalus* and *Trachurusmediterraneus* from Iskenderun Bay, Turkey. *Environmental Research*, *92*(3), 277–281. https://doi.org/10.1016/s0013-9351(02)00082-8

Yu, B., Zhang, Y., Shukla, A., Shukla, S. S. and Dorris, K. L. (2000). The removal of heavy metal from aqueous solutions by sawdust adsorption — removal of copper. *Journal of Hazardous Materials*, *80*(1–3), 33–42. https://doi.org/10.1016/s0304-3894(00)00278-8

Zamboulis, D., Pataroudi, S., Zouboulis, A. and Matis, K. (2004). The application of sorptive flotation for the removal of metal ions. *Desalination*, *162*, 159–168. https://doi.org/10.1016/s0011-9164(04)00039-6

Zodi, S., Potier, O., Lapicque, F. and Leclerc, J. P. (2009). Treatment of the textile wastewaters by electrocoagulation: Effect of operating parameters on the sludge settling characteristics. *Separation and Purification Technology*, *69*(1), 29–36. https://doi.org/10.1016/j.seppur.2009.06.028

# Chapter - 3

# Microplastic Pollution – An Emerging Concern for Freshwaters: Effects and Overcome to Microplastic Pollution

**Inain Jaies[1*], [2]Imtiyaz Qayoom Feroz, A. Shah[1] and [2]Monisa Mehboob**

[1] Division of Aquatic Animal Health Management, Faculty of Fisheries, SKUAST-K, Rangil-190006, Ganderbal, J&K, India
[2] Division of Aquatic Environmental Management, Faculty of Fisheries, SKUAST-K, Rangil-190006, Ganderbal, J&K, India
*Corresponding email: inainjaies@gmail.com

## Abstract

The microplastic footprints in urban agglomerations in water bodies are strongly influenced by anthropogenic factors. A lot of focus is currently being paid to microplastic (MP) pollution. There is growing concern regarding the potential toxicity of MPs to organisms at the morphological and physiological levels. Fish is a major source of protein for humans; hence the existence and ecotoxicological effects of MPs in fish may have an impact on the safety of aquatic food supplies. The present chapter will provide an overview about: 1) potentially harmful effects that MPs may have on aquatic flora and wildlife; 2) the primary methods for eliminating microplastic from water bodies; and 3) provide a case study to demonstrate how this is achieved.

**Keywords:** Microplastics, Wastewater treatment, Deterioration, Filtration methods, Bioaccumulation.

## Introduction

Microplastics are described as the manufactured solid particles or polymeric matrices, of either primary or secondary manufacturing origin, that are insoluble in water, with regular or irregular shape and size ranging from 1 μm to 5 mm (Frias and Nash

2018). The plastics that reach the marine environment could endure for hundreds of thousands or perhaps millions of years until they are broken down by mechanical and photochemical processes and turn into microplastics (Espinosa *et al.*, 2016). There have been several reports of microplastics in a wide variety of habitats around the world. The atmosphere, marine and fresh water ecosystem, agricultural land, food, drinking water resources and biota have shown the presence of microplastics in a wide variety of forms polymers, sizes, and concentrations (Campanale *et al.*, 2020).

There are two categories of microplastics: primary and secondary. Primary MPs are released into the environment directly through a variety of routes, such as product use (for instance, domestic wastewater systems are polluted by personal care items), inadvertent loss from spills during production or transportation, or abrasion during washing (e.g., laundering of clothing made with synthetic textiles). In other words, primary microplastics are made up of plastics that are instantly spread into the environment (Sundt *et al.*, 2014). While bigger plastics are subjected to weathering, such as when they are exposed to things like wave action, wind abrasion, and UV radiation from sunlight, secondary microplastics are produced as a by-product. According to the shape, there are several different varieties of microplastics, including fibre/line, bead/pellet, fragment, film, and foam (Lusher *et al.*, 2014).

Personal care products (such as facial scrub, toothpaste); blasting abrasives; cleaning products; synthetic textile; paint and wood preservatives; synthetic waxes; plastic pellet manufactures; weathering and abrasion of utensils, footwear soles, infrastructure; domestic items and toys; and medical resources especially nappies, sanitary towels, plasters constitute the different sources of microplastics (Mehra *et al.*, 2020). Microplastics get added up into the fresh water bodies by terrestrial and atmospheric routes. Microplastic pollution is impaired by sewage discharges from cities and towns and factories that produce plastic. Important source of plastic pollutants can be found in urban areas (Luo *et al.*, 2019). Numerous microplastics are found in agricultural soil as a result of the fertiliser, plastic film, or plastic greenhouse (Kumar *et al.*, 2020). Land-based input, shipping, marine aquaculture and fishing are some of the sources of microplastics in the marine environment (Jang *et al.*, 2020). They are dispersed throughout the marine ecosystem in the sea surface, water column, sediments, and marine creatures. Microplastics are less prevalent in the water column than in the surface waters (Xiong *et al.*, 2018).

Environmental variables, their interactions with microorganisms, and the type of microplastic (crystallinity, biodegradability, surface properties, oxidation resistance and residual monomers) all affect how quickly they degrade in aquatic ecosystems. Microplastic degradation techniques come in a variety of forms, including biological, physical, and chemical ones (Yuan *et al.*, 2020). Microplastics can degrade physically through photo-, thermal-, and mechanical-degradation processes. Thermal oxidation and hydrolysis degradation are the two basic types of chemical degradation. The process of biological degradation is carried out by microorganisms, primarily bacteria, mould (fungi), and algae. Microbial degradation can be caused by enzyme catalysis or hydrolysis (Yuan *et al.*, 2020).

## Effect of microplastics on flora and fauna

Aquatic plants are poisoned by MPs. Large rooted plants' primary buds may become shorter due to sorbed MPs, which also impeded freshwater phytoplankton's capacity to establish roots and photosynthesize and maintain cell viability. Yu *et al.*, 2020 subjected *Utricularia vulgaris*, an aquatic plant, to 1, 2, and 5m polystyrene fluorescent MP particles at concentrations of 15, 70, and 140 mg/L for 7 days as part of their investigation. On *U. vulgaris* growth rate, morphological makeup, and physiological traits, the harmful effects of MPs were evaluated. The findings demonstrated that, in comparison to the control group, the relative growth rates and functional characteristics of leaves (morphological and photosynthetic) were considerably decreased at a high concentration of MP particles (140 mg/L). The bioaccumulation of MPs in the bladders was most likely the cause of the effects on growth performance. The impact of polyethylene micro-beads on the floating freshwater plant duckweed *Lemna minor* was also examined by Kalciková *et al.* in 2017. Polyethylene micro-beads had no specific effect on the pace of leaf growth or the amount of photosynthetic pigments in duckweed leaves, but they had a strong mechanical blocking effect on root growth. The viability of root cells was likewise decreased by sharp particles, whereas the effect of micro-beads with a smooth surface was ignored. Dovidat *et al.*, (2019) looked at the impact of nano- and microplastics on the vascular plant *Spirodela polyrhiza*, a species of freshwater duckweed. For 120 hours, *S. polyrhiza* was exposed to concentrations of 102 to 106 particles $mL^{-1}$. Confocal microscopy revealed that the duckweed's roots, in particular, had nano-scale plastic particles adsorbed on their surface. Additionally, the transmission of plastic particles to various herbivorous species within the ecosystem may occur as a result of the adsorption of plastic particles to *S. polyrhiza* roots as discovered in the study.

Polypropylene, polyethylene and polystyrene are the primary MP pollutants while polypropylene and polyethylene are also frequently found in fish and macro-invertebrates, these polymers are the most common microplastic pollutants in water and sediments (Cera *et al.*, 2020). Microplastics are ingested by organisms as they swallow their food, and over time, bioaccumulation of microplastics causes a variety of adverse effects. Due to ingesting of microplastics, organ system disruption, digestive system disruption, immune system deterioration, impotence, various respiratory and circulatory difficulties, even organ failure, and in severe situations, organism death, occur (Triebskorn *et al.*, 2019). The microplastics may reduce food intake, impede growth, result in oxidative damage, and cause aberrant behaviour in fish and other aquatic species. Due to their abundance, persistence, and movement across scales, as well as their capacity for extensive distribution and potential for geophysical and biological impact, marine microplastic debris poses a concern on a worldwide scale (Zalasiewicz *et al.*, 2016). Those nano-scale microplastics cross the biological barrier and build up in tissues, which cause the production of ROS, have an impact on lipid metabolism, and possibly have additional molecular effects on life (Li *et al.*, 2021). It has been demonstrated that microplastics clog feeding appendages and restrict food intake, causes physical harm and oxidative stress, impair energy allocation in a variety of aquatic animals, and in some cases cause damage to the alimentary canal (Cole *et al.*, 2015). Some of the compounds linked to microplastics have endocrine disrupting activity, which is what causes organisms' hormonal imbalance (Sazakli and Leotsinidis, 2019). Microplastics can stunt marine life's growth and development and harm its genetic makeup.

**Table 1: Toxicological effects of microplastics in marine fish**

| S. No. | Type of Microplastic | Fish species | Effect |
|---|---|---|---|
| 1 | Polyethylene | *Pomatoshistos microps* | AChE activity decrease |
| 2 | Polyethylene terephthalate | *Acanthochormis polyacanathus* | Growth decrease |
| 3 | Polyethylene | *Dictrarchus labrax* | Mortality increase |
| 4 | Polyvinyl Chloride | *Dictrarchus labrax* | Inflammation |
| 5 | Polymer | *Dictrarchus labrax* | Swimming speed decrease |
| 6 | Polyethylene | *Oryzias lapites* | Abnormal proliferation of sperm cells |
| 7 | Polystyrene | *Carrassius carassius* | Vitality decrease |
| 8 | Polystyrene | *Danio rerio* | Body length decrease |

| 9 | Polyethylene, Polypropylene, Polyamide, PolyvinylChloride | *Danio rerio* | Intestinal injury |
|---|---|---|---|

The importance of attached particles was also demonstrated in the marine crabs *Carcinus maenas* and *Uca rapax*. In addition to causing an accumulation in the hepatopancreas and stomach of affected crabs, exposure to microplastics also caused an accumulation in their gills. The branchial function may be impacted by respiratory absorption and microplastic adherence to the gills that follows (Watts *et al.*, 2014). Corals are not immune to the effects of microplastics because some corals are known to consume microplastics, which have a detrimental effect on their growth, energy level, and the frequency of pathogens in reefs (Lamb *et al.*, 2018). When microplastics are consumed by sea urchins, pelagic planktotrophic pluteus larva frequently change form (Messinetti *et al.*, 2018). Additionally, microplastics can serve as a synthetic substrate for microbes. By providing surfaces for microorganisms and rafting the colonised organisms over great distances, microplastics create biofilms. Additionally, it has been demonstrated that biofilms with potentially harmful microbes can grow on plastic in the marine environment.

Additionally, freshwater biota like fish, crabs, and algae also suffer from the effects of microplastics. Numerous fish species suffer organ damage from the suspended solids of microplastics. These have been discovered to go into epithelial cells, cause lipid peroxidation, disrupt respiratory function, and decrease the tolerance of infection by pathogens (Michel *et al.*, 2013). At lower concentrations of *high-density polyethylene fragments* (50 μg $L^{1}$), Karami *et al.*, 2016 noticed histological changes in the gills of the African catfish (Clarias gariepinus), such as basal cell hyperplasia and necrosis in connective tissue, as well as blood biochemistry parameters, such as plasma cholesterol levels and blood HDL levels. For higher particle concentrations (500 μg $L^{-1}$), more severe effects (epithelial lifting, hyperplasia, and widespread cell sloughing) were described. Additionally, the extent of tissue alteration in the exposed individuals' livers was considerably influenced by concentrations of 500 μg $L^{-1}$. Crustaceans that are exposed to microplastics suffer negative effects on their ability to feed, body mass, metabolic rate, and energy reserves. The marine *Tigriopus japonicus* was shown by Lee *et al.* 2013 to consume 0.05, 0.5, and 6 m PS beads non-selectively. While everyone survived an acute exposure (96 h), a two-generation chronic exposure to 0.05 (>12.5 g $mL^{1}$) and 0.5 g $mL^{1}$ PS beads resulted in death that was size- and concentration-dependent and significantly reduced fecundity by 0.5 and 6 g $mL^{1}$. Studies have shown that the microplastics affect the plankton population as well. The size of the microplastic may have an impact on how it affects plankton, with smaller microplastics being more hazardous life (Li *et al.*, 2021). The maximum concentration of secondary microplastics (105 $PmL^{-1}$) led to increased mortality and somewhat reduced reproduction of daphnids, according to Ogonowski *et al.*, 2016. A general comparison and conclusion are, however, constrained by incoherent exposure regimes (various particle sizes, concentrations, and exposure durations, among others).

**Impacts of microplastics on biofilm:**

In addition to having the ability to transport chemicals, microplastics also have the capacity to act as a substrate for microbes. The microbial colonisation of microplastics and the biopolymer adsorption increases the nutritional value and enhances the "flavour," making them more enticing to biota. In contrast, the presence of pathogens (Kirstein *et al.*, 2016) and toxic algae/bacteria in these may cause infections, chemical toxicity, or the avoidance of microplastics due to "terrible taste." Due to the composition and quantities of additives, such as prooxidants and starch, biofilm formation on synthetic polymers, such as polyethylenes (LDPE, HDPE, LLDPE) and polyvinyl chloride, affects the species composition of biofilm microorganisms. The process of biofilm formation on a microplastic particle is depicted in Figure 1.

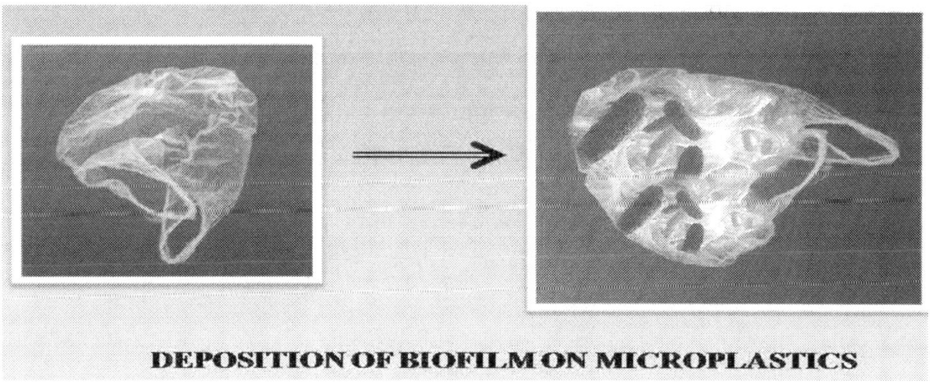


**Figure 1: Process of biofilm formation on a microplastic**

Lifshitz-van der Waals forces and the electrostatic contact between the cell surface and the substrate act indirectly and reversibly during the early stage of bacterial adhesion to the surface, which can result in greater attachment via adhesion receptors. Extracellular polymers, LPS, or proteins are then created in an already irreversible stage of adhesion, following which the micro-colonies grow as a result of the proliferation of the initial adhering cells (Wang *et al.*, 2021). Small microplastics less than 1 m, which are also sometimes referred to as nano-plastics, can easily pierce the biofilm matrix (Drury *et al.*, 1993), impairing the biofilm's ability to function biologically by, for instance, causing oxidative stress (Miao *et al.*, 2019). Larger microplastics, on the other hand, which are unable to pass through the biofilm matrix, have a tendency to assemble at the water-biofilm interface. Microplastics may act as a physical barrier for fluxes of, say, oxygen or nutrients because of this accumulation on the water-biofilm interface.

## Methods of microplastic removal from water bodies:

### » Physical methods:

Physical principles like adsorption, filtration, sedimentation etc. are commonly used for microplastic removal from water bodies. The most popular techniques for treating water are various types of filtrations. Some of these can include membrane filtration, activated carbon filtration, disc filtration, fast sand filtration; etc. Among various filtration methods, the quick sand filter is regarded as an effective filtration technology since it can remove all types of microplastics using the sand grains. Layers of sand consisting of 1 mm of gravel with a grain size of 35 mm and 0.5 mm of quartz with a grain size of 0.1-0.5 mm are used to create rapid sand filters. This technique is a low-cost, effective way for MPs larger than 20 m. Several authors have mentioned different physical methods for removal of microplastics from aquatic ecosystems that have been tabulated in table 2.

**Table 2: Physical methods for microplastic removal**

| S. No. | Principle Method | Media used | Reference |
|---|---|---|---|
| 1 | Adsorption | Magnetic Polyoxometalate Supported Ionic Liquid Phases | Misra *et al.*, 2020 |
| 2 | | Biochar Adsorbents | Siipola *et al.*, 2020 |
| 3 | | Coagulative colloidal gas aphrons | Zhang *et al.*, 2021 |
| 4 | Filtration | Rapid sand filter | Talvitie *et al.*, 2017 |
| 5 | | Zirconium metal-organic frameworks-based foam | Chen *et al.*, 2020 |
| 6 | Retention | Disc filter | Simon *et al.*, 2019 |
| 7 | | Disc filter | Talvitie *et al.*, 2017 |
| 8 | Flocculation | Electrocoagulation | Perren *et al.*, 2018 |

| 9 | Flotation | Dissolved air flotation | Talvitie *et al.*, 2017 |
|---|---|---|---|
| 10 | Gravitational filter | Biofilter | Liu *et al.*, 2020 |

» **Chemical methods:**

Some of the key mechanisms that are frequently examined for MP removal include coagulation and flocculation. The primary goal of the coagulation/flocculation process is to separate the colloidal particles that are already in the solution by neutralising their charges, creating flocs, and then removing them via sedimentation or filtration (Zhou *et al.*, 2021). Researchers utilise a variety of compounds to improve the effectiveness of the coagulation and flocculation process.

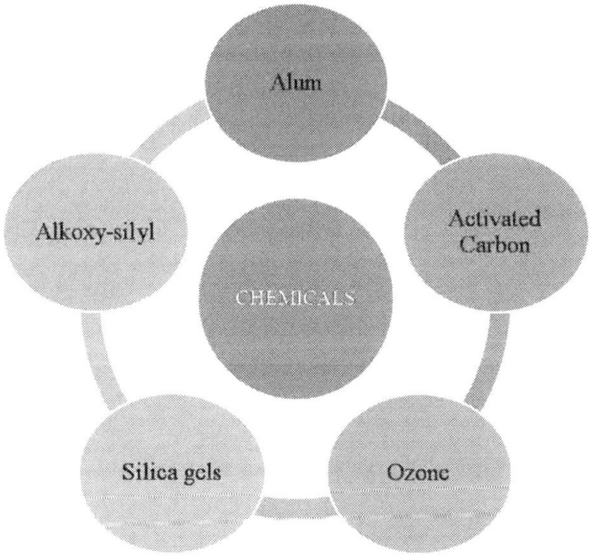


Due to its affordability, great efficiency, and ease, the coagulation sedimentation process (CSP) has emerged as a key technique for treating MPs in water. The removal of microplastic particles (180 nm-125 m) by filtration and a combination coagulation/flocculation/filtration process were compared by Zhang *et al.*, 2020. It was demonstrated that the size of microplastic particles does not affect the removal efficiency linearly; the largest particles in this study were removed with almost absolute efficiency (106-125 m with >99% removal efficiency), but particles with sizes of 10-20 m were removed with less than 90% efficiency. Ozonation is described as a process that causes the exposed substance to deteriorate through the creation of ozone. Ozone molecules attack the bonds in organic molecules as a reactive species (Singh and Sharma 2008). Hidayaturrahman and Lee showed that ozonation on the tertiary stage of wastewater treatment plants was successful in removing microplastics (particles smaller than 1.2 m), and the efficiency was close to 90%, compared to removal by disc-filtration (79%), or quick sand filteration (74%). Although ozonation alone has an impact on the surface characteristics of micro-particles, it frequently increases the hazardous qualities of the treated effluent (Krystynik *et al.*, 2021). In a recent study, Ariza-Tarazona *et al.* (2019) investigated the removal of polyethylene microplastics using ultrafiltration and iron and aluminium salt coagulants. $Al^{3+}$ performs better than $Fe^{3+}$, according to the results of the studies, which were conducted with varying $Al^{3+}$ and $Fe^{3+}$ ion concentrations. Additionally, at a low concentration of Al coagulant source, 0.5 mM, the pH of the solution barely affected the efficacy of removing microplastics, whereas removal efficiency dropped as pH increased, especially for small-sized microplastics with a diameter smaller than 0.5 mm. They discovered that under high Al dose of 5 mM, the boosting coagulation agent polyacrylamide (PAM), enhanced the removal efficiency for small microplastics significantly better than for large particles. Even though alum by itself is a good coagulating agent, it was shown that when cationic polyamine-coated sand was combined with alum, the removal was increased by 26.8%. It was discovered that the removal of microplastics from water also depends on the particle morphology and follows the separation behaviour of elongated-rough>elongated smooth>spherical-rough>spherical-smooth. This pattern demonstrates that spherical-smooth particles are eliminated less effectively than elongated-rough particles (Shahi *et al.*, 2020).

» **Biological methods:**

In biological approaches, organisms break down microplastics that are already present in the environment to combat the problem of microplastic pollution. The microorganisms, primarily bacteria, mould (fungi), and algae, participate in biological deterioration. MPs are broken down by microbes through enzyme catalysis and hydrolysis (Yuan *et al.*, 2020). The ability of several organisms to break down microplastics found in water and wastewater has been investigated. The majority of biological species researched in relation to the capability for degrading microplastics are bacteria (Badola *et al.*, 2022). *Bacillus cereus* and *Bacillus gottheilii,* two types of bacillus bacterial strains isolated from mangrove sediments, were studied by Auta *et al.*, (2017) for their ability to remove various microplastics made of polyethylene, polystyrene, polyethylene terephthalate, and polypropylene. Electron microscopy and FTIR studies were used to scan the morphological and structural changes, and the weight loss of the microplastics was used to calculate the rate of biodegradation. *B. cereus* isolate with polystyrene microplastics provided the fastest mass decrease (0.0019/ day) and the lowest half-life of degradation (363.16 days), whereas *B. gottheilii* on polyethylene provided 0.0016/day and 431.25 days. Based on the mass and size fluctuations of the microplastics in a batch reactor, Paço *et al.* (2017) investigated the potential of the fungus *Zalerion maritimum*, a naturally occurring fungus in maritime habitats, for the biodegradation of polyethylene microplastics. In order to determine the reaction rate order, they measured the contaminant concentration at different time intervals. They discovered that the biodegradation kinetics of the microplastics obey fractional order, which can be explained in terms of the complex mechanism anticipated for such reactions.

An outstanding biological technique for eliminating fine MPs is microalgae (Cunha *et al.*, 2020). Similar to corals, which have been laboratory-tested to be natural sinks for MPs in oceans (Martin *et al.*, 2019), microalgae can be used for the removal of MPs. Arossa *et al.* 2019 have reported on the assessment of *Tridacna maxima*, a gigantic clam from the Red Sea, for the removal of 53-500 m polyethylene microplastics from wastewater. As clam shells removed 66.03% of the microplastics from the wastewater, their findings demonstrated the crucial function they play in the process. In a lab setting, the MPs removal effectiveness of mushroom coral taken from the reef of the island of Magoodhoo, Faafu Atoll, Republic of Maldives, demonstrated efficiency of 97% for virgin + bio-fouled plastic of size 200-1000 m (Corona *et al.*, 2020).

**Case study:**

The functioning of a wastewater treatment facility located in Southwest Europe was examined by Menéndez-Manjón *et al.* in 2022 in relation to the annual occurrence and fate of microplastics. Microplastics were measured and identified using stereomicroscopy and FTIR spectrophotometry from samples taken weekly at various locations throughout the facility (influent, secondary effluent, final effluent, and sludge). This plant was built to handle an average daily flow of 8000 $m^3$, or the equivalent of 85,000 people, and several physical, chemical, and biological processes were used in it in the order shown in figure 2.

In wastewater and sludge samples, the bulk of the microplastics were fibres and fragments. In terms of chemical composition, polyethylene (PE), polyethylene terephthalate (PET), and polypropylene (PP) stood out in the water samples, whereas PET, polyamide (PA), and polystyrene were the primary polymers in the sludge samples (PS). During the wastewater treatment procedures, the microplastics that were the easiest to remove were those larger than 500 m in size.

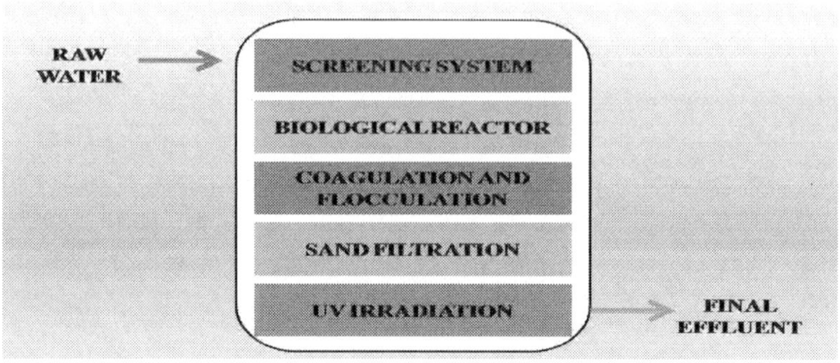


**Figure 2: Wastewater treatment facility**

Results showed that removal efficiencies for microplastics (MPs) were between 89% and 95% for the whole period under study, hence there were no significant monthly fluctuations. Dry sludge samples had MP values between 12 to 39 MPs/g, which was almost 79% of the total MPs eliminated during the wastewater treatment operations.

Similar to this, Akarsu *et al.*, 2021 assessed the effectiveness of membrane filtration and electrocoagulation-electroflotation (EC/EF) methods in eliminating microplastics from wastewaters. To achieve the highest level of treatment success, the effects of several electrode combinations (Fe-Al and Al-Fe), current densities (10–20 A/m$^2$), pH (4.0–10.0), and operating periods (0–120 min) on the removal of two distinct polymer particles in water were examined. Investigations were also done into the impact of pressure (1-3 bar) on the effectiveness of membrane filter removal. The greatest removal efficiencies were 100% for both polymer types when the electrodes were Al-Fe, the initial pH was 7, the current density was 20 A/m$^2$, and the reaction was carried out. Additionally, the membrane filtration approach revealed 100% removal efficiency. Additionally, the one-year average data of a plant being treated with real-scale membranes was compared to these laboratory-scale results. The findings showed that the proposed techniques offered maximal microplastic removal efficiency (100%) in comparison to traditional secondary and tertiary treatment approaches (2-81.6%).

## Conclusion

Different routes, such as air deposition, runoff from contaminated soil, or municipal wastewater, allow microplastics to reach aquatic bodies. Due to their well-documented ubiquity, extended residence period, and propensity to be consumed by biota, microplastics may present a concern to aquatic habitats. Because they can be consumed more easily than larger particles throughout the food web, microplastics are emerging contaminants posing significant threats to the ecosystem. Enhancing the shelf life of plastic products, implementing effective waste management techniques, and raising awareness can significantly reduce the amount of litter that is introduced into the environment, allowing the aquatic ecosystem to recover.

## Bibliography

Akarsu, C., Kumbur, H., & Kideys, A. E. (2021). Removal of microplastics from wastewater through electrocoagulation-electroflotation and membrane filtration processes. *Water Science and Technology*, 84(7), 1648-1662.

Ariza-Tarazona, M. C., Villarreal-Chiu, J. F., Barbieri, V., Siligardi, C., & Cedillo-González, E. I. (2019) New strategy for microplastic degradation: green photocatalysis using a protein-based porous N-TiO2 semiconductor. *Ceram Int*, 45(7), 9618–9624.

Arossa, S., Martin, C., Rossbach, S., & Duarte, C. M. (2019) Microplastic removal by red sea giant clam (*Tridacna maxima*). *Environ Pollut,* 252, 1257–1266.

Auta, H., Emenike, C., & Fauziah, S. (2017) Screening of *Bacillus* strains isolated from mangrove ecosystems in Peninsular Malaysia for microplastic degradation. *Environ Pollut*, 231, 1552–1559.

Badola, N., Bahuguna, A., Sasson, Y., & Chauhan, J. S. (2022). Microplastics removal strategies: A step toward finding the solution. *Frontiers of Environmental Science & Engineering*, 16(1), 1-18.

Campanale, C., Massarelli, C., Savino, I., Locaputo, V., & Uricchio, V. F. (2020). A detailed review study on potential effects of microplastics and additives of concern on human health. *International journal of environmental research and public health*, 17(4), 1212.

Cera, A., Cesarini, G., & Scalici, M. (2020). Microplastics in freshwater: What is the news from the world? *Diversity,* 12, 276.

Chen, Y. J., Chen, Y., Miao, C., Wang, Y. R., Gao, G. K., Yang, R. X., Zhu, H. J., Wang, J. H., Li, S. L., & Lan, Y. Q. (2020). Metal–organic framework-based foams for efficient microplastics removal. *Journal of Materials Chemistry. Materials for Energy and Sustainability*, 8(29), 14644– 14652.

Cole, M., Lindeque, P., Fileman, E., Halsband, C., & Galloway, T.S. (2015). The impact of polystyrene microplastics on feeding, function and fecundity in the marine copepod *Calanus helgolandicus. Environmental science & technology*, 49(2), 1130-7.

Corona, E., Martin, C., Marasco, R., & Duarte, C. M. (2020). Passive and active removal of marine microplastics by a mushroom coral (*Danafungia scruposa*). *Frontiers in Marine Science*, 7, 1 –9.

Cunha, C., Silva, L., Paulo, J., Faria, M., Nogueira, N., & Cordeiro, N. (2020). Microalgal-based biopolymer for nano- and microplastic removal: A possible biosolution for wastewater treatment. *Environmental Pollution*, 263, 114385 1-10.

Dovidat, L. C., Brinkmann, B. W., Vijver, M. G., & Bosker, T. (2020). Plastic particles adsorb to the roots of freshwater vascular plant *Spirodela polyrhiza* but do not impair growth. *Limnology and Oceanography Letters*, 5(1), 37-45.

Drury, W. J., Characklis, W. G., & Stewart, P. S. (1993). Interactions of 1 μm latex particles with Pseudomonas aeruginosa biofilms. *Water Research*, 27(7), 1119–1126.

Espinosa, C., Esteban, M. Á., & Cuesta, A. (2016). Microplastics in aquatic environments and their toxicological implications for fish. *Toxicology–new aspects to this scientific conundrum. InTech, Rijeka*, 113-145.

Frias J., & Nash, R. (2018). Microplastics: Finding a consensus on the definition. *Mar. Pollut. Bull.*, 138, 145–147. doi: 10.1016/j.marpolbul.2018.11.022

Jang, M., Shim, W. J., Cho, Y., Han, G. M., Song, Y. K., & Hong, S. H. (2020). A close relationship between microplastic contamination and coastal area use pattern. *Water Research*, 171, 115400.

Kalčíková, G., Gotvajn, A. Ž., Kladnik, A., & Jemec, A. (2017). Impact of polyethylene microbeads on the floating freshwater plant duckweed Lemna minor. *Environmental Pollution*, 230, 1108-1115

Karami, A., Romano, N., Galloway, T., & Hamzah, H. (2016) Virgin microplastics cause toxicity and modulate the impacts of phenanthrene on biomarker responses in African catfish (*Clarias gariepinus*). *Environ Res*, 151, 58–70.

Kirstein, I. V., Kirmizi, S., Wichels, A., Garin-Fernandez, A., Erler, R., Löder, M., & Gerdts, G. (2016). Dangerous hitchhikers? Evidence for potentially pathogenic *Vibrio* spp. on microplastic particles. *Marine environmental research*, 120, 1-8.

Krystynik, P., Strunakova, K., Syc, M., & Kluson, P. (2021). Notes on Common Misconceptions in Microplastics Removal from Water. *Applied Sciences*, 11(13), 5833.

Kumar, M., Xiong, X., He, M., Tsang, D. C., Gupta, J., Khan, E., Harrad, S., Hou, D., Ok, Y. S., & Bolan, N. S. (2020). Microplastics as pollutants in agricultural soils. *Environmental Pollution*, *265*, 114980.

Lamb, J. B., Willis, B. L., Fiorenza, E. A., Couch, C. S., Howard, R., Rader, D. N., True, J. D., Kelly, L. A., Ahmad, A., Jompa, J., & Harvell, C. D. (2018). Plastic waste associated with disease on coral reefs. *Science*, 359(6374), 460-462.

Lee, K., Shim, W. J., Kwon, O. Y., & Kang, J. (2013). Size-dependent effects of micro polystyrene particles in the marine copepod *Tigriopus japonicus*. *Environ Sci Technol*, 47, 11278–11283.

Li, Y., Sun, Y., Li, J., Tang, R., Miu, Y., & Ma, X. (2021). Research on the Influence of Microplastics on Marine Life. In *IOP Conference Series: Earth and Environmental Science* 631(1), 012006). IOP Publishing.

Liu, F., Nord, N. B., Bester, K., & Vollertsen, J. (2020). Microplastics removal from treated wastewater by a biofilter. *Water*, 12(4), 1085

Luo, W., Su, L., Craig, N. J., Du, F., Wu, C., & Shi, H. (2019). Comparison of microplastic pollution in different water bodies from urban creeks to coastal waters. *Environmental pollution*, 246, 174-182.

Lusher, A. L., Burke, A., O'Connor, I., & Officer, R. (2014). Microplastic pollution in the Northeast Atlantic Ocean: validated and opportunistic sampling. *Marine pollution bulletin*, 88(1-2), 325-333.

Martin, C., Corona, E., Mahadik, G. A., & Duarte, C. M. (2019). Adhesion to coral surface as a potential sink for marine microplastics. *Environmental Pollution*, 255, 113281.

Mehra, S., Sharma, K., Sharma, G., Singh, M., & Chadha, P. (2020). Sources, Fate, and Impact of Microplastics in Aquatic Environment. *Emerging Contaminants*.

Menéndez-Manjón, A., Martínez-Díez, R., Sol, D., Laca, A., Laca, A., Rancaño, A., & Díaz, M. (2022). Long-Term Occurrence and Fate of Microplastics in WWTPs: A Case Study in Southwest Europe. *Applied Sciences*, 12(4), 2133.

Messinetti, S., Mercurio, S., Parolini, M., Sugni, M., & Pennati, R. (2018). Effects of polystyrene microplastics on early stages of two marine invertebrates with different feeding strategies. *Environmental Pollution*, *237*, 1080-1087.

Miao, L., Hou, J., You, G., Liu, Z., Liu, S., Li, T., ... & Qu, H. (2019). Acute effects of nanoplastics and microplastics on periphytic biofilms depending on particle size, concentration and surface modification. *Environmental Pollution*, 255, 113300.

Michel, C., Schmidt-Posthaus, H., & Burkhardt-Holm, P. (2013). Suspended sediment pulse effects in rainbow trout (Oncorhynchus mykiss)—relating apical and systemic responses. *Canadian Journal of Fisheries and Aquatic Sciences*, 70(4), 630-641.

Misra, A., Zambrzycki, C., Kloker, G., Kotyrba, A., Anjass, M. H., Franco Castillo, I., ... & Streb, C. (2020). Water purification and microplastics removal using magnetic polyoxometalate-supported ionic liquid phases (magPOM-SILPs). *Angewandte Chemie International Edition*, 59(4), 1601-1605.

Ogonowski, M., Schür, C., Jarsén, Å., & Gorokhova, E. (2016). The effects of natural and anthropogenic microparticles on individual fitness in *Daphnia magna*. *PloS one*, 11(5), e0155063.

Paço, A., Duarte, K., da Costa, J. P., Santos, P. S., Pereira, R., Pereira, M. E., ... & Rocha-Santos, T. A. (2017). Biodegradation of polyethylene microplastics by the marine fungus *Zalerion maritimum*. *Science of the Total Environment*, 586, 10-15.

Perren, W., Wojtasik, A., & Cai, Q. (2018). Removal of microbeads from wastewater using electrocoagulation. *ACS omega*, 3(3), 3357-3364.

Sazakli, E., & Leotsinidis, M. (2019). Possible effects of microplastics on human health. *Microplastics in Water and Wastewater, Hrissi K. Karapanagioti, Ioannis K. Kalavrouziotis Download citation file: Ris (Zotero) Reference Manager EasyBib Bookends Mendeley Papers EndNote RefWorks BibTex Close Search.*

Shahi, N. K., Maeng, M., Kim, D., & Dockko, S. (2020). Removal behavior of microplastics using alum coagulant and its enhancement using polyamine-coated sand. *Process Safety and Environmental Protection*, 141, 9-17.

Siipola, V., Pflugmacher, S., Romar, H., Wendling, L., & Koukkari, P. (2020). Low-cost biochar adsorbents for water purification including microplastics removal. *Applied sciences*, 10(3), 788.

Simon, M., Vianello, A., & Vollertsen, J. (2019). Removal of> 10 μm microplastic particles from treated wastewater by a disc filter. *Water*, 11(9), 1935.

Singh, B., & Sharma, N. (2008). Mechanistic implications of plastic degradation. *Polymer degradation and stability*, 93(3), 561-584.

Sundt, P., Schulze, P. E., & Syversen, F. (2014). Sources of microplastic-pollution to the marine environment. *Mepex for the Norwegian Environment Agency*, 86, 20.

Talvitie, J., Mikola, A., Koistinen, A., & Setälä, O. (2017). Solutions to microplastic pollution–Removal of microplastics from wastewater effluent with advanced wastewater treatment technologies. *Water research*, 123, 401-407.

Triebskorn, R., Braunbeck, T., Grummt, T., Hanslik, L., Huppertsberg, S., Jekel, M., Knepper, T. P., Krais, S., Müller, Y. K., Pittroff, M., Ruhl, A. S., Schmieg, H., Schür, C., .... & Köhler, H. R. (2019). Relevance of nano-and microplastics for freshwater ecosystems: a critical review. *TrAC Trends in Analytical Chemistry*, 110, 375-392.

Wang, J., Guo, X., & Xue, J. (2021). Biofilm-developed microplastics as vectors of pollutants in aquatic environments. *Environmental Science & Technology*, 55(19), 12780-12790.

Watts, A. J., Lewis, C., Goodhead, R. M., Beckett, S. J., Moger, J., Tyler, C. R., & Galloway, T. S. (2014). Uptake and retention of microplastics by the shore crab Carcinus maenas. *Environmental science & technology*, 48(15), 8823-8830.

Xiong, X., Zhang, K., Chen, X., Shi, H., Luo, Z., & Wu, C. (2018). Sources and distribution of microplastics in China's largest inland lake–Qinghai Lake. *Environmental pollution*, 235, 899-906.

Yu, H., Zhang, X., Hu, J., Peng, J., & Qu, J. (2020). Ecotoxicity of polystyrene microplastics to submerged carnivorous Utricularia vulgaris plants in freshwater ecosystems. *Environmental Pollution*, 265, 114830.

Yuan, J. H., Ma, J., Sun, Y. R., Zhou, T., Zhao, Y. C., & Yu, F. (2020). Microbial degradation and other environmental aspects of microplastics/plastics. *Sci. Total Environ.*, 715, 136968

Zalasiewicz, J., Waters, C. N., Do Sul, J. A. I., Corcoran, P. L., Barnosky, A. D., Cearreta, A., ... & Yonan, Y. (2016). The geological cycle of plastics and their use as a stratigraphic indicator of the Anthropocene. *Anthropocene*, 13, 4-17.

Zhang, M., Yang, J., Kang, Z., Wu, X., Tang, L., Qiang, Z., ... & Pan, X. (2021). Removal of micron-scale microplastic particles from different waters with efficient tool of surface-functionalized microbubbles. *Journal of Hazardous Materials*, 404, 124095.

Zhang, Y., Diehl, A., Lewandowski, A., Gopalakrishnan, K., & Baker, T. (2020). Removal efficiency of micro-and nanoplastics (180 nm–125 μm) during drinking water treatment. *Science of The Total Environment*, 720, 137383.

Zhou, G., Wang, Q., Li, J., Li, Q., Xu, H., Ye, Q., ... & Zhang, J. (2021). Removal of polystyrene and polyethylene microplastics using PAC and FeC13 coagulation: Performance and mechanism. *Science of the Total Environment*, 752, 141837.

# Chapter - 4

# Recent Advances in Techniques for Detection of Aquatic Pollution - Present Scenario and Future Prospective

**[1]Vishva N. Barad, [2*]Asif I. Bhadarka, [2]Vivek Shrivastava and [3]Rohini Kalyani**

[1]College of Fisheries Science, Veraval Kamdhenu University, Gandhinagar
[2]Department of Aquaculture, College of Fisheries Science Himmatnagar, Kamdhenu University Gandhinagar, Gujrat, India
[3]Department of Fish Nutrition and Feed Technology, ICAR -Central Institute of Fisheries Education (CIFE) Mumbai, Maharashtra, India
[*]Corresponding email: asifbhadarka7@gmail.com

## Abstract

Water pollution occurs when harmful substances contaminate a stream, river, lake, ocean, aquifer, or other body of water, degrading water quality and rendering it toxic to humans or the environment. There are two main sources of water pollution; point sources and non-point sources. Point sources include factories, wastewater treatment facilities, septic systems, and other sources that are clearly discharging pollutants into water sources. Non-point sources are more difficult to identify, because they cannot be traced back to a particular location. Non-point sources include runoff including sediment, fertilizer, chemicals and animal wastes from farms, fields, construction sites and mines. Landfills can also be a non-point source of pollution, if substances leach from the landfill into water supplies. This chapter provides an overview of current break throughs in nanotechnologies for water and wastewater treatment processes, including nano-based materials such as nano adsorbents, nanometals, nanomembranes, and photocatalysts, taking into account the influence of these nanomaterials on water treatment. When compared to traditional techniques, the positive features of these materials as well as technical hurdles are discussed.

**Keywords:** Water Pollution, Contamination, Degradation, Effluent, Chemical markers, Waste water.

## Introduction

Water is typically referred as polluted when it is impaired by anthropogenic contaminants. Due to these contaminants, it either does not support a human use, such as drinking water, or undergoes a marked shift in its ability to support its biotic communities, such as fish and other edible plants and animals. Natural phenomena such as volcanoes, algal blooms, storms, and earthquakes also cause major changes in water quality and the ecological status of water. Water pollution is a major global problem which requires ongoing evaluation and revision of water resource policy at all levels (international down to individual aquifers and wells). It has been suggested that water pollution is the leading worldwide cause of death and diseases (Pink, 2006; Savedge, 2019). Water pollution accounted for the deaths of 1.8 million people in 2015 (Staff, 2017). The organization Global Oceanic Environmental Survey (GOES) consider water pollution as one of the main environmental problems that can present a danger for the existence of life on earth in the next decades. One of the main concerns is that water pollution, heart phytoplankton that produces 70% of oxygen and removes a large part of carbon dioxide on earth.

## Water Pollution

Water pollution occurs when harmful substances often chemicals or microorganisms contaminate a stream, river, lake, ocean, aquifer, or other water body, degrading water quality and rendering it toxic to humans or the environment. Water pollution is the contamination of water bodies (like oceans, seas, lakes, rivers, aquifers, and groundwater) usually caused due to human activities. Water pollution is any change in the physical, chemical or biological properties of water that will have a detrimental consequence of any living organism.

The organization proposes a number of measures for fixing the situation, but they should be taken in the next 10 years for being effective. India and China are two countries with high levels of water pollution. An estimated 580 people in India die of water pollution related illness (including waterborne diseases) every day. About 90 percent of the water in the cities of China is polluted. As of 2007, half a billion Chinese had no access to safe drinking water (Kahn and Yardley, 2007). In addition to the acute problems of water pollution in developing countries, developed countries also continue to struggle with pollution problems.

### Causes of Water Pollution

Water is uniquely known as a "universal solvent," is able to dissolve more substances than any other liquid on earth, but vulnerable to pollution. It is the reason we have Kool-Aid and brilliant blue waterfalls. Toxic substances from farms, towns, and factories readily dissolve into and mix with it, causing water pollution.

## Types of Water Pollution

### Groundwater

When rain falls and seeps deep into the earth, filling the cracks, crevices, and porous spaces of an aquifer (basically an underground storehouse of water), it becomes groundwater one of our least visible but most important natural resources. Nearly 40 percent of Americans rely on groundwater, pumped to the earth's surface, for drinking water. For some folks in rural areas, it's their only freshwater source. Groundwater gets polluted when contaminants from pesticides and fertilizers to waste leached from landfills and septic systemsmake their way into an aquifer, rendering it unsafe for human use. Ridding groundwater of contaminants can be difficult to impossible, as well as costly. Once polluted, an aquifer may be unusable for decades, or even thousands of years. Groundwater can also spread contamination far from the original polluting source as it seeps into streams, lakes, and oceans.

### Surface water

Covering about 70 percent of the earth, surface water is what fills our oceans, lakes, rivers, and all those other blue bits on the world map. Surface water from freshwater sources (that is, from sources other than the ocean) accounts for more than 60 percent of the water delivered to American homes. But a significant pool of that water is in peril. According to the most recent surveys on national water quality from the U.S. Environmental Protection Agency, nearly half of our rivers and streams and more than one-third of our lakes are polluted and unfit for swimming, fishing, and drinking. Nutrient pollution, which includes nitrates and phosphates, is the leading type of contamination in these freshwater sources. While plants and animals need these nutrients to grow, they have become a major pollutant due to farm waste and fertilizer runoff. Municipal and industrial waste discharges contribute their fair share of toxins as well. There's also all the random junk that industry and individuals dump directly into waterways.

### Ocean water

Eighty percent of ocean pollution (also called marine pollution) originates on landwhether along the coast or far inland. Contaminants such as chemicals, nutrients, and heavy metals are carried from farms, factories, and cities by streams and rivers into our bays and estuaries; from there they travel out to sea. Meanwhile, marine debris plastics blown in by the wind or washed in via storm drains and sewers. Our seas are also sometimes spoiled by oil spills and leaksbig and smalland are consistently soaking up carbon pollution from the air. The ocean absorbs as much as a quarter of man-made carbon emissions.

## Sources of Water Pollution

### Point sources

When contamination originates from a single source, it's called point source pollution. Examples include wastewater discharged legally or illegally by a manufacturer, oil refinery, or wastewater treatment facility, as well as contamination from leaking septic systems, chemical and oil spills, and illegal dumping. While point source pollution originates from a specific place, it can affect miles of waterways and ocean (Wu and Chen, 2013; Zhou *et al.*, 2016; Callery and Healy, 2019).

### Nonpoint sources

Nonpoint source pollution refers to diffuse contamination that does not originate from a single discrete source. This type of pollution is often the cumulative effect of small amounts of contaminants gathered from a large area. A common example is the leaching out of nitrogen compounds from fertilized agricultural lands. Nutrient runoff in storm water from "sheet flow" over an agricultural field or a forest is also cited as examples of non-point source pollution. Contaminated storm water washed off of parking lots, roads and highways, called urban runoff, is sometimes included under the category of non-point sources. This runoff becomes a point source because it is typically channeled into storm drain systems and discharged through pipes to local surface waters (Hsieh and Yang, 2007; Wu and Chen, 2013; Callery and Healy, 2019).

### Contaminants

The specific contaminants leading to pollution in water include a wide spectrum of chemicals, pathogens, and physical changes such as elevated temperature and discoloration. While many of the chemicals and substances that are regulated may be naturally occurring (calcium, sodium, iron, manganese, etc.) the concentration usually determines what is a natural component of water and what is a contaminant. High concentrations of naturally occurring substances can have negative impacts on aquatic flora and fauna. Oxygen-depleting substances may be natural materials such as plant matter (e.g., leaves and grass) as well as man-made chemicals. Other natural and anthropogenic substances may cause turbidity (cloudiness) which blocks light and disrupts plant growth, and clogs the gills of some fish species. Alteration of water's physical chemistry includes acidity (change in pH), electrical conductivity, temperature, and eutrophication. Eutrophication is an increase in the concentration of chemical nutrients in an ecosystem to an extent that increases the primary productivity of the ecosystem. Depending on the degree of eutrophication, subsequent negative environmental effects such as anoxia (oxygen depletion) and severe reductions in water quality may occur, affecting fish and other animal populations (Harrison, 2001; Xagoraraki and Kuo, 2008; Shaltami *et al.*, 2019).

## Pathogens

Disease-causing microorganisms are referred as pathogens. Pathogens can produce waterborne diseases in either human or animal hosts. Coliform bacteria, which are not an actual cause of disease, are commonly used as a bacterial indicator of water pollution. Other microorganisms sometimes found in contaminated surface waters that have caused human health problems include: *Burkholderia pseudomallei*, *Cryptosporidium parvum*, *Giardia lamblia*, *Salmonella*, *Norovirus* and other viruses, and Parasitic worms including the *Schistosoma* type. High levels of pathogens may result from on-site sanitation systems (septic tanks, pit latrines) or inadequately treated sewage discharges. Older cities with ageing infrastructure may have leaky sewage collection systems (pipes, pumps, valves), which can cause sanitary sewer overflows. Some cities also have combined sewers, which may discharge untreated sewage during rain storms. Silt (sediment) from sewage discharges also pollutes water bodies. Pathogen discharges may also be caused by poorly managed livestock operations (Harrison, 2001; Kong *et al.*, 2009; Reder *et al.*, 2017; Xue *et al.*, 2020).

## Organic, Inorganic and Macroscopic Contaminants

Contaminants may include organic and inorganic substances. Many of the chemical substances are toxic. Organic water pollutants include: Detergents; Disinfection by-products found in chemically disinfected drinking water, such as chloroform; Food processing waste, which can include oxygen-demanding substances, fats and grease; Insecticides and herbicides, a huge range of organohalides and other chemical compounds; Petroleum hydrocarbons, including fuels (gasoline, diesel fuel, jet fuels, and fuel oil) and lubricants (motor oil), and fuel combustion byproducts, from storm water runoff; Volatile organic compounds, such as industrial solvents, from improper storage; Chlorinated solvents, which are dense non-aqueous phase liquids, may fall to the bottom of reservoirs, since they don't mix well with water and are denser (e.g., polychlorinated biphenyl (PCBs) and trichloroethylene); Perchlorate; Various chemical compounds found in personal hygiene and cosmetic products; and Drug pollution involving pharmaceutical drugs and their metabolites, this can include antidepressant drugs or hormonal medicines such as contraceptive pills. These molecules can be small and difficult for treatment plants to remove without expensive upgrades. Inorganic water pollutants include: 1) Acidity caused by industrial discharges (especially sulfur dioxide from power plants), 2) Ammonia from food processing waste, 3) Chemical waste as industrial by-products, 4) Fertilizers containing nutrients- nitrates and phosphates—which are found in storm water runoff from agriculture, as well as commercial and residential use, 5) Heavy metals from motor vehicles (via., urban storm water runoff) and acid mine drainage, 6) Secretion of creosote preservative into the aquatic ecosystem, and 7) Silt (sediment) in runoff from construction sites, logging, slash and burn practices or land clearing sites. Macroscopic pollution – large visible items polluting the water – may be termed floatables in an urban storm water context, or marine debris when found on the open seas, and can include such items as: Trash or garbage (e.g., paper, plastic, or food waste) discarded by people on the ground, along with accidental or intentional dumping of rubbish, that are washed by rainfall into storm drains and eventually discharged into surface waters, Nurdles, small ubiquitous waterborne plastic pellets (Ruan *et al.*, 2013; Manyatshe *et al.*, 2017; Artabe *et al.*, 2020).

## Temperature Change

Thermal pollution is the rise or fall in the temperature of a natural body of water caused by human influence. Thermal pollution, unlike chemical pollution, results in a change in the physical properties of water. A common cause of thermal pollution is the use of water as a coolant by power plants and industrial manufacturers. Elevated water temperatures decrease oxygen levels, which can kill fish and alter food chain composition, reduce species biodiversity, and foster invasion by new thermophilic species. Urban runoff may also elevate temperature in surface waters. Thermal pollution can also be caused by the release of very cold water from the base of reservoirs into warmer rivers (Kennedy, 2004).

## Measurement

Water pollution may be analyzed through several broad categories of methods: physical, chemical and biological. Most involve collection of samples, followed by specialized analytical tests. Some methods may be conducted in situ, without sampling, such as temperature. Government agencies and research organizations have published standardized, validated analytical test methods to facilitate the comparability of results from different testing events. Sampling of water for physical or chemical testing can be done by several methods, depending on the accuracy needed and the characteristics of the contaminants. Many contamination events are sharply restricted in time, most commonly in association with rain events. For this reason, grab samples are often inadequate for fully quantifying contaminant levels. Scientists gathering this type of data often employ auto-sampler devices that pump increments of water at either time or discharge intervals. Sampling for biological testing involves collection of plants and animals from the surface water body. Depending on the type of assessment, the organisms may be identified for bio-surveys (population counts) and returned to the water body, or they may be dissected for bioassays to determine toxicity. Common physical tests of water include temperature, solids concentrations and turbidity. Water samples may be examined using the principles of analytical chemistry. Many published test methods are available for both organic and inorganic compounds. Frequently used methods include pH, biochemical oxygen demand (BOD), chemical oxygen demand (COD), nutrients (nitrate and phosphorus compounds), metals (including copper, zinc, cadmium, lead and mercury), oil and grease, total petroleum hydrocarbons (TPH), and pesticides. Biological testing involves the use of plant, animal or microbial indicators to monitor the health of an aquatic ecosystem. They may be any biological species or group of species whose function, population, or status can reveal what degree of ecosystem or environmental integrity is present. One example of a group of bio-indicators is the copepods and other small water crustaceans that are present in many water bodies. Such organisms can be monitored for changes (biochemical, physiological, or behavioral) that may indicate a problem within their ecosystem.

## Nonpoint source

Nonpoint source pollution is contamination derived from diffuse sources. These may include agricultural or stormwater runoff or debris blown into waterways from land. Nonpoint source pollution is the leading cause of water pollution in U.S. waters, but it is difficult to regulate, since there's no single, identifiable culprit.

### Transboundary

It goes without saying that water pollution can't be contained by a line on a map. Transboundary pollution is the result of contaminated water from one country spilling into the waters of another. Contamination can result from a disasterlike an oil spillor the slow, downriver creep of industrial, agricultural, or municipal discharge.

## Most Common Types of Water Contamination

### Agricultural

Not only is the agricultural sector the biggest consumer of global freshwater resources, with farming and livestock production using about 70 percent of the earth's surface water supplies, but it's also a serious water polluter. Around the world, agriculture is the leading cause of water degradation. In the United States, agricultural pollution is the top source of contamination in rivers and streams, the second-biggest source in wetlands, and the third main source in lakes. It's also a major contributor of contamination to estuaries and groundwater. Every time it rains, fertilizers, pesticides, and animal waste from farms and livestock operations wash nutrients and pathogens such as bacteria and virusesinto our waterways. Nutrient pollution, caused by excess nitrogen and phosphorus in water or air, is the number-one threat to water quality worldwide and can cause algal blooms, a toxic soup of blue-green algae that can be harmful to people and wildlife.

### Sewage and wastewater

Used water is wastewater. It comes from our sinks, showers, and toilets (think sewage) and from commercial, industrial, and agricultural activities (metals, solvents, and toxic sludge). The term also includes stormwater runoff, which occurs when rainfall carries road salts, oil, grease, chemicals, and debris from impermeable surfaces into our waterways.

More than 80 percent of the world's wastewater flows back into the environment without being treated or reused, according to the United Nations; in some least-developed countries, the figure tops 95 percent. In the United States, wastewater treatment facilities process about 34 billion gallons of wastewater per day. These facilities reduce the number of pollutants such as pathogens, phosphorus, and nitrogen in sewage, as well as heavy metals and toxic chemicals in industrial waste, before discharging the treated waters back into waterways. That's when all goes well. But according to EPA estimates, our nation's aging and easily overwhelmed sewage treatment systems also release more than 850 billion gallons of untreated wastewater each year.

### Oil pollution

Big spills may dominate headlines, but consumers account for the vast majority of oil pollution in our seas, including oil and gasoline that drips from millions of cars and trucks every day. Moreover, nearly half of the estimated 1million tons of oil that makes its way into marine environments each year comes not from tanker spills but from land-based sources such as factories, farms, and cities. At sea, tanker spills account for about 10 percent of the oil in waters around the world, while regular operations of the shipping industry through both legal and illegal discharges contribute about one-third. Oil is also naturally released from under the ocean floor through fractures known as seeps.

### Radioactive substances

Radioactive waste is any pollution that emits radiation beyond what is naturally released by the environment. It is generated by uranium mining, nuclear power plants, and the production and testing of military weapons, as well as by universities and hospitals that use radioactive materials for research and medicine. Radioactive waste can persist in the environment for thousands of years, making disposal a major challenge. Consider the decommissioned Hanford nuclear weapons production site in Washington, where the cleanup of 56 million gallons of radioactive waste is expected to cost more than $100 billion and last through 2060. Accidentally released or improperly disposed of contaminants threaten groundwater, surface water, and marine resources.

## Marine Pollution

Marine pollution has been an ever-present problem since the advent of large-scale agricultural activity and industrialization. However, significant laws and regulations at an international level to tackle the problem came only in the mid-twentieth century. During United Nations Conventions on the Law of the Sea in the early 1950s, the various stakeholders come together to deliberate and formulate laws pertaining to marine pollution. Till mid-twentieth century the majority of the scientists maintained that oceans were vast enough to be able to dilute the amount of pollution being drained into them, thus, considering pollution harmless to the marine life.

## Causes of Marine Pollution

The marine environment becomes polluted and contaminated through various sources and forms. Major sources of marine pollution are the inflow of chemicals, solid waste, discharge of radioactive elements, industrial and agricultural effluents, man-made sedimentation, oil spills, and many such factors. The majority portion of the marine pollution comes from the land that contributes to 80 percent of the marine pollution; air pollution also carries pesticides from farms and dust into the marine waters. Air and land pollution is a major contributor to the growing marine pollution that is not only hampering the aquatic ecology but also affecting the life on land. The non-point sources like wind-blown debris, agricultural runoff, and dust become the major source of pollution. Apart from these, factors like land runoff, direct discharge, atmospheric pollution, pollution caused by ships, and deep-sea mining of natural resources contribute heavily.

## Types of Marine Pollution

*Eutrophication*: there is an excess of chemical nutrients mainly nitrates and phosphates in the water, it leads to eutrophication or nutrient pollution. Eutrophication decreases the level of oxygen, reduces the quality of water, makes the water inhabitable for fish, affects the breeding process within the marine life and increases the primary productivity of the marine ecosystem.

*Acidification*: Oceans act as a natural reservoir for absorbing the carbon dioxide from the Earth's atmosphere. But, due to rising level of carbon dioxide in the atmosphere, the oceans across the world are becoming acidic in nature, as a consequence, it leads to acidification of oceans. Researches and scientists have not been able to uncover the potential damage ocean acidification may have on the Earth's atmosphere. But there is a strong concern that acidification might lead to dissolution of calcium carbonate structures, that can affect the shell formation in shellfish and also the corals.

*Toxins*: There are persistent toxins that do not get dissolved or disintegrate with the marine ecosystem rapidly. Toxins such as pesticides, DDT, PCBs, furans, TBT, radioactive waste, phenols, and dioxins get accumulated in the tissue cells of the marine life forms and lead to bioaccumulation hampering the life underwater and sometimes lead to a mutation in aquatic life forms.

*Plastics:* The ever-growing dependence of human population on plastic has filled the oceans and the land; it consists of 80 percent of the debris found in the oceans. Plastic dumped and found in the oceans are dangerous for the marine life forms and wildlife, as sometimes it strangles and chokes them to death. The rising levels of plastic dumps found in the oceans are suffocating, ingesting, and entangling the life underwater as well as above it.

## Effects of Marine Pollution

The contamination of water by excessive nutrients is known as nutrient pollution, a type of water pollution that affects the life under water. When excess nutrients like nitrates or phosphates get dissolved with the water it causes the eutrophication of surface waters, as it stimulates the growth of algae due to excess nutrients. Most of Benthic animals and plankton are either filter feeders or deposit feeders, take up the tiny particles that adhere to potentially toxic chemicals. In the ocean food chains, such toxins get concentrated upward. This makes estuaries anoxic as many particles combine chemically depletive of oxygen.

When the marine ecosystem absorbs the pesticides, they are incorporated into the food webs of the marine ecosystem. After getting dissolved in the marine food webs, these harmful pesticides cause mutations, and also results in diseases, which can damage the entire food web and cause harm to the humans. When toxic metals are dumped or flown into the oceans through drains, it engulfs within the marine food webs. These can cause a change to tissue matter, biochemistry, behavior, reproduction, and suppress and alter the marine life's growth. Marine toxins can be transferred to several animals feeding on the fish or fish hydrolysate as a meal, toxins are then transferred to dairy products and meat of these affected land animals.

## Steps to Prevent Marine Pollution

» Stop using plastic and littering garbage as they not only choke up the drains but also releases into the ocean. Ensure that chemicals mentioned above are not used anywhere near the streams of water and try cutting down on the usage of such chemicals.

» For farmers, they need to switch from chemical fertilizers and pesticides towards the usage of organic farming methods. Use public transport and reduce the carbon footprint by taking small and substantial measures that will not help in reducing the pollution from the environment but will ensure a safe and healthy future for the Upcoming generations.

» Prevent from any oil or chemical spill in the oceans and if in case there is an oil or chemical spill near you volunteer and help in cleaning outthe ocean water. Volunteer or initiate beach cleanup activities and spread awareness about the same in the nearby vicinity.

» Use public transport and reduce the carbon footprint by taking small and substantial measures that will not help in reducing the pollution from the environment but will ensure a safe and healthy future for the upcoming generations.

## Current Status of Aquatic Pollution

India has more than 1.2 billion people, but only a little over 30% of its population lives on less than $1.25 a day. Around 80% of India's water is severely polluted because people dump raw sewage, silt and garbage into the country's rivers and lakes. This has led to water being undrinkable and the population having to rely on illegal and expensive sources. Each year, more than 1.5 million Indian children die from diarrhea. Out of the entire Indian population, experts predict that 40% of people may not have a connection to a clean water source by 2030. So, how did water pollution in India get so out of hand and how can Indians reverse the damage?

## Water Pollution in India

India suffers from increased urbanization, unauthorized slums and the absence of pipe planning. Estimations suggest that by 2030, almost 600 million Indians might live in slums due to the ever-growing population. Because of that, tanker mafias are prominent. Tanker mafias are business owners that hold septic tanks that illegally sell water from lakes, wells and groundwater. They charge around $50 per 1,000 liters, and for most Indians, it is unaffordable.

Between the years of 2001 and 2012, 3,245 hectares of lakes dissipated in the city of Hyderabad. The water recedes by nine feet a year on average in southern New Delhi all because of tanker mafias. Oil leaks, inadequate treatment of waste, poor sanitation and open defecation are the leading causes of water pollution in India. By drinking dirty water, the human digestive system suffers from harmful bacteria that disrupt the balance of the gut, causing diarrhea and other diseases.

## Consequences of India's Poor Water Quality

Around 70% of wastewater goes untreated and each day, more than 40 million liters of wastewater flows directly into India's lakes, rivers and ocean. Eventually, contaminated water also enters the groundwater. Because of this, proper waste management and sewage pollution cannot occur, upsetting the irrigation system. The crops are not able to grow properly because of the infectious bacteria and disease in the water. Because of the poor infrastructure and absence of sewage control, 38 million Indians suffer from waterborne diseases like typhoid, cholera and hepatitis every year. Over the last decade, the frequency of these illnesses remained at the same level.

Worldwide, waterborne diseases cause more deaths than AIDS, tuberculosis and measles combined in children under 5-years-old. Water pollution in India not only harms people's health and food security, but it also contributes to the decrease in India's GDP and economic stagnation. Not only does GDP growth reduce by one-third when the pollution in the country's waterbodies exceeds a certain limit, but agricultural revenues lower by 9% in the districts that are close to industrial territories. The degradation of the environment, including water pollution in India, leads to a loss of $80 billion annually. Meanwhile, estimates determine that the health costs to treat waterborne diseases are almost $9 billion per year.

## India's Action Steps to Eradicate Water Pollution

India is taking several steps to rebalance the quality of its water source, from flocculation and reuse of industrial water to the contributions that local Indian startups are making. In Chennai, a city in Eastern India, industrial water reuse rose from 36,000 to 80,000 cubic meters in 3 years, from 2016 to 2019. VA Tech Wabag, a water company quartered in Chennai, also built numerous water reuses plants all across India. As of 2020, VA Tech Wabag contributed immensely to the production of more than 18 million cubic meters of clean water every day, which has positively impacted almost 100 million people globally. It also specifically focused on installing a water treatment plant in Panjrapur, Maharashtra, through implementing a combination of techniques like flocculation, skimming and filtration. To reduce the turbidity and colloids in the water, which can cause *E. coli*, dysentery, cholera and salmonella; it uses flocculants to clean the water. Its project in Maharashtra has a large capacity of 455,000 cubic meters of water per day. As a result, it will enable more than 112 million people to have access to clean water.

In Gujarat, a state of more than 70 million citizens, the government launched its Reuse of Treated Waste Water Policy, which aims to drastically decrease the use of the Narmada River. It will install 161 sewage treatment plants all across Gujarat in order for industrial and construction sectors to use the treated water. Evaluations offer that in 2015, the Indian government installed almost 16,000 reverse osmosis systems in Karnataka and 281 solar electrolytic defluoridation plants in Madhya Pradesh.

## Detection Techniques of Aquatic Pollution

Analytical methods used for the detection of pollutants in wastewater

In most of the time, the methods utilized for the identification of organic compounds are unique about those utilized for heavy metal contaminants. As already chromatographic techniques are typically utilized for the examination of organic compounds while atomic spectroscopic strategies are utilized for heavy metal investigation. To be noted, electrochemical methods can be utilized for the two kinds of pollutants.

» Electrochemical.
» Chromatography.
» Atomic Spectroscopy.
» Fluorescence-Based Detection of Water Pollutants.
» Colorimetric Detection of Water Pollutants.

### Use of chemical markers for water pollution detection

A wide range of chemicals has been utilized to detect wastewater contamination in surface water and groundwater. In recent years, numerous studies have successfully utilized polypropylene copolymers (PPCPs), artificial sweeteners (ASs), fluorescent whitening agents (FWAs) and fecal sterols/stanols (SSs) to detect contamination of wastewater tosurface water and groundwater (Glassmeyer *et al.*, 2005). However, it remains highly challenging to accurately predict the extent of wastewater contamination using these developed chemicalmarkers. Currently, there is no single chemical that could serve asmarker for wastewater contaminationfor all sites accurately. Understanding of land use pattern, types and levels of contaminants inwastewater, fate and transport of chemicals are needed in order to select suitable marker(s).

Chemical markers can be divided into three main categories: (a) those produced by humans, e.g., fecal sterol; (b) those that can pass through human bodies, e.g., PPCPs; and (c) those that are associated with sewage contaminated waste system, e.g., detergents. The use of chemical markers has the advantages of more rapid and reliable detection, source specific and stable compared to microbial markers. The disadvantages of utilizing chemical markers include: (a) the presence of chemical marker does notabsolute represent contamination by wastewater; and (b) could lack specificity and can be impacted byenvironmental factors such as degradation and sorption.

One of the key considerations for selecting thesuitable chemical markers is that they should be associated with wastewater sources andoccurring in receiving water bodies at specific study area. The specific chemicals that are ubiquitously present and detected in high concentration in raw wastewater could be potentially selected as markersin the studied area. The proposed chemical marker(s) should be specific to wastewater contamination sources. For instance, markers should be absent or detected in significantly lower concentrationsin background receiving bodies with no wastewater contamination sources. The use of markers issite-specific and may not be applicable from one location to another.

## Considerations of the Use of Chemical Markers to Detect Wastewater Contamination

The selection of potential chemical markers is dependent on various factors such as land use, population, geology, type and level of pollutants in the sewage, hydroclimatic and environmental conditions. In order to select realistic wastewater markers, it is useful to understand the characteristics of markers. It is generally agreed that the use of single marker is not sufficient for better results. For instance, ratio of coprostanol/cholesterol can be used to assess the degree of sewage pollution. Higher ratio of stigmastanol to other SSs is an indication of presence of fecal matter from herbivorous animals such as cow and sheep (Vane *et al.*, 2010). In addition, measurement of related physicochemical parameters is recommended. The data may be useful to support findings on the source of contamination. In a recentstudy by De Sousa *et al.* (2014), measurement of electrical conductivity was not only useful for screening of pollution level, but could also be used to support findings on the contamination sources with the aid of statistical analysis (Principal component analysis (PCA) andcorrelation). Figure 1 summarizes the approach in selecting the appropriate chemical markers

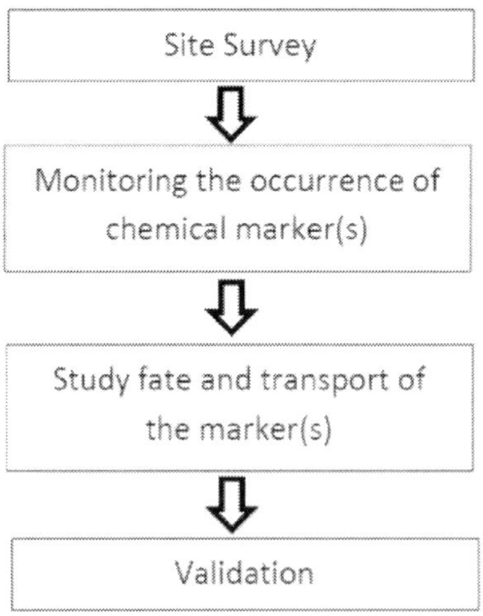


**Figure 1: Approach for detecting sewer contamination using chemical markers**

## Site Survey

Chemicals found in wastewater are siteand sourcespecific. The occurrence, types, transportation and fate of pollutants are largely dependent on human activity, environmental conditions and land use pattern. Thus, selection and consideration of chemical marker(s) in a new site begins with site survey and data collection. Characterization of the specified site/catchment and the potential fecal input are important steps in selecting potential markers. Characterization of site with the evaluation and analysis of the land use, hydrology, population, weather conditions, soil, wastewater treatment processes and other related information could help to identify appropriate potential marker(s) for selection. In addition, types and concentration level of pollutants could be assessed and predicted based on the data collection such as PPCPs consumption data, biodegradation and sorption rates, stability during transportation and fate in soil, and average removal capacity in wastewater treatment plants. Site characterization data should be analyzed and interpreted for a comprehensive assessment such of potential source. For example, food and beverages wastes from food courts and restaurant can be asource of ASs. This information suggests that ASs could be used to track the pollution as artificial sweeteners (ASSs) were reported to be persistent and poorly removed in conventional WWTPs (Tran *et al.*, 2014a).

## Occurrence and Detection of Chemical Markers

Detecting chemical marker(s) in both wastewater source and receiving bodies can be a challengingtask due to dilution during transportation and interference of matrix constitutes in the water samples. The method quantification limits for the markers are recommended to be subprecipitate levels, so that it candetect the target analytes in receiving bodies. Recently, there have been significant improvement and development in chemical markers quantitation methods to provide high through put capability and to enhance detection sensitivity. Advances in analytical techniques allow chemical markers to bedetected in wastewater.

The sample should be collected at source (wastewater), during transport (canal), and receiving bodies (reservoir, groundwater, surface water) and, moreover, the sampling schedule should cover different weather conditions such as wet and dry periods, different seasonal periods, as well as temporal and spatial distribution. Background samples (without impact bywastewater source) should be collected for comparison. Statistical analysis and indices could beperformed and used to identify potential markers. For instance, PCA was used to evaluate the spatial and temporal variation of the chemical parameters and emerging contaminants to select promising markers. Several researchers studied the relationship between caffeine and other water quality parameters (Peeler *et al.*, 2006; Nodler *et al.*, 2013). It was found that the measurement of caffeine, nutrients and indicator bacteria was useful to distinguish human versus non-human sources of wastewater contamination. In addition, detectionratio and detection frequency are widely used indices to identify chemical markers for wastewater contamination. Fenech *et al.* in2012 proposed the use of acetaminophen, caffeine and carbamazepineas chemical markers at Tullow site as their detection frequencies (>83%) are the highest.

In another study, five compounds (benzophenone, ethyl citrate, galaxolide (HHCB), tributyl phosphate andtriclosan) with detection frequencies of 50%–70% were proposed as suitable chemical markers for human fecal contamination. The concentration ratio of ephemeral compound and a recalcitrant compound should decrease with the increased time in a water body (Buser *et al.*, 1999). Glassmeyer *et al.*, 2005 applied this ratio in their study and found that it was a useful tool in predictingthe composition of compounds in lake as compared to lake's hydraulic residence time and alsotemporal and spatial distance from a known source in a river.

## Fate and Transport of Chemical Markers

An important factor for consideration in selecting the appropriate markers for wastewater contamination would be to understand their fate and transport in the environment. Transport of chemical markers may be affected by various environmental factors such as dilution and natural attenuation. It was found that adsorption and biodegradation processes were the main transportmechanisms of sewage molecular markers in a subsurface area (Kagle *et al.*, 2009). Yamaji *et al.*, 2010 showed that photodegradation affected the fate of FWAs in the lacustrine system significantly, with the DSBP and DAS-1 removal rates of o 93%–99% and 44%–63%, respectively. They demonstrated that FWAs and DSBP/DAS-1 ratio were good indicators for studying the role of direct photodegradation inbehavior of water-soluble chemicals in aquatic environments. Carbamazepine, benzotriazole andprimidone were determined to be suitable markers based on their low sorption affinity to soil, whilethe use of sulfamethoxazole as a marker was limited by the formation of non-extractable residues atlower pH (Yu *et al.*, 2009). Yu et al. also highlighted that organic carbon could impact sorption process (Badruzzaman *et al.*, 2013) found that sucralose was the most recalcitrant marker in the environment based on their low removal rate (15%) by adsorption, biodegradation or photolysis. Markers with high attenuation were iohexol (90% by photolysis) and carbamazepine (50% by photolysis). They also notedthat atenolol was highly biodegradable (60%–80% by biodegradation and adsorption).

Foolad *et al.* (2015), studied the fate of nine target sewage markers and found that acetaminophen (ACT), caffeine (CF) and diethyl ltoluamide (DEET) were removed completely through biotic soil column due to adsorption and biodegradation processes. It is well-known that the transport of chemical markers from wastewater source to receiving bodies are highly affected by the characteristics of wastewater, degree of dilution, type of sediment, temperature, pH and degradation during transport. Limited study was performed to determine the synergistic and antagonistic effects of markers and their metabolites during transport. Indeed, comprehensive study into the fate and transport of chemicals should be conducted for selecting appropriate markers for wastewater contamination.

**Validation**

It is essential that markers should be absent or detected in significantly lower concentration in the background samples (no known wastewater sources), higher concentrations should be detected in samples with wastewater sources contribution. Glassmeyer *et al.*, 2005, collected a wide range of samples which consisted of samples from upstream, downstream, WWTPs, and further downstream in US. The samples collected represent a variety of climatic conditions, population, streams size and treatment capacities of WWTPs. They demonstrated that many compounds present in WWTP effluent were not found in the upstream sample. In contrast, the concentrations of markers were similar (no statistically different) in WWTP effluent and downstream samples, which indicated that contamination of wastewater to river water quality was caused by incomplete pollutant removal in WWTP. The concentrations of markers were found to be diminished with greater distances from WWTP. This may be due toinstream processes such as dilution, degradation, sorption, etc. In addition, the authors showed the targeted markers (acetaminophen, caffeine, methyl salicylate) were not found at Michigan (a reference site). This further suggested these targeted markers could potentially serve as chemical markers. The absence of target compounds at the reference site clearly indicates that the target compounds are not detected in unpolluted streams, and therefore could be proposed as chemical markers. Tran *et al.*, 2014b, found that none of the targeted ASs (ACE, CYC and SAC) was detected in thebackground groundwater samples. However, these compounds were found in all groundwater samples with known leaking sewer segment. In the same study, ACE and SAC were not found in all the background surface water samples. In contrast, CYC was detected in a relatively low concentration compared to those in potentially impacted surface water samples. The authors suggested that the presence of CYC, SAC and ACE in wastewater and impacted surface water was affected by wastewater contamination. Suitable markers that can be used to detect wastewater contamination should be absent (or detected at relatively lower concentration) in reference site (background samples) and detectable at higher concentration in both wastewater source and contaminated surface water. Hence, samples should be collected at different sites to better evaluate and validate suitability of the markers.

## Ability to Distinguish Sources

Identification of pollution sources will help in the management of wastewater contamination. Selection of appropriate markers enables pollution source tracing, which could distinguish between human and non-human, and discernment among different animals' origins. Most of the studies to differentiate pollution sources were based on the use of SSs markers. Source identification and apportionment could be achieved through marker ratio analysis and statistical tools such as (PCA) and multiple linear regressions (MLR).

The use of SSs as marker provides additional advantage which can be used to discriminatebetween human and non-human contamination. In addition, a ratio of cholestanonewas proposed (Grimalt *et al.*, 1990). This ratio indicates human fecal pollution if the ratio is greater than 0.7. The ratio of COP5 stigmasterol for source identification, where ratio above 1.5 indicates human fecal pollution (Isobe *et al.*, 2002), while as Mudge and Seguel, 1999 proposed a ratio of ECOPCOP, where the ratio lower than 0.2 indicates untreated sewage contamination. This index has been used later on to study Barigui river watershed (Froehner *et al.*, 2009). The ratios were below 0.20 at most stations andit was concluded that sewage discharged into the river did not undergo prior treatment or incomplete removal of the measured pollutants during treatment. This finding has been confirmed using ratios of isomeric forms of sterols and ketones. A consistent result was obtained for the source identification by the SSs based PCA method and microbial markers in river waters contaminated by fecal matter of unknown origin (Derrien *et al.*, 2012). The approach developed in this study could be applied to differentiate sources of fecal contaminations in water body where fecal contaminations from human, bovine and porcineco-present. This study also validated the use of PCA method to detect the wastewater contamination sources. McCalley *et al.*, 1981 proposed ratio of COP/CSE to elucidate the contributions from biogenic and sewage sources. For instance, ratio greater than 1 indicates sewage source while ratio less than1 represents biogenic sources. Leeming *et al.*, 1996 applied the various ratios of sterols to differentiate the feces sources between herbivore and omnivore. In another study, Leeming *et al.*, 1996, successfully distinguish feces from cows and sheep, humans, hens, dogs and cats with the aid of PCA analysis. Kuroda *et al.*, 2014 utilized PPCP analysis, tritium analysis; principal component analysis-multiple linearregression (PCA-MLR), perfluoroalkyl carboxylic acid ratio analysis (PFCAR) and their combined useto identify the sources of perfluoroalkyl acids (PFAAs) in Japan. They noted that the primary source ofPFAAs in polluted groundwater were sewage (44.6%) and point sources (45.7%) whilst minor sourcewas street runoff (9.7%). Gregor *et al.*, 2002 modeled the selected chemical profiles of different fecal sources using classification and regression tree (CART) analysis. The authors concluded that effluents from human wastes could be identified by the use of FWAs.

It is widely accepted that utilizing multiple chemical marker(s) are useful to distinguish sourcesof pollution and to distinguish "fresh" vs. "aged" wastewater. For instance, Zirlewagen *et al.*, 2016 utilized CYC/ACE ratios to identify fresh untreated wastewater. The authors noted that using ACE as indicator alone could not distinguish between "fresh" and "old" wastewater contamination. Hence, they suggested the combined use of ACE and CYC as marker set could provide additional information of the age of wastewater contamination. Low CYC/ACE ratios could be related to bothtreated wastewater and untreated wastewater with a minimum residence time of several weekswithin the karst system. Higher CYC/ACE ratio was due to recent contamination by untreated wastewater. The presence of both biodegradable markers (CYCand SAC) and persistent marker (ACE) could be useful to detect fresh wastewater contamination e.g., sewer leakage or overflows discharge (Tran *et al.*, 2014b). This finding had been validated in few sampling locations in Singapore. The ratio of valsartan and valsartan acid was recommended to identify the age of contamination within the study area, e.g., >1 related to fresh untreated sewage and 1 for wastewater of higher relative treatment level or residence time. However, it is worth mentioning that the occurrence of valsartan and valsartan acid is site-specific and depends on the wastewater characteristics. Sidhu *et al.*, 2013 suggested that the presence of caffeine, paracetamol and salicylic acid could explain recent contamination from raw sewage due to high prevalence of human adenovirus and human poly-omavirus found in their study. Further studies on the use of multiple markers to assess "fresh" vs "aged" wastewater contamination are needed in order to better understand and manage the wastewater contamination issues. It is recommended that combination of few chemical markers (both recalcitrant and biodegradable) and/or microbial markers would be useful to assess the wastewater contamination.

## A. Nanotechnological Approach for Tracing of Aquatic Pollution

Access to clean; microorganism-free water is a significant concern that necessitates intensive investigation. To secure a supply of drinking water and prevent global water pollution, significant difficulties in the global water situation, mostly coming from global population expansion and climate change, necessitate unique revolutionary water solutions. In this sense, the use of cutting-edge nanotechnology to classical process engineering opens up new avenues for technical advancements in improved water and wastewater treatment. Nanotechnology has enormous benefits in water treatment because it allows for the research and fabrication of nanomaterials with unique features in the range of 1-100 nm. These nanoparticles are effective in water treatment and save money, time, and effort compared to standard methods. The relevance of nano-material synthesis in numerous sectors, as well as top-down and bottom-up methods to nanomaterial synthesis.

Nanotechnological wastewater technology is still in the research phase, and only a few products are available in the market as of now. More rigorous research is needed to overcome the short comings of this technology to utilize its full capabilities for environmental protection. The uses of nanotechnology are numerous. As a result, it's going to be a crucial piece of technology in the future. Both developing and developed countries are interested in using it for water and wastewater purification because of its wide range of potential applications in these environments. The usage of nanometals (silver and gold) and nanomembranes can aid in the improvement of water treatment technologies that can be used to tackle water-related problems such as waterborne pathogens, biofouling, and the removal of hazardous metals (lead, arsenic, and chromium) among others. Nanotechnology can also be utilized to increase the effectiveness of waterfilters. Sedimentation, filtration, and chemical or biological degradation are not totally capable of eradicating new pollutants or some types of bacteria in drinking water. Nanotechnology may be able to help with the problems caused by different pollutants and bacteria. This chapter discusses advanced nanomaterials and approaches for removing pollutants and contaminants from aqueous solutions, functionalized nanostructured materials andprocesses in water and wastewater treatment technologies, heavy metal removal/recovery using amorphous metal oxide nanoparticles, nanoparticle assisted nanofiltration, and the development of innovative and natural-based nano-sorbents for water treatment (Watlington, 2005; Ealias and Saravanakumar, 2017).

## Main nanotechnological processes for wastewater treatment

### Nanofiltration

Membrane filtration plays an important role in removing various types of contamination and enables high level of water purification. Until recently, its biggest problem was a substantial investment cost (about 70 % of the total investment costrefers to membranes). As the price is lowering, the membrane wastewater treatment process becomes more and more popular in the market, mainly due to its high efficiency in the removal of solid waste materials, monovalent and divalent ions, various pathogens, etc. Nanofiltration (with reverse osmosis, RO) is a high-pressure membrane treatment process. But unlike the RO, it requires amuch lower drive pressure (7 to 14 bar), and so allows lower energy consumption. Centrifugal pumps are most oftenused for the pressure and circulation of wastewater within the nanomembrane. The plant consists of a large number of modules, with different membrane configurations within each module. In nanofiltration, the usual length of the module variesfrom 0.9 to 5.5m and the diameter ranges from 100 to 300mm (Tchobanoglous *et al.*, 2014). The modules are installed on the stand and can be arranged either horizontally or vertically. For vertical installation, a smaller number of connecting pipes and fittings are required, and the footprint is smaller. Nanofiltration produces water that meets highly stringent requirements in terms of water reuse. Since this process is highly efficient in the removal of organic and inorganic substances, bacteria and viruses, the need for subsequent disinfection of water is minimal.

Nanotechnology offers a wide range of solutions for membrane materials (Lens *et al.*, 2013) including

Ceramic membranes for nanofiltration

New polymeric membranes with anti-fouling coating (Organic brush-like coating, membrane impregnated with nanoparticles)

New composite membranes

Thin film composite membranes

Metal/metal oxides + polymer

Carbon nanotube + polymer

Zeolites + polymer

Aquaporin + polymer

According to Pendergast and Hoek (2011), all membrane types used in nanofiltration can be grouped in three categories:

Nanostructured ceramic organic-inorganic membranes

Biologically inspired membranes

They also develop methods for evaluating certain nanomaterial properties, such as permeability, strength, practical application, etc. They concluded that biologically inspired membranes possess the greatest potential for improvement but are far from commercial use. On the otherhand, zeolite membranes have limited possibilities for further development but are closest to commercial use (Lens *et al.*, 2013).

**Nanomaterials for catalysis and photocatalysis**

Metal nanoparticles and metal oxides have proven to be very good catalysts in oxidation reactions. They exhibit a strong catalytic activity through which pollution molecules are oxidized forming less-toxic substances, or converted into ecologically acceptable final products (Lens *et al.*, 2013). The main reasons for theseproperties of nanoparticles are: very small particle size, i.e., a large surface to volume ratio, and high reactivity directly related to nanoparticle size.

Nano-catalysts can effectively be used for chemical oxidation of organic and inorganic pollutants in water in advanced oxidation processes (AOP). These processes are based on formation of highly reactive radicals that react easily with pollutant molecules. The application of this process is often limited because of the extremely high costs of providing required energy (UV lamps, ozonators, ultrasonicators, etc.) (Lens *et al.*, 2013). Photocatalysis is the most significant oxidation process. This is a chemical reaction change that is induced by adsorption ofa photon whose energy is greater than the energy needed to overcome the interstitial of two electron shells (valentine and conductive) of a semiconductor. When the photon illuminates the catalytic surface, the electron (negatively charged particle) is transferred from the valentine shell to the empty conduction shell and leaves a "hole" behind it with a positive charge. This "e-h" pair ("electron-hole") creates highly reactive radicals that bind the molecules of pollution and thus break them down (Bora and Dutta, 2014).

However, there are several technical challenges that have to be met to enable broader practical application of this process, including

Optimization of catalysts in the exploitation of available light energy,

More efficient separation of nano-catalysts after treatmentand re-application,

Improvement of selective properties during chemical reactions.

However, the biggest drawback of this technology is the high operating cost of providing the required light energy (UVradiation), which is why this technology is still not considered to be economically viable. The research has therefore, concentrated on the exploitation of (natural) solar energy for photocatalytic processes. Apart from being free, the advantage of solar energy is that it can also be used in open air processes. The exploitation of renewable energy in these processes opens up new opportunities for the development of technologically efficient and economically viable wastewater treatment technologies (Bora and Dutta, 2014).

## Nanomaterials for water disinfection

In addition to having excellent adsorption and catalytic properties, some nanomaterials have proven to have great antimicrobial activity as well. Such materials include chitosan, silver nanoparticles, titanium dioxide, fullerene nanoparticles, carbon nanotubes, etc. All these nanomaterials are mild oxidants and are relatively inert in water, and are therefore not expected to create harmful by-products. There are several ways of applying the nanomaterials in water disinfection processes: direct action on (bacterial) cells in the sense of preventing electron passage through the membrane, break through the cell membrane, oxidation of some cellular components, hydroxyl radicals (within the action of nanoparticles as photocatalysts), and the formation of dissolved metal ions that can cause damage to cellular components as shown in Figure 2 (Li *et al.*, 2008).

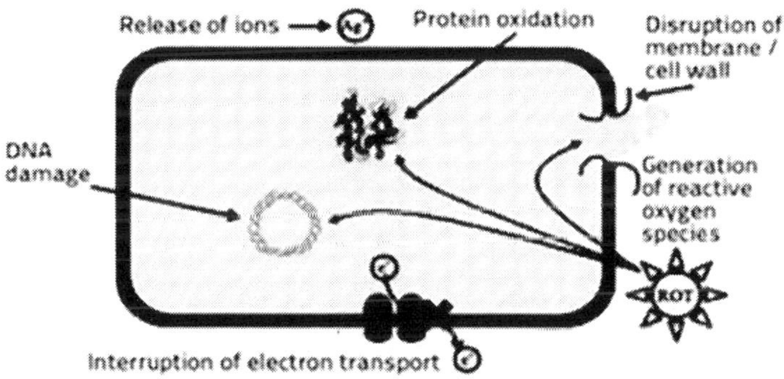


**Figure 2: Various mechanisms of antimicrobial activities exerted by nanomaterials**

However, there are some limitations regarding the use of nanotechnology in wastewater disinfection processes. For some nanomaterials to be effective in removing different types of microorganisms, they must be in direct contact with the cell membrane of bacteria, viruses, etc. Therefore, some nanomaterials (carbon nanotubes) need to be strongly connected to the reactive surface. Also, the deficiency of nanotechnology in disinfection processes is that there is no residual, i.e., subsequent antimicrobial activity in wastewater (such as in the case of chlorine use) (Lens *et al.*,

2013). The great advantage of conventional disinfection procedures is currently low cost. Nanotechnological processes will become more competitive after standardization of the production of nanomaterials and following reduction of total labor costs.

**Nanomaterials for adsorption of pollutants**

Nanoparticles possess two important characteristics that make them very good adsorbents. These are the large specific surface of nanomaterials and surface multifunctionality or the ability to easily chemically react and bind to different adjacent atoms and molecules as shown in Figure 3. These characteristics make nanoparticles not only effective adsorbents for various contaminants in wastewater but also allow for long-term stability, as this also results in adsorbent degradation (with the addition of catalytic properties of nanoparticles) and improves the adsorption efficiency. Compared to the best known such material - activated carbon - carbon nanotubes possess approximately the same large specific surface, but their great advantage lies in the structure of nanomaterials and a much better arrangement of carbon atoms. In addition, nanomaterials possess unique mechanical, electrical, chemical, optical and many other characteristics that allow them to have much better adsorption properties for some contaminants (heavy metals and organic pollutants). This is why they are called the "material of the 21st century" (Ren *et al.*, 2011).

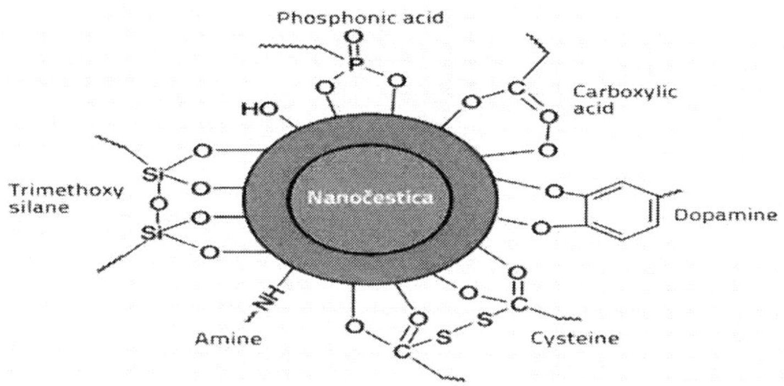


**Figure 3: Multifunctionalities of metal oxide nanoparticle surface**

Besides carbon nanotubes, metal-based nanoparticles also have adsorption characteristics. The most common metal oxides used as adsorbents are iron oxides ($Fe_xO_y$), silicon (Si), titanium (Ti) and tungsten (W). They are mainly used for adsorption of heavy metals and radionuclides (unstable nuclides). The adsorptionprocess is based on the electrostatic interaction of dissolved metals in wastewater and the nanoadsorbent surface. Changing the pH of the solution can significantly affect the strength of this interaction. Thus, the surface of the nanoadsorbent may be:

- » acidic, with positive charge attracting anions
- » basic, with a negative charge attracting cations from waste water (Qu *et al.*, 2013).

**Nanoscale zerovalent iron (nZVI)**

Remediation of wastewater using nanotechnology is also gaining a greater interest. Compared to conventional materials and technologies, nanoparticles can provide great material and energy savings due to their prominent properties. Their nanostructure allows them to act as colloids, and so they can easily be injected into the soil and associated with contamination, regardless of location and depth. Key characteristics that nanoparticles must have in terms of remediation of contaminated groundwater are:

-high reactivity with contaminants

-high mobility within porous medium

-appropriate life span

-negligible harmful effects.

According to available research results, the zerovalent iron (nZVI) is considered to be the most interesting nanomaterial due to its low production costs, positive environmental effects, and high reactivity with contaminants. This nanotechnology has found extensive application in contaminated groundwater treatment, and has already reached the commercial level ofapplication in the world. In addition, it has already proven tobe a highly efficient technology for the removal of various organic and inorganic pollutants, including chlorinated solvents, pesticides, nitroamines and nitroaromatics, organophosphates, inorganic anions, arsenic, uranium, numerous metals, etc (Crane and Scott, 2012).

Depending on the mobility of the contaminant, the nZVIinjection method can be tailored for the treatment of:

-mobile contamination

-immobile contamination.

Thus, for the treatment of mobile contaminant plumes, low mobility nZVI is typically used for the formation of a "reactive treatment zone". The nanoparticles are sequentially injected and adsorbed to native aquifer material, forming a physicochemical filter. Mobile nZVI is typically selected for the treatment of a static contaminant body and is injected upstream for direct treatment (Crane and Scott, 2012).

A legal framework has to be put in place for the application of the described nanotechnology which has already been defined in North America and in some European countries. A broader application of this technology relies on clear understanding of all processes and behaviour of the nZVIin the underground water system. Great emphasis is placed on its impact on the environment. The same experts who proposed this technology as an efficient method of water treatment also point to possible negative environmental effects. So far, little research has been done on the subject of its potentially harmful influence. Additional research and knowledge in thisfield is needed to accurately predict behaviour of nanoparticles in natural environment. This detailed knowledge combined with the expected reductionin total costs will enable this nanotechnology to become more competitive, as compared to other conventional methods (Crane and Scott, 2012).

### D. Wireless Sensor Technology for Aquatic Pollution Tracing

Nowadays, water is considered as one of the scarcest natural resources on our planet (Bakker, 2012). It is important to human kind animals and plants. Depending on the quality of water, it may either be a source of life and good health or a source of diseases and deaths. The growing environmental degradation in recent years brought about by development, population increase and climate change increases the need for researchers to look into its negative impact in the environment, especially in water sources and its implication. Increasing water pollution in oceans, lake, and river triggers worldwide demand more advanced methods in environmental monitoring systems particularly in the field of waterquality monitoring.

Wireless sensor networks have been widely adopted to monitor different phenomena in the environment. An application area that has received attention in recent years has been water quality monitoring (Chung and Yoo, 2015; Alippi *et al.*, 2011; Faustine *et al.*, 2014) which presents many advantages like its portability (Rashid and Rehmani, 2016), near real-time data acquisition and data logging capability (Can and Demirbas, 2015), water chemical monitoring (Hayes *et al.*, 2008), hydrodynamic performance monitoring (Perez *et al.*, 2011), irrigation (Dinh *et al.*, 2007), and agriculture (Kotamaki *et al.*, 2009). This Wireless Sensor Network (WSN) is suitable for monitoring physical and chemical water characteristics in remote areas at lower cost and reduce manpower requirement (De Marziani *et al.*, 2011). Other studies are focused on the enhancement of the transmission, power harvesting and management and environment adaptability in the past few years of applications. Hence, this particularly cover large aquatic area monitoring, focusing on DO, pH, and temperatureas water quality indicators by using WSN technologies. Also, Global System for Mobile Communication (GSM) will be used to transmit sensed data from the sink nodeIts aim to deploy a system which is reproducible, scalable and can operate independently for a long duration, and can be applied inremote areas.

For many years, the procedure for testing the quality of water followed a simple work flow that involved manually collecting samples of the water and then transporting these samples to a laboratory for analysis to detect chemicals and microbial contaminants. In recent years, low-power sensor devices with wireless communication capabilities have become widely used for environmental sensing and monitoring, as they can capture data from distributed sensors in real time; perform local processing of the data; are robust; and are able to self-configure, self-power and operate autonomously. A main advantage of using WSNs for environmental monitoring is that they do not require human presence or intervention to operate. Murphy *et al.* (2015) designed a low-cost autonomous optical sensor formonitoring a range of water quality parameters. The optical sensor designed by the authors is robust, easily deployable, and simple to operate. This sensor consists of a multi wave length light source with two photodiode detectors. This sensor is capable of measuring the transmission and side scattering of the light in the detector head, which contains five LEDs of different wavelengths. The detector head forms part of the sensor head, which covers the optics and controls the detection abilities of the sensor.

## Wireless Sensor Nodes

WSNs extend the capabilities of *in-situ* monitoring systems. Typically, a wireless sensor node consists of the sensor unit, the interface circuit, a processor, a trans receiver system, and a power supply unit (Yang *et al.*, 2002; Yang and Pan 2010; Wang *et al.*, 2010). These wireless sensor nodes are designed to accommodate a variety of situations ranging from short-term or spot sampling of the water quality parameters to long-term, unattended monitoring and analysis. Additionally, there are other sensors that are designed to float on water, such as buoys that come with flow controls and mini wet labs to collect water samples for analysis (Burke and Allenby, 2014). The sensors are classified into single and multiple parameters, depending on the number of water quality parameters that the sensor is able to detect. Several studies also proposed the use of specialized hardware devices to support the wireless sensor nodes for measuring water quality parameters. These include, for example, autonomous underwater vehicles (AUVs) (Zhan *et al.*, 2009), fish robots (Shin *et al.*, 2007), unmanned airboats (Kaizu *et al.*, 2011), and digital cameras (Goddijn and White, 2006).

**Specialized Hardware Devices for Supporting Wireless Sensor Nodes:**

a. Digital videos
b. Automatic underwater vehicles
c. Fish robots
d. Battery operated mini-boat
e. Buoysmounted with probes
f. Air boat

**Sensors:** The availability, affordability, and compatibility with Arduino-based architecture were the main reason for the selection of Atlas Scientific brand of water quality sensors.

**System design:** The system is designed and implemented with its maingoal of monitoring water quality (DO, temperature, andpH) with the consideration that the system: 1) is suitablein a large aquatic area; 2) can measure and store informationin a database in a real-time scenario; and 3) has a mechanism system to deliver a timely notification to the managers or authorities and users of the system. Therefore, the system was designed to provide thefollowing: 1) data collection and storage; 2) data disseminationand display (charts, graphs, and datasheets) (Randhawa *et al.*, 2016; Roadknight *et al.*, 2004); 3) power management for longer measurement periods (Gao *et al.*, 2015); and 4) maintenance interface for administrators. Software configuration on the system for the operation of the proposed water quality monitoring system. Upon implementation of a node, the system will initialize the SD card and the real-time clock (RTC). It will also ready the GSM module for data transmission. After a few seconds, the system will now start collecting data through the on-board electrodes (pH, DO and temperature). Data will be sent using GSM transceiver to the base station orthe pre-identified cell phone number. Also, data will bestored in the RTC as backup data storage with the same format with the base station. Signal, data will be automatically saved to the RTC and ready to be transferred to the database located in the base station. After the transmission and data backup through the RTC, the system will automatically turn-off and wait for the 30 min time interval to conserve the energy. A separate timer circuit will awake the sensor ifit reaches the required time interval of 30 min. The cycle will go on until the required tasked is achieve. It provides graphs for easy monitoring of abnormalities in the covered area. It could also provide data sheets and pdf file for research purposes. Real-time values for the 3 parameters will also be sent using SMS into a predefined cell phone number of some concerned persons.

**Wireless communication:** The sensed data from the sensor node are relayed wirelessly to a sink node or directly to the server. Low power consumption is the main reason for choosing Zigbee over other short distance communication systems protocol which is the main concerns for this study. It used the simplest ZigBee setup (sensor nodes to sink node) for less installation: and operational complication.

**Sensor node:** A sensor node collects and pre-process data from theelectrodes and sends data to the sink node making itas the main building block of the water quality monitoring system (WQMS) prototype.

**Sink node:** Sink node is basically a sensor node with the capability of sending data to long distance using GSM module.

**Data storage and visualization:** The system was designed to store data in three possible ways: a) in a pre-assigned cellular phone number for short period data storage; b) at the base station which houses the database and control center, and c) using an onboard Secure Digital (SD) card memory module which serves as backup storage for all individual sensor units. The Bathy is a new integrated sensor. This dependable, precise, and rugged device offers enhanced functionality ideal for specific operations of up to 6000m in depth. Valeport's highly accurate sensor technology can be combined for multiple profiles in one drop; the SWiFT CTD includes a new rapid-response temperature probe and functions reliably down to 500m as standard. SWiFT CTD's operational battery life of up to five days and option of charging via USB, make these profilers ideal for offshore, coastal, harbor and inland environmental and hydrographic survey use. A CTD plus version is also available with turbidity measuring capabilities. It couples CTD and turbidity measurements, while providing computed salinity, density and sound velocity.

**Wireless Network Technologies:**

a. (Local Communication) ZigBee, Wi-Fi (a/b/g/n), Wi-Fi Direct

b. (Remote Communication) RF Module, GSM, WiMAX LTE

**Challenges and recommendations**

» Weather conditions like the constant variation of temperature and water intrusion could affect some functions of the electronic component, thus, it is very important to consider it in the design stage of the entire system.

» Energy source issues can be addressed through design of energy management and utilization of high-capacity battery.

» Cost and affordability are one of the main problems for a widespread implementation of automated WQMS in developing countries.

**Open issues and future research directions**

1. Data Computation, Analysis, and Reporting
2. Data Communication and Transmission
3. Security, Privacy, and Data Confidentiality
4. Energy Harvesting
5. Bio fouling
6. Sensor Drift and Calibration
7. Underwater Propagation

## Future Prospective of Recent Techniques Use for Detection of Aquatic Pollution

In the future of chemical marker selection, it would be beneficial to evaluate the other emerging contaminants as potential markers for wastewater contamination. Relationship between fecal input and the markers should be checked periodically to assess the specificity and sensitivity of the use of the markers. It is recommended to use appropriate statistical tools to develop indices/ratio for the identification of wastewater contamination sources. Application of statistics techniques such as PCA, CA and DA would be beneficial for the interpretation of experimental data. Furthermore, analyzing data with these statistical techniques assists in the identification of potential chemical markers and could provide valuable information related to specific source of wastewater contamination.

Another research gap to address in future study is the influence of natural source which are not originated from wastewater, e.g., caffeine from Cola acuminate. There is limited study about background levels of chemical markers in aquatic systems and the wastewater sources. Further research should be designed to address this research gap to provide more accurate assessment of wastewater contamination and its origin.

## Conclusion

Various classes of chemicals have been proposed as chemical markers to detect wastewater contamination. The advantages of using chemical marker(s) are relatively fast and sensitive detection methods, spatially and temporally stable, and sourcespecific compared with microbial markers. PPCPs and ASs have been shown to be specific to wastewater contamination. Numerous PPCPs and ASs compounds are selected as markers because they are persistent, not effectively removed by WWTPs and stable during transport. The persistent nature of FWAs and their widespread distribution in aquatic environment made them suitable to use as markers for sewage contamination. The limitations of using FWAs as markers include the measurement by fluorometer may be interfered by organic matters in the water samples, and some studies noted lack of relationship between FWAs and fecal bacteria. The use of all chemical markers is region and temporally specific.

## Bibliography


Alippi, C., Camplani, R., Galperti, C. and Roveri, M. A. 2011. Robust, adaptive, solar poweredWSN framework for aquatic environmental monitoring. *IEEESensorsJournal, 11*, 45–55.

Artabe, A. E., Cunha-Silva, H. and Barranco, A. (2020). Enzymatic assays for the assessment of toxic effects of halogenated organic contaminants in water and food. A review. *Food and Chemical Toxicology, 145*: 111677.

Badruzzaman, M., Oppenheimer, J. A. and Jacangelo, J. G. 2013. Impact of environmental conditions on the suitability of microconstituents as markers for determining nutrient loading from reclaimed water. *Water Resources, 47*, 6198–6210.

Bakker, K. 2012. Water Security: Research Challenges and Opportunities. *Science, 337*(6097), 914-915.

Bora, T. and Dutta, J. 2014. Applications of nanotechnology in wastewater treatment – A review, *Journal of Nanoscience and Nanotechnology, 14*, 613-626, https://doi.org/10.1166/jnn.2014.8898

Burke, D. G. and Allenby, J. (2013). Low-cost water quality monitoring needs assessment. In *Chesapeake Conservancy, Technical Reports*, 3.

Buser, H. R., Poiger, T. and Muller, M. D. 1999. Occurrence and environmental behavior of the chiral pharmaceutical drug ibuprofen in surface waters and wastewater. *Environmental Science & Technology, 33*, 2529–2535.

Callery, O. and Healy, M. G. (2019). A novel method to rapidly assess the suitability of water treatment residual and crushed concrete for the mitigation of point and nonpoint source nutrient pollution. Resources. *Conservation and Recycling, 10*(2): 100010.

Can, Z. and Demirbas, M. 2015. Smartphone-based data collection from wireless sensornetworks in an urban environment. *Journal of Network and Computer Applications, 58*, 208–16.

Chung, W. Y, Yoo, J. H. 2015. Remote water quality monitoring in wide area. *Sensor andActuators B: Chemical, 217*, 51–7.

Crane, R. A. and Scott, T. B. 2012. Nanoscale zero-valent iron: Future prospects for an emerging water treatment technology, *Journal of Hazardous Materials, 211-212*, 112-125, https://doi.org/10.1016/j.jhazmat.2011.11.073

De Marziani C, Alcoleas R, Colombo F, Costa N, Pujana F, Colombo A, et al. 2011. A lowcost reconfigurable sensor network for coastal monitoring. In: OCEANS IEEE.6-9.

De Sousa, D. N. R., Mozeto, A. A., Carneiro, R. L. and Fadini, P. S. 2014. Electrical conductivity and emerging contaminant as marker of surface freshwater contamination by wastewater. *Science of The Total Environment, 484*, 19–26.

Derrien, M., Jarde, E., Gruau, G., Pourcher, A. M., Gourmelon, M., Jadas-Hecart, A. and Wickmann, A. C. P. 2012. Origin of fecal contamination in waters from contrasted areas: Stanols as microbial source tracking markers. *Water Resource, 46*, 4009–4016.

Dinh, T. L., Hu, W., Sikka, P., Corke, P., Overs, L. and Brosnan, S. 2007. Design and deployment of a remote robust sensor network: experiences from an outdoor water quality monitoring network. In: 32nd IEEE Conference on Local Computer Networks. 15-18.

Ealias, A. M. and Saravanakumar, M. P. 2017. A Review on the Classification, Characterisation, Synthesis of nanoparticles and their application. *IOP Conference Series: Materials Science and Engineering. 263*, 032019.

Fenech, C., Rock, L., Nolan, K., Tobin, J., Morrissey, A. 2012. The potential for a suite of isotope and chemical markers to differentiate sources of nitrate contamination: A review. *Water Resources, 46*, 2023–2041.

Foolad, M., Ong, S. L. and Hu, J. 2015. Transport of sewage molecular markers through saturated soil column and effect of easily biodegradable primary substrate on their removal. *Chemosphere, 138*, 553–559.

Froehner, S., Martins, R. F. and Errera, M. R. 2009. Assessment of fecal sterols in Barigui River sediments in Curitiba, Brazil. *Environmental Monitoring and Assessment, 157*, 591–600.

Gao, H., Chen, Q. and Yang, Q. 2015. Power management system for microbial fuel cellpowered water quality monitoring system. In: 2015 IEEE 16th Workshop onControl and Modeling for Power Electronics. Vancouver; 12-15.

Glassmeyer, S. T., Furlong, E. T., Kolpin, D. W., Cahill, J. D., Zaugg, S. D., Werner, S. L., Meyer, M. T. and Kryak, D. D. (2005). Transport of chemical and microbial compounds from known wastewater discharges: Potential for use as indicators of human fecal contamination. *Environmental Science & Technology 39*, 5157–5169.

Goddijn, L. M., and White, M. (2006). Using a digital camera for water quality measurements in Galway Bay. *Estuarine, coastal and shelf science, 66*(3-4), 429-436.

Gregor, J., Garrett, N., Gilpin, B., Randall, C. and Saunders, D. 2002. Use of classification and regression tree (CART) analysis with chemical faecal indicators to determine source of contamination. *New Zealand Journal of Marine and Freshwater Research, 36*, 387–398.

Grimalt, J. O., Fernandez, P., Bayona, J. M. and Albaiges, J. 1990. Assessment of fecal sterols and ketones as indicators of urban sewage inputs to coastal waters. *Environmental Science & Technology, 24*, 357–363.

Harrison, R. M. (2001). Pollution: Causes, effects and control. 4th edition, Royal Society of Chemistry, 580p.

Hayes, J., Beirne, S., Lau, K. T. and Diamond, D. 2008. Evaluation of a lowcost wirelesschemical sensor network for environmental monitoring. In: IEEE Sensors. 26-29.

Hsieh, C. D. and Yang, W. F. (2007). Optimal nonpoint source pollution control strategies for a reservoir watershed in Taiwan. *Journal of Environmental Management, 85*(4): 908-917.

Isobe, K. O., Tarao, M., Zakaria, M. P., Chiem, N. H., Minh, L. Y. and Takada, H. 2002. Quantitative application of fecal sterols using gas chromatography-mass spectrometry to investigate fecal pollution in tropical waters: Western Malaysia and Mekong Delta, Vietnam. *Environmental Science & Technology,* 36, 4497–4507.

Kagle, J., Porter, A. W., Murdoch, R. W., Rivera-Cancel, G. and Hay, A. G. 2009. Biodegradation of Pharmaceutical and personal care products. In Advances in Applied Microbiology; Allen, I., Laskin, S.S., Geoffrey, M.G., Eds.; Academic Press: Cambridge, MA, USA, pp. 65–108.

Kahn, J. and Yardley, J. (2007). As China roars, pollution reaches deadly extremes (available at https://www. nytimes.com/2007/08/26/world/asia/26china.html).

Kaizu, M., Iio, H., Yamada, N. and Noguchi. 2011. Development of unmanned airboat for water-qualitymapping. Biosystems Engineering, 109(4), 338–347.

Kennedy, V. S. (2004). Thermal Pollution. In *Encyclopedia of Energy*; Cleveland, C. J., Ed.; Springer: Boston, MA, USA, 6: 79–89.

Kong, R. Y. C., Mak, M. M. H. and Wu, R. S. S. (2009). DNA technologies for monitoring waterborne pathogens: A revolution in water pollution monitoring. *Ocean and Coastal Management, 52*(7), 355 358.

Kotamaki, N., Thessler, S., Koskiaho, J., Hannukkala, A. O., Huitu, H., Huttula, T., et al. 2009. Wireless in-situ sensor network for agriculture and water monitoring on a river basin scale in southern Finland: evaluation from a data user's perspective. *Sensors Basel 9*, 2862–83

Kuroda, K., Murakami, M., Oguma, K., Takada, H. and Takizawa, S. 2014. Investigating sources and pathways of perfluoroalkyl acids (PFAAs) in aquifers in Tokyo using multiple tracers. *Science of The Total Environment, 488–489*,51–60.

Leeming, R., Ball, A., Ashbolt, N. and Nichols, P. 1996. Using faecal sterols from humans and animals to distinguish faecal pollution in receiving waters. *Water Resources, 30*, 2893–2900.

Leeming, R., Latham, V., Rayner, M. and Nichols, P. 1997. Detecting and distinguishing sources of sewage pollution in Australian Island and coastal waters and sediments. *ACS Symp. Series, 671*, 306–319.

Lens, P. N. L., Virkutye, J., Jegatheesan, V., Kim, S. H., Al-Abed, S. 2013. Nanotechnology for water and wastewater treatment, IWA Publishing.

Li, Q., Mahendra, S., Lyon, D. Y., Brunet, L., Liga, M. V., Li, D., Alvarez, P. J. J. 2008. Antimicrobial nanomaterials for water disinfection and microbial control: Potential applications and implications, *Water research, 42*, 4591-4602,

Manyatshe, A., Fosso-Kankeu, E., van der Berg, D., Lemmer, N., Waanders, F. and Tutu, H. (2017). Dispersion of inorganic contaminants in surface water in the vicinity of Potchefstroom. *Physics and Chemistry of the Earth, Parts A/B/C, 100*: 86-93.

McCalley, D. V., Cooke, M. and Nickless, G. 1981.Effect of sewage treatment on faecal sterols. *Water Resources, 15*, 1019–1025.

Mudge, S. M. and Seguel, C. G. 1999. Organic contamination of San Vicente Bay, Chile. *Marine Pollution Bulletin, 38*, 1011–1021.

Murphy, K., Heery, B., Sullivan, T., Zhang, D., Paludetti, L., Lau, K. T., Diamond, D., Costa, E., O'Connor, N. and Regan, F. 2015. A low-cost autonomous optical sensor for water quality monitoring. *Talanta, 132*: 520-527.

Nodler, K., Hillebrand, O., Idzik, K., Strathmann, M., Schiperski, F., Zirlewagen, J. and Licha, T. 2013. Occurrence and fate of the angiotensin II receptor antagonist transformation product valsartan acid in the water cycle—A comparative study with selected b-blockers and the persistent anthropogenic wastewater indicators carbamazepine and acesulfame. *Water Research, 47*, 6650–6659.

Peeler, K. A., Opsahl, S. P. and Chanton, J. P. 2006. Tracking anthropogenic inputs using caffeine, indicator bacteria and nutrients in rural freshwater and urban marine systems. *Environmental Science & Technology, 40*, 7616–7622.

Pendergast, M. T. M. and Hoek, E. M. V. 2011. A review of water treatment membrane nanotechnologies, *Energy and environmental science, 4* (6), 1946-1971.

Perez, C. A., Jimenez, M., Soto, F., Torres, R., Lopez, J. A. and Iborra, A. 2011. A system formonitoring marine environments based on wireless sensor networks. In: OCEANS IEEE, 6-9.

Pink, D. H. (2006). Investing in tomorrow's liquid gold (available at https://finance.yahoo.com/columnist/article/ trenddesk/3748).

Qu, X., Alvarez, P. J. J. and Li, Q. 2013. Applications of nanotechnology in water and wastewater treatment. *Water research, 47*, 3931-3946.

Randhawa, S., Sandha, S. S. and Srivastava, B. 2016. A multi-sensor process for in-situ monitoring of water pollution in rivers or lakes for high-resolutionquantitative and qualitative water quality data. In: 2016 IEEE InternationalConference on Computational Science and Engineering and IEEEInternational Conference on Embedded and Ubiquitous Computing and15th International Symposium on Distributed Computing and Applicationsfor Business Engineering. 24-26.

Rashid, B. and Rehmani, M. H. 2016. Applications of wireless sensor networks for urbanareas: a survey. *Journal of Network and Computer Applications, 60*, 192–219.

Reder, K., Alcamo, J. and Flörke, M. (2017). A sensitivity and uncertainty analysis of a continental-scale water quality model of pathogen pollution in African rivers. *Ecological Modelling, 351*: 129-139.

Ren, X., Chen, C., Nagatsu, M. and Wang, X. 2011. Carbon nanotubes asadsorbents in envioromental pollution mamagement: A review. *Chemical Engineering Journal, 170*, 395-410, https://doi.org/10.1016/j.cej.2010.08.045

Roadknight, C., Parrott, L., Boyd, N. and Marshall, I. W. 2004. A layered approach to in situdata management on a wireless sensor network. In: 2004 IntelligentSensors, Sensor Networks and Information Processing Conference.Melbourne, 14-17.

Ruan, X., Huang, S., Chen, H. and Qian, G. (2013). Sorption of aqueous organic contaminants onto dodecyl sulfate intercalated magnesium iron layered double hydroxide. *Applied Clay Science, 72*: 96-103.

Savedge, J. (2019). Water pollution: Causes, effects, and solutions (available at https://www.thoughtco.com/waterpollution-causes-effects-and-solutions-1140786).

Shaltami, O. R., Fares, F. F., Errishi, H., EL Oshebi, F. M., Elabbar, F. A., Eltaboni, F. B., Elshelmani, N. M., Elghazal, R. and Bustany, I. (2019). Geochemical evaluation of water in the Benghazi City, NE Libya. 4th International Conference on Science and Natural Resources (ICSNR4), Kota Kinabalu, Malaysia, Proceeding Book; pp. 13 22.

Shin, D., Na, S. Y., Kim, J. Y., and Baek, S. J. (2007). Sonar localization using ubiquitous sensor network for water pollution monitoring fish robots. In *2007 IEEE International Symposium on Signal Processing and Information Technology* (pp. 80-85). IEEE.

Sidhu, J. P. S., Ahmed, W., Gernjak, W., Aryal, R., McCarthy, D., Palmer, A., Kolotelo, P. and Toze, S. 2013. Sewage pollution in urban stormwater runoff as evident from the widespread presence of multiple microbial and chemical source tracking markers. *Science of The Total Environment, 463–464*, 488–496.

Staff, R. (2017). Study links pollution to millions of deaths worldwide (available at https://www.reuters.com/article/us-health-pollution/study-links-pollution-to-millions-of-deaths-worldwide-idUSKBN1CO39B).

Tchobanoglous, G., Stensel, H. D., Tsuchihashi, R. and Burton, F.2014. Wastewater engineering: Treatment and resource recovery, Fifth Edition, New York, McGraw-Hill.

Tran, N. H., Hu, J., Li, J. and Ong, S. L. 2014b. Suitability of artificial sweeteners as indicators of raw wastewater contamination in surface water and groundwater. *Water Resources, 48*, 443–456.

Tran, N. H., Li, J., Hu, J. and Ong, S. L. (2014a). Occurrence and suitability of pharmaceuticals and personal care products as molecular markers for raw wastewater contamination in surface water and groundwater. *Environmental Science & Pollution Resources, 21*, 4727–4740.

Vane, C. H., Kim, A. W., McGowan, S., Leng, M. J., Heaton, T. H. E., Kendrick, C. P., ... & Swann, G. E. A. (2010). Sedimentary records of sewage pollution using faecal markers in contrasting peri-urban shallow lakes. *Science of the Total Environment, 409*(2), 345-356.

Wang, J., Ren, X. L., Shen, Y. L., and Liu, S. Y. (2010). A remote wireless sensor networks for water quality monitoring. In *2010 International Conference on Innovative Computing and Communication and 2010 Asia-Pacific Conference on Information Technology and Ocean Engineering* (pp. 7-12). IEEE.

Watlington, K. 2005. Emerging Nanotechnologies for Site Remediation and Wastewater Treatment. *National Network for Environmental Management Studies Fellow North Carolina State University.*

Wu, Y. and Chen, J. (2013). Investigating the effects of point source and nonpoint source pollution on the water quality of the East River (Dongjiang) in South China. *Ecological Indicators, 32*: 294-304.

Xagoraraki, I. and Kuo, D. (2008). Water pollution: Emerging contaminants associated with drinking water. *International Encyclopedia of Public Health,* 539-550.

Xue, N., Wang, L., Li, W., Wang, S., Pan, X. and Zhang, D. (2020). Increased inheritance of structure and function of bacterial communities and pathogen propagation in plastisphere along a river with increasing antibiotics pollution gradient. *Environmental Pollution, 265*(A): 114641.

Yamaji, N., Hayakawa, K. and Takada, H. 2010. Role of photodegradation in the fate of fluorescent whitening agents (FWAs) in lacustrine environments. *Environmental Science & Technology, 44,* 7796–7801.

Yang, S. C., and Pan, Y. (2010). The application of the wireless sensor network (WSN) in the monitoring of Fushun Reach River in China. In *2010 Second International Conference on Computer and Network Technology* (pp. 331-333). IEEE.

Yang, X., Ong, K. G., Dreschel, W. R., Zeng, K., Mungle, C. S., & Grimes, C. A. (2002). Design of a wireless sensor network for long-term, in-situ monitoring of an aqueous environment. *Sensors, 2*(11), 455-472.

Yu, L., Fink, G., Wintgens, T., Melin, T., & Ternes, T. A. (2009). Sorption behavior of potential organic wastewater indicators with soils. Water Research, 43(4), 951-960.

Zhan, A., Chen, G., and Wang, W. (2009). Utilizing automatic underwater vehicles to prolong the lifetime of underwater sensor networks. In *2009 Proceedings of 18th International Conference on Computer Communications and Networks* (pp. 1-6). IEEE.

Zhou, P., Huang, J., Jr, R. G. P. and Hong, H. (2016). New insight into the correlations between land use and water quality in a coastal watershed of China: Does point source pollution weaken it? *Science of The Total Environment, 543*(A): 591-600.

Zirlewagen, J., Licha, T., Schiperski, F., Nodler, K. and Scheytt, T. 2016. Use of two artificial sweeteners, cyclamate and asesulfame, to identify and quantify wastewater contributions in a karst spring. *Science of The Total Environment, 547*, 356–365.

# Chapter - 5

# Nanoplastic Pollutants Affecting Fisheries Sector All Over the World

**Kafeela M. Qureshi[1*], Feroz A. Shah[1], Imtiyaz Qayoom[2], Irtifa Fayaz[1] and Kaiser Farooq[1]**

Division of Aquatic Animal Health Management, Faculty of Fisheries, SKUAST-K, Rangil-190006, Ganderbal, J&K, India

[2] Division of Aquatic Environmental Management, Faculty of Fisheries, SKUAST-K, Rangil-190006, Ganderbal, J&K, India

[*]Corresponding mail: qureshik982@gmail.com

## Abstract

Future environmental hazards from developing pollutants like nanoplastics cannot be disregarded. There is general agreement that these are not only present in the environment - either directly released or as a result of weathering of larger fragments - but that they pose a serious threat to the environment and human health, as well. However, their presence has been difficult to adequately ascertain due to the inherent technical difficulties for isolating and quantifying them. These particles are suitable for consumption by species at the bottom of the food chain due to their smaller size. Furthermore, because nanoparticles often have a high surface area to volume ratio, additional pollutants, like persistent organic pollutants, may be adsorbed and subjected to bioaccumulation and bio-amplification events. Recent discussions among scientists, policymakers, and the general public have focused on nanoplastics since the weathering of macroplastics could result in a significant load.of nanoplastics in different ecosystems. This chapter will put light on the adverse effects caused by nanoplastics accumulation in the aquatic environment that is the subject of concern today and some of the remediation processes with special reference to microbial remediation.

**Keywords:** Nanoplastics, Bioaccumulation, Heteroaggregates, Degradation, Remediation

## Introduction

Plastic wastes have been acknowledged as prevalent, largely non-biodegradable, and widely dispersed over the earth's various ecosystems (Amobonye*et al.*, 2020). Plastic manufacturing has increased quickly, from hundreds of millions of tonnes in 2016 to an anticipated double by 2025 and triple by 2050. Due to the longevity and resistance to biodegradation of plastics, the rates of accumulation in the environment are high, but there has not yet been a significant release of plastic garbage. Since most moldable wastes end up in the ocean, which receives roughly 12 million tonnes of plastic every year, the ocean is their final resting place. As a result, there is rising concern about the consequences of microplastics (MPs) and nanoplastics (NPs) in the environment. Moreover, the monitoring of the plastic characteristics should be continuously observed for both produced and disposed plastics (Gaylarde, 2021). Since micro (defined here as 100 nm to 5 mm in size) and nano (defined as 100 nm in one dimension) sized particles are small enough to be ingested by many animals, concerns about potential bioaccumulation and biomagnification are raised by their widespread presence in the environment. The risk posed by nanoplastics is greater than that of microplastics because they can penetrate through biological membranes (Ng *et al.*, 2018). Even though it has been shown that the majority of polymeric bulk materials used to make micro and nano plastics (MNPs) are relatively biochemically inert, changes to MNPs physicochemical qualities have increased their bioavailability and toxicity (Ma *et al.*, 2020). Due to their widespread distribution and potential for contamination, MNPs have thus been the subject of several scientific investigations in both aquatic and terrestrial ecosystems. Plastic waste has been discovered in a variety of aquatic biota, including small algae and large mammals, in addition to abiotic compartments.

Three qualities should be present in an organism employed as a bioindicator of plastic-polluted areas: universal occurrence, including relatively simple collection, extensive contact with the environment through particle-feeding and impartial ingestion of the majority of plastic particles to which it is exposed (Ward *et al.*, 2019). It is anticipated that nanoplastics may naturally aggregate together as well and exist as huge colloids a few hundred nanometers in size, in contrast to nanotechnology, where manufactured nanoparticles typically aggregate into clusters up to several micrometers in size. Then, it is conceivable that smaller plastic particles would bind. The reactivity, toxicity, fate, transport, and risk to the environment and organisms are largely determined by the dynamic processes of spontaneous heteroaggregation and disaggregation of nanoplastics (Ter Halle and Ghiglione, 2021). The least well-known component of marine litter, NPs may also be the most dangerous. There are many potential sources for NPs, including product release or "Nanofragmentation," which is the breakdown of bigger particles. In addition to physiological impacts, NPs may also have chemical ones. Since NPs have a large surface area, they may have extraordinarily high sorption affinities for poisonous substances, which result in chemical and particle toxicity consequences once NPs pass cell membranes. Additionally, if the process of nanofragmentation is relevant, the release of non-polymer nanoscale additives from the product fragments may increase the total risk (Nowack *et al.,* 2012; Koelmans *et al.,* 2015). Nano-scale plastic materials behave and interact with living things in ways that are very different from nanoparticles. It has been demonstrated that the surface area of nano scale particles has an effect on the biotoxicity of polystyrene nanoparticles, and the high surface area improves their potential for heightened reactivity. Important parameters include, among others, chemical composition, purity, doping, hydrodynamic size, morphology, size heterogeneity, redox parameters, tendency for aggregation, nature and composition of the shell or coating material, surface modifications and surfactants, chemical and colloidal stability, solubility, biodegradability, concentration, duration of exposure, and unknown interactions with other nanoparticles or behavior under electromagnetic field exposure (Mattsson *et al.,*2018).

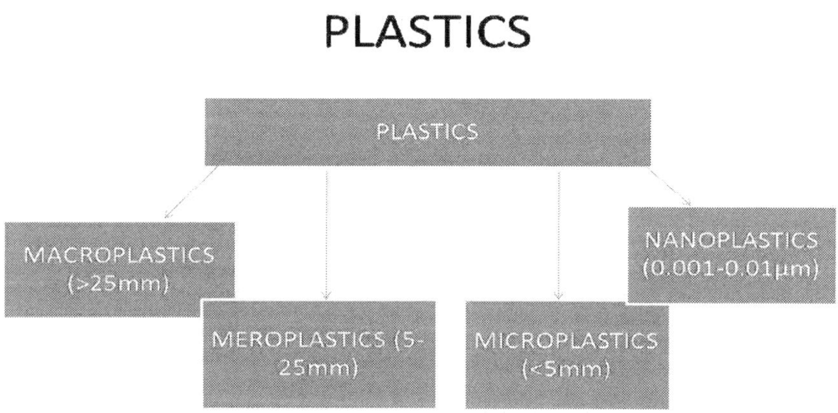


**Figure 1: Classification of plastics modified from (Alprol *et al.,* 2021)**

## Nanoplastics and Manufactured Plastics

Nanoplastics and other "nanomaterials" cannot be compared because they have various physical and chemical characteristics and separate production processes. A manufactured nanomaterial is one that has been "deliberately developed for commercial reasons to have defined qualities or precise composition," according to the International Standard Organization (ISO/TS 80004e1:2015). Size, shape, surface, and composition are typically the specific physical and chemical features, and they cannot be extended to nanoplastics. Numerous researchers have thought about produced nanomaterials and their potential release and impact on the environment in the Nano-Environmental Health and Safety programmes. On the other hand, unlike produced nanomaterials, nanoplastics do not have a purpose for size selection; rather, they are the end product of the erosion or breakdown of bigger plastic trash. As a result, the physical characteristics and composition of nanoplastics are highly heterogeneous and polydisperse. Indeed, it is very likely that nanoplastics will aggregate with other natural and/or man-made materials because they are unintentionally created by the decomposition of microscale plastic waste. Colloids are typically heteroaggregates made up of clay and organic compounds in environmental systems. Consequently, the physical and chemical characteristics of the system (pH, salinity, natural organic matter, etc.) usually have little control over and influence on their surface and structure (Gigault *et al.*, 2018).

## Sources of Nanoplastics in aquatic systems

The abiotic and biotic breakdown of microplastics results in the production of nanoplastics. For instance, it has been demonstrated that microplastics can become nanoplastics when exposed to UV light, and gastric fragmentation has been suggested as a way to produce nanoplastics from microplastics (Wang *et al.*, 2020a). The removal of microplastics by current wastewater treatment techniques has been demonstrated to be effective (Carr *et al.*, 2016), but this may not hold true for smaller particles like nanoplastics. Additionally, it also relies on the procedures used, as others have pointed out that wastewater treatment facilities struggle to remove tiny pieces of plastic effectively (Browne *et al.*, 2011). It has been proposed that household items like cosmetics and cleaning supplies, which are disposed of in domestic wastewater, are one of the primary sources of polymeric micro- and nano-sized pollutants in aquatic systems (Da Costa *et al.*, 2016). These aren't the only places where plastics are found in the environment, additional sources include those with an industrial origin, such as raw materials used to make plastic items and from the spilling of powdered or pelleted plastic resin used for air blasting. Whether occurring before entering the environment, as is the case with the fragmentation of synthetic fibers during the washing of clothes, or after, as a result of the wear-and-tear of the plastic items exposed to the elements, the breakdown of larger plastic items constitutes an additional source of micro- and nanoplastics (Browne *et al.*, 2011). This fragmentation process has been described by (Shim *et al.*, 2014). They were able to create micro- and nanoplastics from expanded polystyrene using an accelerated mechanical abrasion system, simulating conditions found on rivers or beaches (Corcoran *et al.*, 2009), where prolonged abrasion by sand particles may result in the formation of these particulates. It should be emphasized that other elements were not taken into account that could have speed up the degradation of bigger plastics, like UV radiation and microbial activity. Other significant sources of plastics in aquatic systems include ship-generated waste; trash dumped after recreational activities, misplaced, or carelessly handled fishing fleet's plastic fishing gear (Da Costa*et al.*, 2016). Agricultural polyethylene foils, polystyrene foam thermal cutting, and drying garments are a few examples of NP sources. The development of 3D printing, which is now widely used for quick prototypes and small projects, has been named as one of the sources of ultrafine particles. Ultrafine particles may also be found in the environment as a result of using therapeutic items made from polymeric nanoparticles for biological purposes (Mofijur *et al.*, 2021). Engineered plastics are electro-pinned to create mats with fibers that could eventually deteriorate to the nanoscale. MPs can degrade to a size of less than 100 nm, which is another source of NPs.

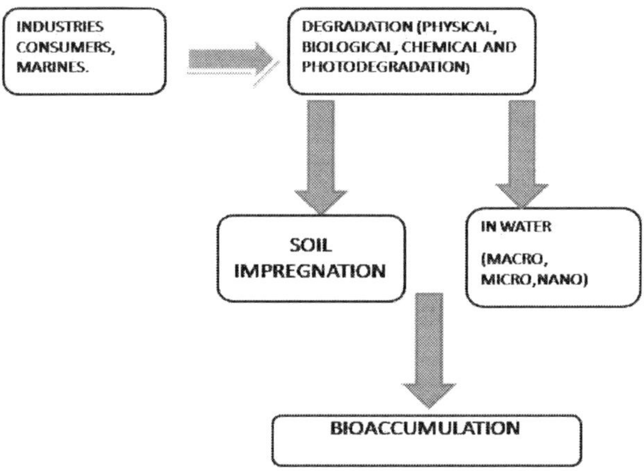


**Figure 2: Sources of Nanoplastics in the environment (Gagadoo *et al.*, 2020)**

## Toxicity of Nanoplastics

The intrinsic toxicity of nanoscale polymers differs little from that of microscale plastics. The toxicities of Polystyrene nanoplastics (100 nm) on *Mytilus eludes* and *Crassostrea virginica* after a 45-min exposure were examined by Ward and Kach, (2009) in one ecotoxicity research in a marine setting, recording ingestion and egestion of both species. The study found that nano plastics can get to the digestive gland and stay in the body longer than micro plastics. In a different study, Chen *et al.*, (2011) examined the extraction of exopolymeric compounds by three phytoplankton species and the accumulation of these substances using 10 and 100 ppb of non-fluorescent nanoparticles and fluorescent PS modified nano particles (both 23 nm). In order to evaluate the effects of nanoplastics on the formation of pseudo feces and the weight of faeces and pseudo feces in *M. edulis*, Wegner *et al.*, (2012) also utilized PS nanoplastics (30 nm). Nanoplastics were removed from the water by *M. edulis*, which also shows a modest decline in filtering capabilities when algae were present (food). PS nanoplastics with carboxyl (40 nm) and amine (50 nm) functional groups were employed by Della Torre *et al.*, (2014) in the sea urchins*Paracentrotus lividus.* The organisms had aberrant gene expressionand were harmful to develop embryos. Additionally, Bergami *et al.*, (2016) examined *Artemia franciscana* mortality and sub-lethal effects using carboxylated and amino-modified PS nanoplastics (40–50 nm). Eating, motility, and multiple molting were negatively impacted in *A.franciscana*by nanoplastics. These findings suggest that ecosystems as well as organisms and populations are impacted by nanoplastics. Kashiwada (2006) exposed *Oryzias latipes* to 39.4 nm PS nanoplastics and measured adsorption, accumulation, distribution, and death to examine the effects of nanoplastics in a freshwater ecosystem. *Scenedesmus obliquus* and *Daphnia magna* were tested for toxicity by Besseling *et al.*, (2014) using PS nanoplastics functionalized with carboxylic acid groups. Nanoplastics also had an impact on the survival, reproduction, body size, and physical development of *D. magna* neonates, in addition to the photosynthetic capability and biomass of algae. These investigations evaluated hazardous values and looked into how nanoplastics affected growth, survival, and immobility. Nanoplastics have been employed in numerous in vitro tests with fish (Casado *et al.*, 2013; Greven *et al.*, 2016) and mussels (Canesi *et al.*, 2015; Canesi *et al.*, 2016) due to their smaller size and similarity to nanoparticles, which have been the subject of extensive research. Nanoplastics were tested for trophic transfer by Cedervall *et al.*, (2012) and Mattsson *et al.*, (2014). Using the same food chain, Mattsson *et al.*, (2014) examined fish behaviour and carried out NMR spectroscopy. Nanoplastics have a significant likelihood of penetrating organisms´ body and tissues because of their small volume and sizable surface area. One of the main ways people are exposed to pollutants is through trophic transfer, which needs more research (Chae and An, 2017).

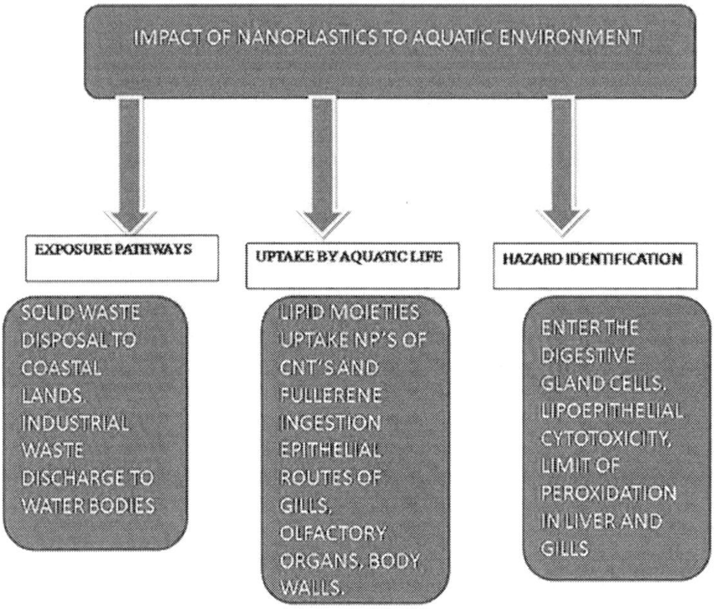


**Figure 3: Impact of nanoplastics to the aquatic environment modified from (Sajid *et al.*, 2014)**

## Controlling Measures of Nanoplastics

To more fully comprehend NP dangers, more research is needed. The first task, however, is to gather reliable data on exposure to marine, freshwater, and terrestrial environments (Trevail *et al.*, 2015). Environmental NP concentrations are currently only estimates because of the inadequate data and the lack of precise analytical procedures (Schwaferts *et al.*, 2019; Trevail *et al.*, 2015). Emerging remediation technology, public policy, education, and awareness are the main NP mitigation strategies. Though there hasn't been much advancement in the development of NP remediation techniques, recent studies have offered various possible directions. The increasing substitution of conventional non-biodegradable NPs with ecologically friendly biodegradable NPs is made possible by advances in biotechnology. In addition, specialized enzymes, bacteria, and fungi may be used to facilitate an effective and efficient disposal process. Reducing NPs in wastewater treatment plants is the most efficient way to limit NP transmission into the marine environment.

It has been demonstrated that pre-treatment methods (coagulation, density, and membrane separation) are successful in eliminating MPs from drinkable water; however, further research is needed to determine whether or not they are also successful in removing NPs. According to reports, altering the soil (by adding biochar, for instance) can lessen the amount of nitrogen that moves between particles in porous medium, stabilizing the risk. In future research, pH regulation and the addition of inorganic or organic material may be employed to dramatically increase the accumulation of NP in a wastewater treatment process, resulting in the successful separation of NP (Mofijur *et al.*, 2021). Extrapolating cleanup techniques for other pollutants to NPs is advised. Adding "green" remedial materials like biochar (Rajapaksha*et al.*, 2016; Wang *et al.*, 2020b) and clay minerals (Simoes *et al.*, 2019) is one example.

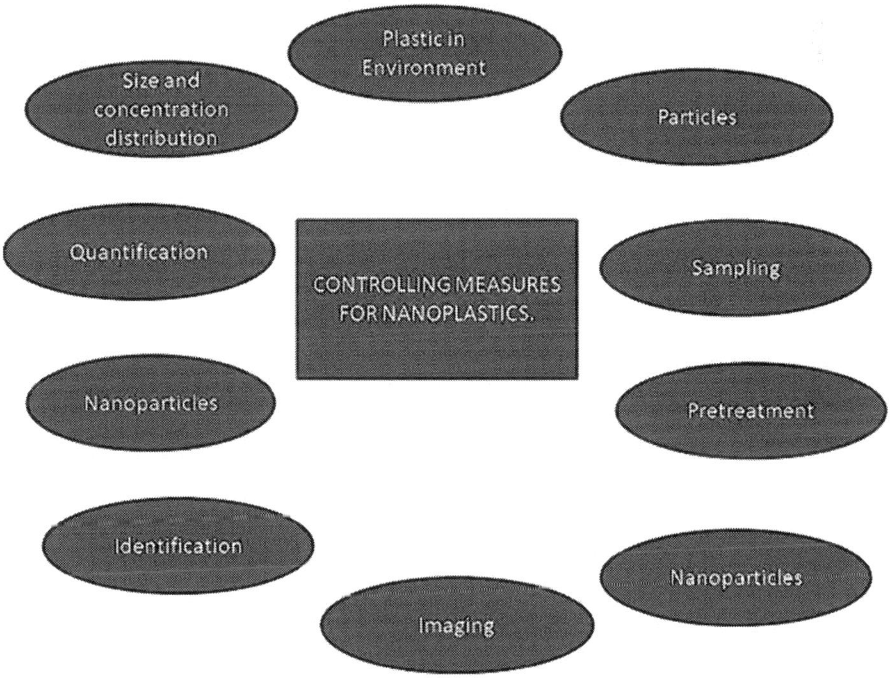


**Figure 4: Controlling measures for nanoplastics (Cai *et al.*, 2021)**

## Microbial Remediation Techniques of Nanoplastics

Micro-nano plastics pollutants can be removed by microbial decomposition, which is practicable, clean, and inexpensive. The microbial deterioration of micro-nanoplastics takes place along a number of biochemical reaction pathways. MNPs' different chemical structures also affect how they degrade, and this needs to be fully understood if we want to protect our environment. Two key procedures are required to increase the rate of biodegradation of pollutants in nutrient-limited environments: bio-stimulation (the provision of limiting nutrients to stimulate microbial growth) and bio-augmentation (the addition of living cells capable of breakdown). Numerous factors influence the microbial breakdown of polymers, which affects how the materials behave physico-chemically. Surface morphology, topography, surface hydrophobicity, electric charge distribution, and other environmental elements all have an impact on the degradation process. Biotic factors include metabolic activity, the release of acids, and enzymatic activity (such as temperature, pH, salinity, oxygen level). The crucial MNP properties, such as additiveness, hydrophobicity, and the persistence of organic contaminants on the surface, are important because they have an impact on remediation. Even so, it might be challenging to comprehend the ways in which MNPs degrade. The subsequent sequential phases are used to describe how microbial plastic breakdown happens. Assimilation is the incorporation of the molecules by microorganisms. Biodeterioration is the change in the physical and chemical properties of the polymer caused by biological agents. Bio-fragmentation is the lytic cleavage of complex polymers into simpler forms (Oxidized metabolites produced by degradation such as $CO_2$, $CH_4$, and $H_2O$). According to reports, abiotic variables including UV radiation and photo-oxidation cause a rise in microbial deterioration. Due to the existence of large molecular fragments that are challenging for cellular absorption, higher molecular weight plastic polymers impede the process of microbial breakdown. Microorganisms use two important mechanisms called intracellular and extracellular degradation to depolymerize such big particles. Microbes gather on the surface of MNPs as a result of intracellular breakdown and hydrolyze the plastic into short chains. Bacteria release extracellular enzymes like hydrolases that break down complex polymers into simpler pieces so they can be digested by various metabolic pathways during extracellular breakdown. Additionally, through aerobic or anaerobic metabolism, bacteria transform these short chains into end products ($CO_2$, $H_2O$, or $CH_4$) in a process known as mineralization. Biological natural attenuation refers to the process by which these final products are used as sources of carbon and energy (Tiwari *et al.*, 2020).

## Conclusion

The interaction of pollutants with MNPs has been the subject of extensive investigation, although most of these studies only used one sort of pollutant and one form of plastic particle. The concentration of contaminants in the water column or sediment can be considerably increased by NPs, although not all pollutants interact with NPs with the same affinity. Therefore, in a complex mixture scenario that occurs in real life, NPs may result in the differential enrichment of contaminants already present in the environment. This differential enrichment will vary based on the physico-chemical properties of each NP particle. The way that NPs interact with additional environmental pollutants will also depend on the pollutants that have previously been adsorbed to them. Other significant issues in this area include (i) how different environmental factors, such as temperature, pH, salinity, and organic matter, will affect these interactions; (ii) how such complex and variable enrichment of co-contaminants on NPs' surfaces will impact NPs' toxicity; and (iii) how weathering might affect the physico-chemical properties of both the NPs and their adsorbed pollutants. As opposed to environmental pollutants, these concerns relate to how NPs interact with microbial communities. These are but a few of the urgent demands and difficulties in the dynamic and multifaceted field of plastic's ecotoxicity, which is without a doubt a major environmental concern. Therefore, there is a need for more research on the impact and ecotoxicity caused by NP's.

## Bibliography

Alprol, A. E., Gaballah, M. S. and Hassaan, M. A. (2021). Micro and Nanoplastics analysis: Focus on their classification, sources, and impacts in marine environment. *Regional Studies in Marine Science*, *42*: 101625.

Amobonye, A., Bhagwat, P., Singh, S. and Pillai, S. (2021). Plastic biodegradation: Frontline microbes and their enzymes. *Science of the Total Environment, 759*: 143536.

Bergami, E., Bocci, E., Vannuccini, M. L., Monopoli, M., Salvati, A., Dawson, K. A. and Corsi, I. (2016). Nano-sized Polystyrene affects feeding, behavior and physiology of brine shrimp *Artemia franciscana* larvae. *Ecotoxicology and Environmental Safety, 123*: 18-25.

Besseling, E., Wang, B., Lürling, M. and Koelmans, A. A. (2014). Nanoplastic affects growth of *S. Obliquus* and reproduction of *D. magna*. *Environmental Science & Technology, 48*(20): 12336-12343.

Browne, M. A., Crump, P., Niven, S. J., Teuten, E., Tonkin, A., Galloway, T. and Thompson, R. (2011). Accumulation of microplastic on shorelines worldwide: sources and sinks. *Environmental science & technology*, *45*(21): 9175-9179.

Cai, H., Xu, E. G., Du, F., Li, R., Liu, J., & Shi, H. (2021). Analysis of environmental nanoplastics: Progress and challenges. *Chemical Engineering Journal*, *410*: 128208.

Canesi, L., Ciacci, C., Bergami, E., Monopoli, M. P., Dawson, K. A., Papa, S., Canonica, B. and Corsi, I. 2015. Evidence for immunomodulation and apoptotic processes induced by cationic polystyrene nanoparticles in the hemocytes of the marine bivalve *Mytilus*, *Marine Environmental Research, 111:* 34-40.

Canesi, L., Ciacci, C., Fabbri, R., Balbi, T., Salis, A., Damonte, G., Cortese, K., Caratto,

V., Monopoli, M. P., Dawson, K., Bergami, E. and Corsi, I. (2016). Interactions of cationic polystyrene nanoparticles with marine bivalve hemocytes in a physiological environment: role of soluble hemolymph proteins. *Environmental Research, 150*: 73-81.

Carr, S. A., Liu, J. and Tesoro, A. G. (2016). Transport and fate of microplastic particles in wastewater treatment plants. *Water Research, 91*(15): 174-182.

Casado, M. P., Macken, A. and Byrne, H. J. (2013). Ecotoxicological assessment of silica and polystyrene nanoparticles assessed by a multitrophic-test battery. *Environment International, 51*: 97–105.

Cedervall, T., Hansson, L. A., Lard, M., Frohm, B. and Linse, S. (2012). Food chain transport of nanoparticles affects behaviour and fat metabolism in fish. *PLoS ONE, 7*(2): e32254.

Chae, Y. and An, Y. J. (2017). Effects of micro- and nanoplastics on aquatic ecosystems: Current research trends and perspectives. *Marine Pollution Bulletin, 124*(2): 624-632.

Chen, C. S., Anaya, J. M., Zhang, S., Spurgin, J., Chuang, C. Y., Xu, C., Miao, A. J., Chen, E. Y. T., Schwehr, K. A., Jiang, Y., Quigg, A., Santschi, P. H. and Chin, W. C. (2011). Effects of engineered nanoparticles on the assembly of exopolymeric substances from phytoplankton. *PLOS ONE, 6*(7): e21865.

Corcoran, P. L., Biesinger, M. C., Grifi, M. (2009). Plastics and beaches: a degrading relationship. *Marine Pollution Bulletin,* 58: 80–84.

Da Costa, J. P., Santos, P. S., Duarte, A. C. and Rocha-Santos, T. (2016). (Nano) plastics in the environment–sources, fates and effects. *Science of the total environment, 566*: 15-26.

Della Torre, C., Bergami, E., Salvati, A., Faleri, C., Cirino, P., Dawson, K.A. and Corsi, I. (2014). Accumulation and embryotoxicity of polystyrene nanoparticles at early stage of development of sea urchin embryos *Paracentrotuslividus*. *Environmental Science and Technology,* 48(20): 12302–12311.

Gangadoo, S., Owen, S., Rajapaksha, P., Plaisted, K., Cheeseman, S., Haddara, H. and Chapman, J. (2020). Nano-plastics and their analytical characterization and fate in the marine environment: From source to sea. *Science of the Total Environment, 732*: 138792.

Gaylarde, C. C., Neto, J. A. B. and da Fonseca, E. M. (2021). Nanoplastics in aquatic systems-are they more hazardous than microplastics? *Environmental Pollution, 272*: 115950.

Gigault, J., Ter Halle, A., Baudrimont, M., Pascal, P. Y., Gauffre, F., Phi, T. L. and Reynaud, S. (2018). Current opinion: what is a nanoplastic? *Environmental pollution, 235*: 1030-1034.

Greven, A. C., Merk, T., Karagöz, F., Mohr, K., Klapper, M., Jovanović, B. and Palić, D. (2016). Polycarbonate and Polystyrene nanoplastic particles act as stressors to the innate immune system of fathead minnow (*Pimephalespromelas*). *Environmental Toxicology & Chemistry, 35*(12): 3093-3100.

Kashiwada, S. (2006). Distribution of nanoparticles in the see-through medaka (*Oryzias latipes*). *Environmental Health Perspectives,* 1697–1702.

Koelmans, A. A., Besseling, E. and Shim, W. J. (2015). Nanoplastics in the aquatic environment. Critical review. *Marine anthropogenic litter*, 325-340.

Ma, H., Pu, S., Liu, S., Bai, Y., Mandal, S. and Xing, B. (2020). Microplastics in aquatic environments: toxicity to trigger ecological consequences. *Environmental Pollution, 261*: 114089.

Mattsson, K., Ekvall, M. T., Hansson, L. A., Linse, S., Malmendal, A. and Cedervall, T. (2014). Altered behavior, physiology, and metabolism in fish exposed to polystyrene nanoparticles. *Environmental Science Technology, 49*(1): 553–561.

Mattsson, K., Jocic, S., Doverbratt, I. and Hansson, L. A. (2018). Nanoplastics in the aquatic environment. *Microplastic contamination in aquatic environments*, 379-399.

Mofijur, M., Ahmed, S. F., Rahman, S. A., Siddiki, S. Y. A., Islam, A. S., Shahabuddin, M., ... & Show, P. L. (2021). Source, distribution and emerging threat of micro-and nanoplastics to marine organism and human health: Socio-economic impact and management strategies. *Environmental research, 195*: 110857.

Ng, E. L., Lwanga, E. H., Eldridge, S. M., Johnston, P., Hu, H. W., Geissen, V. and Chen, D. (2018). An overview of microplastic and nanoplastic pollution in agroecosystems. *Science of the total environment, 627*: 1377-1388.

Nowack, B., Ranville, J. F., Diamond, S., Gallego-Urrea, J. A., Metcalfe, C., Rose, J., Horne, N., Koelmans, A. A. and Klaine, S. J. (2012). Potential scenarios for nanomaterial release and subsequent alteration in the environment. *Environmental Toxicology and Chemistry,31*(1): 50-59.

Rajapaksha, A. U., Chen, S. S., Tsang, D. C., Zhang, M., Vithanage, M., Mandal, S. and Ok, Y. S. (2016). Engineered/designer biochar for contaminant removal/immobilization from soil and water: potential and implication of biochar modification. *Chemosphere, 148*: 276-291.

Sajid, M., Ilyas, M., Basheer, C., Tariq, M., Daud, M., Baig, N. and Shehzad, F. (2014). Impact of nanoparticles on human and environment: review of toxicity factors, exposures, control strategies, and future prospects. *Environmental science and pollution research,22*: 4122-4143.

Schwaferts, C., Niessner, R., Elsner, M. and Ivleva, N. P. (2019). Methods for the analysis of submicrometer-and nanoplastic particles in the environment. *Trends in Analytical Chemistry, 112*: 52-65.

Shim, W., Hong, S. and Jang, M. G. H. (2014). Producing fragmented micro- and nano-sized expanded polystyrene particles with an accelerated mechanical abrasion experiment. 23rd Annual Meeting SETAC. Basel, Switzerland, p. 238.

Simoes, A., Stringfellow, A. M., Smallman, D., Beaven, R., Marshall, J., Powrie, W. and Potter, H. A. B. (2011). Sorption of organic contaminants by Oxford Clay and Mercia Mudstone landfill liners. *Quarterly Journal of Engineering Geology and Hydrogeology, 44*(3): 345-360.

Ter Halle, A. and Ghiglione, J. F. (2021). Nanoplastics: A complex, polluting terra incognita. *Environmental Science & Technology*, *55*(21): 14466-14469.

Tiwari, N., Santhiya, D. and Sharma, J. G. (2020). Microbial remediation of micronano plastics: Current knowledge and future trends. *Environmental Pollution, 265*: 115044.

Trevail, A. M., Gabrielsen, G. W., Kühn, S. and Van Franeker, J. A. (2015). Elevated levels of ingested plastic in a high Arctic seabird, the northern fulmar (Fulmarus glacialis). *Polar Biology*, *38*(7): 975-981.

Wang, L., Ok, Y. S., Tsang, D. C., Alessi, D. S., Rinklebe, J., Wang, H. and Hou, D. (2020). New trends in biochar pyrolysis and modification strategies: feedstock, pyrolysis conditions, sustainability concerns and implications for soil amendment. *Soil Use and Management*, *36*(3): 358-386.

Wang, Y. L., Lee, Y. H., Chiu, I. J., Lin, Y. F. and Chiu, H. W. (2020b). Potent impact of plastic nanomaterials and micromaterials on the food chain and human health. *International journal of molecular sciences*, *21*(5): 1727.

Ward, J. E. and Kach, D. J. (2009). Marine aggregates facilitate ingestion of nanoparticles by suspension-feeding bivalves. *Marine environmental research*, *68*(3): 137-142.

Ward, J. E., Zhao, S., Holohan, B. A., Mladinich, K. M., Griffin, T. W., Wozniak, J. and Shumway, S. E. (2019). Selective ingestion and egestion of plastic particles by the blue mussel (*Mytilus edulis*) and eastern oyster (*Crassostrea virginica*): implications for using bivalves as bioindicators of microplastic pollution. *Environmental science & technology*, *53*(15): 8776-8784.

Wegner, A., Besseling, E., Foekema, E. M., Kamermans, P., Koelmans, A. A. (2012). Effects of nanopolystyrene on the feeding behavior of the blue mussel (Mytilus edulis L.). Environmental Toxicology and Chemistry, 31(11): 2490-2497.

# Chapter - 6

## Freshwater Floral Diversities as Pollution Indicators

**Kaiser Farooq[1*], Feroz Ahmad Shah[1], Imtiyaz Qayoom[2], Kafeela Qureshi[1] and Irtifa Fayaz[1]**

[1] Division of Aquatic Animal Health Management, Faculty of Fisheries, SKUAST-K, Rangil-190006, Ganderbal, J&K, India

[2] Division of Aquatic Environmental Management, Faculty of Fisheries, SKUAST-K, Rangil-190006, Ganderbal, J&K, India

*Corresponding email: Kaiserfarooq798@gmail.com

### Abstract

Water quality changes are well reflected by changes in the distribution, number, and variety of organisms as well as by the bioaccumulation of toxin in them. Phytoplankton, zooplankton, macrophytes, molluscs, and other macro invertebrates are examples of some of the indicators that range from plankton to benthic invertebrates. This chapter focuses on the floral species viz., Cyanophyta, microalgae, phytoplankton (*Anacystis* sp., *Chlamydomonas* sp., *Navicula*)of fresh water that act as pollution indicators.

**Keywords:** Pollution indicator, Biomarkers, Phytoaccumulation, Industrialization.

### Introduction

Pollution is an unfavourable alteration in the physical, chemical, or biological properties of the air, water, or soil that may adversely influence the life or pose threat to human health or any living organism.Environmental pollution is a result of significant environmental damage brought on by human progress, civilization, and rapid industrialization. There has been a significant increase in concern about a number of issues, including cities suffocating on impure air, trees wilting, fish decaying on seashores, dangerous chemicals being cycled into food, and disease epidemics occurring so regularly.The biotic communities, including bacteria, plants, animals, and microorganisms, exhibit varying degrees of sensitivity and can be successfully used as indicators to quickly detect and foresee environmental degradation.

## Pollution indicators:

Pollution may be indicated by any structural and functional changes in species, populations, communities, relative densities, frequencies, and abundance. A species' presence or absence may signal favourable or unfavourable changes to its eco-system. Pollution indicators are organism(s) or community characteristics that can be used to provide information about the state of the environment and deviations from the normal condition (Geelani *et al.*, 2012). Pollution indicators are any organisms (microbes, plants, insects, bacteria, etc.) that serve as indicators of environmental pollution and provide information about dangerous substances or the level of environmental pollution of water, air, soil, etc. by having the capacity to accumulate pollutants.

## Pollution indicators in fresh water ecosystem:

About 6% of all known species are found in freshwater habitats, which make up only around 0.8% of the planet's surface. Aquatic ecosystems' biota is frequently subject to a variety of stressors, including anthropogenic disturbances like the release of contaminants into water supplies and natural environmental changes (Strayer and Dudgeon, 2010). Urbanization is a widespread and quickly expanding land use modification (Paul and Meyer, 2001). The degradation of freshwater sources, such as ponds, lakes, marshes, and so on, as a result of runoff water that continuously adds up contaminants from various places, is the most serious (Parmar *et al.*, 2016). The degradation of freshwater resources alters its physico-chemical and microbial structure and poses a significant threat to aquatic life.

Somatic coliphages and bacterial indicators (*E. coli*, total coliforms (TC), and faecal coliforms (FC)) serve as bacterial markers of bathing water and extensively utilised as indicator micro-organisms in aquatic systems. Any organism employed in this way must represent a faecal source, be alien to the soil environment, and have traits that allow it to be distinguished from any closely related species. Such criteria have been applied to certain faecal streptococcal biotypesas indicators of soil pollution, as indicator of faecal pollution of water due to their abundance in faeces and they are generally found only in polluted waters. *Cladophora glomerata* acts as bioindicators of heavy metal pollution likeManganese (Mn), Copper (Cu),

Zinc (Zn), Arsenic (As), Cadmium (Cd) and Lead (Pb) in streams (Prakash and Awasthi, 2013).

The aquatic environment and its water quality are thought to be the primary determinants of health and disease in both humans and animals. The most dangerous chemical contamination in the world today is caused by the growing usage of waste chemical and agricultural drainage systems. Zn, Cu, Pb, Cd, Hg, Ni, and Cr are the most significant heavy metals in terms of water pollution. Some of these metals, such as Cu, Ni, Cr, and Zn, are trace metals that are necessary for living things but are hazardous at higher amounts. Others, like Pb and Cd, are poisonous elements but their biological purpose is unknown. Biomarkers are crucial in the effort to define and quantify the impacts and presence of contaminants on aquatic systems (Doğan and Kizilkaya, 2010). The fish, shellfish, oyster, mussels, clams, aquatic animals, aquatic plants, and algae *Angustifola*, *Potamogeton pectinatus*, *Ranunculus sphaerosphermus*, and *Groenlandia densa* act as bioindicators of pollution in marshes and lakes, respectively. Various aquatic organisms are found in rivers, lakes, seas, and marines and may be useful as biomarkers of metal pollutants.

**Floral indicators of aquatic pollution:**

Algae and phytoplankton are helpful indicators of water quality. For example, in addition to being an invasive and dangerous plant, *Eichhornia crassipes* is also a valuable species with exceptional metal pollution phytoaccumulation capabilities. Toxic metals like Cr, Cu, Co, Ni, Zn, Pb, As and Cd can be bioconcentrated by *E. crassipes* in its root system (Laet*et al.*, 2019).Similarly, different phytoplankton are also used as indicator of eutrophic waters e.g., nutrient enriched water body can easily be traced using diatoms, *Microcystis aeruginosa*, *Chlorella vulgaris*, *Cladophora crispate* and many more. According to Yusuf, (2020), the organic pollution indicators are *Closterium* sp., Navicula sp., Nitzschia sp., Synedra sp., Chlamydomonas sp., Cyclotella sp., and *Anacystis* sp.

**Macro flora of water as pollution indicators:**

Higher plants are more vulnerable to different types of ecosystem contaminants due to their immovability and prolonged survival, and as a result, they frequently appear to be worthy biological indicators. Pollutants have an effect on plants in a number of ways, including morphological changes, biochemical changes, and/or cellular changes. These effects are frequently seen in how plants perform, grow, and have a high population density. In general, the first biological sign is thought to be an outward vegetative symptom. Macrophytes have also emerged as bioindicator in water body. Laet *et al.*, (2019), showed bioaccumulation capacities of organic pollutants by *Eichhornia crassipes* and described it as potential bioindicator. Some other examples of plants considered as potent pollution indicators are:

### *i. Myriophyllum spicatum*

*Myriophyllum spicatum* (Eurasian watermilfoil), a submerged aquatic plant present in fresh water streams, is considered a threat for aquatic environment. *M. spicatum,* a scarcely studied submerged plant, has been found to have efficiency towards accumulation of non-essential elements such as Caesium and Cobalt. *Hence it can be used for the t*reatment and purification of simulated wastewater contaminated by radioactive or non-radioactive Cobalt and Caesium isotopes (Saleh *et al.*, 2020).

### *ii. Typha angustifolia:*

*Typha angustifolia* is a perennial macrophyte that may grow quickly and produce significant amounts of biomass (Bah *et al.*, 2011). The ability of *T. angustifolia* to store metals in its structure for the rhizofilteration of As, Cd, Cr, Cu, Fe, Mn, Ni, Pb, and Zn (Chandra and Yadav, 2010) from water bodies has been demonstrated (Bah *et al.*, 2011; Ganjo and Khwakaram, 2010; Chandra and Yadav, 2011). Like many other stem plants, this plant can grow both emerged and submerged. Furthermore, Duman *et al.*, (2015) has also studied the temporal variation in heavy metal accumulation in narrow-leaved cattail (*Typha angustifolia L.*).

### *iii. Eichhornia crassipes:*

*Water hyacinth, also known as Eichhornia crassipes, is one of the aquatic plants that has garnered the most attention from scientists in the past ten years.* Originally from South America, its ornamental appeal led to its introduction into Africa, Asia, the South Pacific, North America and Europe, where it has become invasive (Jones *et al.*, 2018). *E. crassipes* is well known as an invasive aquatic plant. It is also used very effectively in phytoremediation, particularly for the rhizofiltration of effluents contaminated by heavy metals. (Kriticos and Brunul, 2016). Laet *et al.*, (2019) showed bioaccumulation capacities of organic pollutants by *E. crassipes* and described it as potential bioindicator. *E. crassipes* is capable of bioconcentrating toxic metals such as Cr, Cu, Co, Ni, Zn, As, Pb and Cd in its root system.

### *iv. Potamogeton:*

Different Potamogeton species, mainly *P. filliforms* and *P. polygonifolius*, have been reported as sensors of water contamination. Numerous additional species, such as *P.*

*alpinus*, *P. lucens*, *P. praelongus*, *P. zooterifolius*, *P. aqualifoliu*, *P. colorates (P. coloratus)*, *P. densus* and *P. ratilus*, are slow-growing but also serve as markers of the water quality (Ansari *et al.*, 2017). These macrophytes have the potential to be used as heavy metal bioindicators of pollution because they reflect spatial changes in metals in water and sediments of freshwater bodies. In marshes and lakes, they serve as bioindicators of iron and manganese pollution. These are submerged macrophytes that are commonly seen in freshwater bodies especially rivers. These plants can occasionally be discovered with floating rhizomes and roots on the sandy substrate of river banks with moderate currents.

### v. Some other examples of macrophytes with pollution indication property:

Changes in pH in an aquatic ecosystem, is recognized by the growth of some acidophytes e.g., common ling (*Calluna vulgaris*), hair grass (*Deschampsia flexuose*), or sunflower (*Drosera rotundifolia*) whereas wild barley (*Hordeum murinum*), French mercury (*Mercurialis annua*), or large nettle (*Urtica dioica*) are the indicators of change in nitrite content. Changes in total soluble salts content are governed by the growth of lead grass (*Salicornia europeae*), tripoly (*Aster tripolium*), and wild march beet (*Statice limonium*) (Zaghloul *et al.*, 2020). The different macro floral indicators of water pollution as presented by Ansari *et al.*, 2017 are given in table 1.

**Table 1. Macro Floral indicators of water pollution**

| S.no | Species | Taxonomic group/Family | Ecological habitat |
|---|---|---|---|
| 1 | *Azolla filiculoides sp.* | Pteridophyta/Salvinaceae | Free Floating |
| 2 | *Hydrilla* sp. | Angiosperm/ Hydrocharitaceae | Rooted Submerged |
| 3 | *Myrophyllum sp.* | Angiosperm/Haloragaceae | Rooted Submerged |
| 4 | *Lemna sp.* | Angiosperm/Araceae | Free Floating |
| 5 | *Wolffia sp.* | Angiosperm/Araceae | Free Floating |
| 6 | *Potamogeton sp.* | Angiosperm/ Potamogetonacea | Rooted Submerged |

(Ansari *et al.*, 2017)

## Micro flora of fresh water bodies as pollution indicators:

Variations in phytoplankton species diversity are a clear sign of contamination in aquatic habitats. Cadmium pollution is indicated by morphological alterations, cellular malformation, chlorosis, and a notable rise in heterocyst frequency in *Anabaena cylindrica*. Cadmium contamination has been linked to the development of excessively long filaments and the loss of cellular contents in *A. inequalis* (Stratton and Corke, 1979). Chlamydomonas zygospore germination has also been shown to noticeably decrease in response to several herbicides and pesticides (Ansari *et al.*, 2017). Due to their sensitivity towards any change in their habitat, phytoplanktons are considered as potential indicators of aquatic pollution. Anthropogenic activities, industrial effluent contamination, and other sources that introduced eutrophication into water bodies have been proven to affect phytoplankton's reactions to environmental pollution circumstances. The community's presence or absence reveals changes in the physico-chemical environment of the water bodies.

### i. Cyanophyta

The Phylum Cyanophyta (Myxophyceae, blue-green algae) differs from other algae in having a prokaryotic cell organisation, i.e., it lacks organised double membraned nuclei, chromatophores, and mitochondria, and possesses distinctive photosynthetic pigments such as biliproteins, myxoxanthin, and myxoxanthophyll in addition to chlorophyll-a and ß-carotene (Kumar and Singh, 1979). The rapid eutrophication of water bodies, such as reservoirs, lakes, etc., is recognised to be indicated by these very potent pollution indicators by the appearance of bloom formations. Due to their immobility and ability to quickly regain equilibrium with their natural habitat in the face of changes like an increase in pollution, aquatic plants are an invaluable resource for predicting the state of the water environment.

### ii. Phytoplankton as pollution indicators:

Aquatic ecosystem pollution has been successfully observed using phytoplankton as a valuable biological indicator. Phytoplanktons are found in the upper portion of aquatic ecosystem. Pollutants, however, may have an impact on the relationship between the growth rate and each of those factors. If, for instance, a colourful, suspended solid light-containing industrial effluent was filtered or absorbed, the growth rates would be reduced (Malik and Bharti, 2012). From the upper tropics to the ocean, phytoplankton is a significant source of pollutants that are transferred to humans. This group of biological indicators is also referred as microalgae; because they have chlorophyll comparable to that found in terrestrial plants, they need sunlight to survive and grow (Hosmani, 2014).

### iii. Bacterial indicators

Bacterial indicators are bacterial populations or species whose existence in an ecosystem, above a given threshold, signals exposure to contaminants. Indeed, bacteria are useful for analysing pollution issues in many habitats. *Escherichia coli* and enterococci are two species of bacteria that are frequently used to detect and gauge the degree of pollution in a variety of habitats. These bacteria are included in the category of "bacterial indicators. *Salmonella typhimurium* and *Clostridium* spp. could also be used as biological indicators for the presence of aquatic logging in a specific habitat, together with anaerobic mesophilic bacterial populations. Coliforms, which are a diverse group of facultative aerobic and anaerobic bacteria that can digest lactose with gas production in less than 48 hours at 37°C, are frequently utilised as biological indicators. The coliforms known as faecal coliforms are those that can thrive and ferment lactose at temperatures above average (44-45°C). They serve as useful biological indicators for contamination of relatively recent faecal origin since they mostly contain large concentrations of *E. coli* types I and II (Saber *et al.*, 2015).

### iv. Microalgae

Microalgae are used as bioindicators to assess the health of natural ecosystems such as pond water. They are utilized to evaluate the media in which a biological system operates, as well as the molecules within it. They serve as a symbol of the high quality of its surroundings. Microalgae respond quickly to changes in the environment, making them an excellent biomarker for evaluating the quality of pond and water pollutants (Okpashi, 2022). *Cladophora glomerata* acts as bioindicators of heavy metal pollution like Manganese (Mn), Copper (Cu), Zinc (Zn), Arsenic (As), Cadmium(Cd) and Lead (Pb)) in streams. *Anacystis* sp., *Chlamydomonas* sp., *Navicula* sp., *Nitzschia* sp., *Synedra* sp., and *Closterium* sp. act as indicators of the presence of organic pollution (Kumari and Paul, 2020). Rheophilic moss *Platyhypnidium aquaticum* exposed to polluted waters can bioaccumulate greater amounts of metals, and a metalloid, than *P. aquaticum* exposed to pollution-free water (Benítez*et al.*, 2020). This proves the pollution indicating characteristic of *P. aquaticum*. Marine algae e.g., *A. curicuatum*, *C. gracilis* and *P. capillacea* are important pollution indicators of heavy metals (Co, Cr, Cu, Fe, Mn, Ni and Zn) in seas. The different micro floral indicators of water pollution as presented by Ansari *et al.*, 2017 are given in table 2.

**Table 2. Micro Floral indicators of water pollution**

| S.no | Species | Taxonomic group/Family | Ecological habitat |
|---|---|---|---|
| 1 | *Microcystis* sp. | Cyanobacteria/Microcystaceae | Free Floating |
| 2 | *Chara* sp. | Chlorophyta/Characeae | Rooted Submerged |
| 3 | *Nitella* sp. | Chlorophyta/Characeae | Rooted Submerged |
| 4 | *Scendesmus* sp. | Chlorophyta/Scendesmaceae | Free Floating |
| 5 | *Chlorella* sp. | Chlorophyta/Chlorellaceae | Free Floating |

(Ansari *et al.*, 2017)

## Conclusion

The current chapter discusses the floral species which are potential indicators of pollution in fresh waters. The significance of such pollution indicators is intense and could be used for the assessment of water quality. This chapter has mentioned microflora, macrophytes and different phytoplankton as pollution indicators of fresh water ecosystem which can be used to obtain quantitative information on pollution levels in aquatic ecosystems.

## Bibliography

Ansari, A.A., Shalini Saggu, S.S., Al-Ghanim, S.M., Abbas, Z.K., Gill, S.S., Khan, F.A., ... and Khan, A.A. (2017). Aquatic plant biodiversity: a biological indicator for the monitoring and assessment of water quality. In *Plant biodiversity: monitoring, assessment and conservation* (pp. 218-227). Wallingford UK: CABI.

Bah, A.M., Dai, H., Zhao, J., Sun, H., Cao, F., Zhang, G. and Wu, F. (2011). Effects of cadmium, chromium and lead on growth, metal uptake and antioxidative capacity in *Typha angustifolia. Biological trace element research*, 142(1), 77-92.

Benítez, Á., Torres, S., Morocho, R., Carrillo, W., Donoso, D. A. and Calva, J. (2020). *Platyhypnidium aquaticum* as Bioindicator of Metal and Metalloid Contamination of River Water in a Neotropical Mountain City. *Plants*, 9(8), 974.

Chandra, R. and Yadav, S. (2010). Potential of *Typha angustifolia* for phytoremediation of heavy metals from aqueous solution of phenol and melanoidin. *EcolEng* 36, 1277–1284.

Chandra, R. & Yadav, S. (2011). Phytoremediation of Cd, Cr, Cu, Mn, Fe, Ni, Pb and Zn from aqueous solution using *Phragmites cummunis*, *Typha angustifolia* and *Cyperus esculentus*. *International journal of phytoremediation*, 13(6), 580-591.

Doğan, M. and Kizilkaya, B. (2010). Some air and water pollution indicators in and around the Lake Abant, Turkey. *Journal of Black Sea/Mediterranean Environment*, 16(1), 53-74.

Duman, F., Urey, E. and Koca, F.D. (2015). Temporal variation of heavy metal accumulation and translocation characteristics of narrow-leaved cattail (*Typha angustifolia* L.). *Environmental Science and Pollution Research*, 22(22), 17886-17896.

Ganjo, D.G. and Khwakaram, A.I. (2010). Phytoremediation of wastewater using some of aquatic macrophytes as biological purifiers for irrigation purposes. *InternationalJournal of Environmental and Ecological Engineering*, 4(6), 222-245.

Geelani, M.S., Bhat, S.J.A., Geelani, H.S. and Haq, S. (2012). Pollution indicators and their detection. *The Journal of Plant Science Research*, 28(2), 193.

Hosmani, S. (2014). Freshwater Plankton Ecology: A Review. *Journal of Research In Management & Technology*, 3, 1-10.

Jones, R.W., Hill, J.M., Coetzee, J.A. and Hill, M.P. (2018). The contributions of biological control to reduced plant size and biomass of water hyacinth populations. *Hydrobiologia*, 807(1), 377-388.

Kriticos, D.J. and Brunel, S. (2016). Assessing and managing the current and future pest risk from water hyacinth, (*Eichhornia crassipes*), an invasive aquatic plant threatening the environment and water security. *PloS one*, 11(8), e0120054.

Kumar, H.D. and Singh, H.N. (1979). Cyanophyta. In *A Textbook on Algae* (pp. 51-68). Palgrave, London.

Kumari, D. and Paul, D.K. (2020). Assessing the role of bioindicators in freshwater ecosystem. *Journal of Interdisciplinary Cycle Research,* 12(9), 58- 74.

Laet, C., Matringe, T., Petit, E. and Grison, C. (2019). *Eichhornia crassipes*: a powerful bio-indicator for water pollution by emerging pollutants. *Scientific reports*, 9(1), 1-10.

Malik, D.S. and Bharti, U. (2012). Status of plankton diversity and biological productivity of Sahastradhara stream at Uttarakhand, India. *Journal of Applied and Natural Science*, 4(1), 96-103.

Okpashi, V.E. (2022). Microalgae (*Selenastrum Capricornutum*) Growth Inhibition Following Exposure to Pond Water: Monitoring Pond Toxicity. *Research Square* 1-10.

Parmar, T.K., Deepak, R. and Agrawal, Y.K. (2016). Bioindicators: The Natural Indicator of Environmental Pollution. *Frontiers in Life Science* 9(2), 110-118

Paul, M.J. and Meyer, J.L. (2001). Streams in the urban landscape. *Annual review of Ecology and Systematics*, 32(1), 333-365.

Prakash, J. and Awasthi, G. (2013). Accumulation of Heavy Metals in Different Water Bodies by Biological Source Algae. *Advancements in Life Sciences*, 2(1), 10-14.

Saber, M., Abouziena, H., Hoballah, E., El-Ashry, S. and Zaghloul, A. (2015, September). Phytoremediation of potential toxic elements in contaminated sewaged soils by sunflower (*Helianthus annuus*) and corn (Zea mays L.) plants. In *12th international phytotechnology conference, Manhattan, Kansas, USA* (pp. 27-30).

Saleh, H.M., Moussa, H.R., Mahmoud, H.H., El-Saied, F.A., Dawoud, M. and Wahed, R.S.A. (2020). Potential of the submerged plant *Myriophyllum spicatum* for treatment of aquatic environments contaminated with stable or radioactive cobalt and caesium. *Progress in Nuclear Energy*, 118, 103147.

Stratton, G.W. and Corke, C.T. (1979). The effect of cadmium ion on the growth, photosynthesis, and nitrogenase activity of Anabaenainaequalis. *Chemosphere*, 8(5), 277-282.

Strayer, D.L. and Dudgeon, D. (2010). Freshwater biodiversity conservation: recent progress and future challenges. *Journal of the North American Benthological Society*, 29(1), 344-358.

Yusuf, Z.H. (2020). Phytoplankton as bioindicators of water quality in Nasarawa reservoir, Katsina State Nigeria. *Acta Limnologica Brasiliensia*, 32.

Zaghloul, A., Saber, M., Gadow, S. and Awad, F. (2020). Biological indicators for pollution detection in terrestrial and aquatic ecosystems. *Bulletin of the National Research Centre*, 44(1), 1-11.

# Chapter - 7

# Sewage Pollution as a Major Cause for Deterioration of Water Quality in Lake Ecosystems

**[1]Maqsooda Akhtar, [1]Bakhtiyar and [2]Imtiyaz Qayoom**

[1]Department of Zoology, University of Kahmir, Hazratbal, Srinagar - 190006, J&K, India
[2]Division of Aquatic Environmental Management, Faculty of Fisheries, SKUAST-K, Rangil-190006, Ganderbal, J&K, India
[3]Department of Zoology, DSB Campus, Kumaun University, Nainital-263001
*Corresponding email: drsheikhimtiyaz@gmail.com

## Abstract

Water supplies are running out on a global scale. The water crisis is a serious problem, especially in crowded urban areas. Poor sanitation as a result of water constraint has made the issue worse in most places with expanding urban populations. Additionally, home sewage and industrial effluents in enormous quantities are produced as a result of urbanization and industrialization and dumped into water bodies. India has a plentiful supply of water bodies, such as reservoirs, lakes, and wetlands, which have long met the needs of the country's rapidly expanding population. Their sustainability has been greatly influenced by the wise management of these waterbodies. Environmental protection demands the deployment of suitable purification systems with high microbiological agent removal efficiencies since wastewater enters the environment without restriction and microbial pollutants are transported to people and other living things.

**Keywords:** Sewage, Pollution, Urbanization, Heavy metal toxicity, Eutrophication.

## Introduction

The waste liquids and solids carried away by sewers are called sewage. The sewage from industries, domestic sources and agricultural activities as runoff is discharged into water bodies as wastewater. Global sanitation crisis produces huge loads of sewage, which are ultimately discharged into water bodies. Globally, sewage pollution is considered of very high importance. The inadequate human sanitation is the biggest challenge in controlling sewage pollution. It has been reported that about 4.5 billion people do not possess adequate sanitation (Bill and Melinda Gates Foundation, 2020). Discharge of widespread and heavy human waste into waterbodies leads to contamination of surface water (Kiulia *et al.*, 2015). The indicators of sewage pollution include contamination of water system with fecal viruses, heavy metals, endocrine disruptors, pharmaceuticals, and pathogens (Wear and Thurber, 2015).

## Composition of Sewage

The domestic effluents include black water (excreta, urine, and fecal sludge) and gray water (used water from washing and bathing); water from institutions (hospitals; industrial effluent, stormwater), commercial establishments and urban agricultural, horticultural and aquaculture runoff (Raschid-Sally & Jayakody, 2009). The composition of sewage includes 99.9% of water and 0.1% of solids. The solids include both dissolved and suspended materials (organic and inorganic solids). The major portion of sewage consists of dissolved solids than suspended solids. The organic constituents of sewage include fraction fats, proteins, carbohydrates, lignin, and their decomposition products, while the inorganic composition of sewage includes industrial and domestic derivatives, which include heavy metals (cadmium, zinc, mercury, copper, arsenic, etc.). In addition, a diversity of microbes from domestic waste (Washing, bathing, sweat, urine and feces) can be observed in sewage (Ajonina *et al.*, 2015). The microbes in sewage are received from the human body and are present in the skin, respiratory tract, oral cavity, urogenital tract, and gastrointestinal tract. Wastewater is an important source of pathogen such as Fecal Coliform, Salmonella, E coli, *Vibrio cholera*, Shigella, parasitic eggs and cysts, fungi and viruses (Grandclément *et al.*, 2017), intestinal nematodes (hookworm (*Ancylostoma duodenale*), roundworm (*Ascaris lumbricoides*), and whipworms (*Trichuris*). Further, sewage contains nitrogen (N) and phosphorus (P). Among domestic wastes, human urine is a rich source of nitrogen and phosphorus than feces (Kirchmann and Pettersson, 1994), and it amounts for 50% of the phosphorus and 80% of nitrogen in the sewage.

### Sewage pollution in water bodies

Globally, the sewage is the first factor responsible for water pollution as domestic sewage, industrial effluents and other polluting agents are directly discharged into water bodies including ponds, lakes and reservoirs. These water bodies are polluted to such an extent that sometimes even drinking for one time is very difficult. During 20th century, population explosion, urbanization, industrialization, agriculture extensiveness and water intensive lifestyles have led to enormous degradation of lakes and other waterbodies. The pollution of water bodies due to sewage from domestic sources and/or industrial toxic effluents, agricultural chemical runoff has affected the diversity and distribution of their flora and fauna.

Though industrial revolution has paved way for the development of countries' economy globally, they are considered as the biggest source of water pollution. Waste waters from manufacturing plants are ejected from these industries in the form of organic and inorganic matters or toxic components that are harmful to human health and to the environment. The organic and inorganic sewage from industries include heavy metals, dyes, polyphenols, textile dyes etc., which are toxic. The toxicity of heavy metals depends on metal dose and the time exposition. Some of the metals such as arsenic, cadmium, chromium, lead, and mercury are capable of damaging multiple organs at lower exposure dosage and are considered carcinogenic. Furthermore, textile dyes are also highly toxic and potentially carcinogenic and can cause animal and human diseases, and environmental degradation (Al-Tohamy *et al.*, 2022). Thus, the removal of recalcitrant substances from wastewaters is very critical as they can harm the environment and human health.

Fresh water bodies are considered as best sink for the deposition of sewages (Das and Acharya, 2003). About 73% of contamination of water bodies is contributed by sewage that is directly discharged from urban areas as wastewater (58%) and by 81% of industrial sewage, which are not treated (Vargas-Gonzalez *et al.*, 2014). Sewage discharge into water bodies leads to increased water pollution and reduction in quality of water resources. Densely populated cities release huge loads of sewage, which find their way through drainage systems and finally end up into water bodies including rivers (Tukura *et al.*, 2009). The continuous discharge of sewage into water bodies has caused an extensive ecological degradation, which is manifested by reduced water quality and availability, ecological imbalance in biotic components, changes in species diversity and distribution and increased. The composition and concentration of sewage constituents, their quantity and frequency of discharge determines the negative impact on a water body (Akpor andMuchie, 2011). The organic nature of sewage reduces the oxygen concentration of water bodies, thus, resulting in a high biological oxygen demand (BOD). The preservation of the biotic elements in an aquatic environment depends heavily on the oxygen balance of the system (Momba *et al.*, 2006). Reduced dissolved oxygen levels in a water ecosystem due to sewage discharge affects fish activity and ultimately causes death of fish population (Igbinosa and Okoh, 2009).

Water is used in industrial processes for the discharge of effluents and other discharges from the industry (Demirak *et al.*, 2006). The wastewater from industries often goes to freshwater ecosystems, which leads to the water pollution. The freshwater ecosystems and other aquatic ecosystems are exposed to supplementary contamination than other environs, as water is used in various industrial practices as well as release of discharges commencing from industry. About 80% of municipal wastewater gets discharged into water bodies, and industries dump millions of tons of heavy metals, solvents, toxic sludge and other wastes into these water bodies annually (Mateo-Sagasta *et al.*, 2017). Further, agriculture runoff and other processes related to agriculture is also a key factor of water pollution (Mateo-Sagasta *et al.*, 2017). The agrochemical industries discharge huge amounts of pesticides, organic matter, sediments, and drug residues into water bodies. Thus, agrochemical industries and saline runoff from agricultural land that goes into lakes pose significant threats to human health and aquatic environments. Severe contamination of water bodies affects flora and fauna of the community (MateoSagasta *et al.*, 2017). Water pollutants can cause primary damage with direct negative effects on the environment or secondary damage by disturbing the trophic level of a food web, whose impact is detected over a long period of time. The water quality in an aquatic ecosystem is one of the most critical challenges (Hampel *et al.*, 2015). Different physiochemical factors of a lake that are influenced by sewage impact the biotic diversity and distribution of the lake.

**Temperature:** Sewage discharge in a lake affects its temperature, which in turn impacts both physical and chemical composition of a lake. The temperature of a lake is important for sustenance of its biotic flora and fauna and an increase in the average temperature of a lake has an ecological impact due to thermal enhancement (Horner *et al.*, 1994). The change in temperature affects the diversity and distribution of its flora and fauna (Gupta *et al.*, 2009). The reduction in the dissolved oxygen as a result of change in temperature occurs due to less dissolution of oxygen. This leads to the less availability of oxygen for the bacteria to degrade the wastes in water, thus increasing its demand. The higher lethal temperatures can even cause tissue anoxiain aquatic animals. Further, the primary production and litter decomposition of the lake is affected due to low oxygen concentrations and reduced energetic processes (Lecerf *et al.*, 2007). In addition, high temperature in water ecosystem including lakes impairs the osmoregulatory system of aquatic organisms, reduces the production of cellular enzymes and membrane lipids, and causes protein denaturation, which in turn leads to thermal death of these organisms. The growth of macrophytes, epiphytes, and phytoplankton is altered by the increased temperatures in a lake (Whitehead and Hornberger, 1984).

**Dissolved oxygen:** Sewage discharge from domestic sources, industriesand agricultural runoff into lakes reduced the availability of oxygen in lakes leading to high BOD. The water quality and aquatic ecosystem health depends on BODand determines the diversity and distribution of flora and fauna (Connolly *et al.*, 2004). Fish depletion and other aquatic life forms face threat in lakes with high BOD. Oxygen from lakes is depleted by the sewage that consumes dissolved oxygen in water. The hypoxic, anoxic and euxinic conditions are developed in lakes due to alteration in oxygen levels on account of sewage discharge such as metal and phosphate, sulphide etc. (Frodge *et al.*, 1990). High BOD due to the discharge of wastes will lead to accumulation of toxic substances in lakes, increased nutrient levels and biomagnification in organisms. Aquatic forms of a high BOD lake become susceptible to diseases, suffer retarded growth, altered feeding habits, and in severe cases, death of the organisms. Severe reduction in oxygen levels will alter the biotic composition of the lake (Chambers and Mill, 1996).

**Total suspended solids:** Suspended solids are fine water borne particles with particle size of >2 microns. Suspended solid particles include inorganic materials; however, can also include algae and bacteria. These particles get stuck in the fish gill, filter feeding apparatus of zooplankton and clog the gills of like epibenthos, which live on the sediment. Apart from the effect on aquatic organisms, suspended solids reduce light penetration into lakes, which in turn, reduces photosynthesis, which may lead to death of aquatic flora (Horner *et al.*, 1994). Further, total suspended solids lead to the reduced dissolved oxygen and increased temperature in lakes, which directly affects its biotic flora and fauna. The suspended solids such as bacteria and other microorganisms will have harmful effects on human health. The organisms may cause gastrointestinal issues, while other suspended solids including metals may result in death.

## Sewage and eutrophication

Accumulation of large amounts of plant nutrients in lakes and estuaries is called eutrophication (from the Greek words 'eu' meaning "well" and 'trophe' meaning "nourishment"). Eutrophication leads to the excessive growth of phytoplankton in water ecosystem, which causes an imbalanced primary and secondary productivity. This results in high rate of succession in seral stages due to nutrient enrichment. Though eutrophication occurs naturally in an aquatic ecosystem over centuries, anthropogenic activities have increased eutrophication to a greater extent and at a very high rate through point-source discharges and non-point loading nutrients including nitrogen and phosphorus (i.e., cultural eutrophication). This drastically affects the drinking water quality, fisheries, and recreational water bodies in an aquatic ecosystem (Bhat *et al.*, 2017). Algal bloom, which is a main outcome of eutrophication in water bodies, has been linked to nutrient enrichment resulting from accelerated human activities such as agriculture, industry, and sewage disposal (Schindler, 1974). The nutrient enrichment occurs mainly due to runoff with overused fertilizers from agroecosystems and/or human wastes discharged from settlements. Though eutrophication is a very slow, natural process, it is accelerated by increase in the rate of nutrient entry in a water ecosystem as a result of human activities. Eutrophication is a serious water pollution, which directly affects the fauna of a water body by reducing the levels of dissolved oxygen. Rapid upwelling of eutrophic lakes can be seen. Faster fish mortality can be seen in eutrophic lakes due to polluted water. Further, growth of plants in a eutrophic lake makes fishing and navigation difficult. Indirectly, hydroelectric generation from eutrophic water storages is affected due to chemical effect of nutrient rich waters on the turbines. The algal blooms in eutrophic waters reduce the water quality and alter the characteristics of water, which lead to the growth of disease-causing bacteria. Eutrophication reduces the quality of surface waters for the industrial and recreational uses. The dense blooms in autotrophic lakes are noxious, smelling foul phytoplankton, which affects water clarity and quality. The light penetration in eutrophic lake is reduced by algal blooms, which causes reduced growth and death of plants in littoral zones. The water contaminated with algal bloom toxins in eutrophic lakes cause fish mortality. Further, predator activity is reduced due to limited light (Lehtiniemi *et al.*, 2005). Additionally, algal blooms make eutrophic water unfit for swimming and boating. In a eutrophic lake, high turbidity, increased biotic biomass, "high sedimentation rate, reduced species distribution and diversity and anoxic conditions are common (Edokpayi, 2016).

## Effects of sewage on human health

For human health and survival, the quality of drinking water comes from freshwater ecosystem such as rivers, lakes, streams. These water bodies are highly susceptible to sewage pollution as humans live near them and use them for waste dumping. Sewage pollution is an extremely serious issue in nations with large population. For example, In India, the increase in human population has increased to one billion, which has directly affected several rivers (Central Pollution Control Board (CPCB), 2015). This is due to the discharge of huge untreated sewage from human population (CPCB, 2015). Not only in developing countries, the developed countries including U.S. also face brunt of sewage pollution in freshwater bodies in cities like Chicago, New York City, Los Angeles and San Francisco (Kiulia *et al.*, 2015). As per Environmental Protection Agency (EPA), in U.S., 46% rivers and streams, and

35% lakes are polluted by sewage and/or agricultural run-off (EPA, 2012, 2016). The transmission of human pathogens by sewage polluted freshwater ecosystems has been found to cause about 829,000 deaths per year (Prüss-Ustün *et al.*, 2019). Additionally, heavy metals in sewage polluted waterbodies results in metal toxicity in fish and humans (Ade *et al.*, 2014). Further, huge discharge of sewage leads to low oxygen levels in water, bacterial decomposition, thus, organic matter load increase in the waterbodies, resulting in high BOD. The dangerously low oxygen levels cause dead zones in freshwater ecosystem, in which fish and shellfish become highly susceptible to disease due to the stress caused by low or no oxygen (Petrovic *et al.*, 2002). Furthermore, in sewage polluted waters, aquatic organisms including fish, mussels, insects, and other aquatic animals exhibit high amounts of endocrine disruptors (Petrovic *et al.*, 2002). The increased endocrine disruptors in fish lead to reduced reproductive activity, malformed reproductive organs, thereby decreased population (Wu *et al.*, 2003). Similarly, amphibians suffer skewed sex ratios in sewage polluted waterbodies (Lambert *et al.*, 2015).

Pathogens such as bacteria, fungi, viruses and parasites in sewage and wastewater cause intestinal, lung, and other infections. Main diseases caused by bacteria include diarrhea, fever, cramps, which are accompanied with vomiting, headache, weakness, or loss of appetite. The sewage and wastewater contain *E. coli*, shigellosis, typhoid fever, salmonella, and cholera causing microorganisms. *Aspergillus* and other fungi that grow in compost cause allergic symptoms, however, can also cause lung infection or make asthma worse. Persons with prior health problems, when exposed to microorganisms in sewage and wastewater are more likely to get sick due to *Aspergillus*. The parasites that generally occur in sewage and wastewater include Cryptosporidium and *Giardia lamblia*. They cause diseases such as diarrhea and stomach cramps, and even nausea or a slight fever. Further, roundworm (*Ascariasis*) is common in sewage; they cause coughing, trouble breathing and/or pain and sometimes even block intestines. Hepatitis A virus in sewage causes liver disease, the symptoms of which include feeling tiredness, pains, nausea, jaundice, diarrhea, or loss of appetite.

## Treatment of sewage

Sewage produces undesirable and harmful effects on environment including human health. The effluent treatment is highly important before the waste waters are released out of the industry. Wastewater treatment includes biological and physicochemical processes or both depending on the presence of recalcitrant pollutants and biodegradability effluents. Thus, it is the prime importance now-a-days to dispose the sewage properly to avoid harmful effects on environment especially water ecosystem. Generally, sewage is disposed-off into water reservoirs without looking at the consequences, partly, because we are indifferent to the consequences and it is assumed that the water reservoirs are large enough to withstand the dilution of sewage without harmful effects. The industrial effluents are one of the most critical sewage components, which are discharged into water bodies. Thus, treatment of the toxic substances in both domestic and industrial sewage is highly important before they are discharged into natural environment including water ecosystems to reduce their pathogenicity and pollution potentiality. Various physicochemical and biological processes are involved in treatment of sewage pollutants. The use of the treatment techniques depends on the sewage load, toxic substances, and production of byproducts. The physicochemical processes are generally used to treat industrial sewage pollutants, especially recalcitrant pollutants, but are very expensive. Biological treatment processes are cost effective, more efficient, and eco-friendly.

The physiochemical and biological methods can be applied wither independently or in combination where synergy could be observed.

### 1. Physicochemical treatments

Several physiochemical treatments are involved in wastewater treatments which include membrane filtration, reverse osmosis, coagulation-flocculation, electrocoagulation, and adsorption.

**(a) Membrane filtration**: In this method, the wastewater is physically filtered through porous membranes and the separation of pollutants occurs due to hydraulic pressure. It is used for the separation of industrial sewage from textile, pharmaceutical, paper effluents etc., (Ravikumar *et al.*, 2014), and works on the principal of adsorption, electrostatic phenomenon, and sieving. Membrane filtration removes microorganisms, heavy metals, dyes, odors and turbidity from industrial sewages. Depending on pore size of the membrane, there can be microfiltration, where pollutants with a size greater than 10μm are separated through a membrane having pores between 0.1–10μm and are used for the purification of sewage by removing microorganisms such as bacteria. The separation occurs under pressure gradient of 1–3 bars, ultrafiltration, where the pore size of the membranes are between 0.001 and 0.1 μm, and water and small ions are filtered through this membrane and molecules such as polymers, proteins, bacteria, and viruses are separated, and nanofiltration, where pore size of the membrane is below 1 nm and is generally used for treatment of industrial sewage, especially, for the removal of heavy metals like chromium and nickel ions (Basaran *et al.*, 2016). The organic substance with molecular weight less than 300 Daltons are retained in nanofiltration. The structure of membranes can be uniform (isotropic) or composite (anisotropic). In membrane filtration, mostly low-cost organic membranes are used, but mineral membranes are used for extreme conditions of temperature, pH etc. Reverse osmosis is an important membrane filtration process, in which the pollutants are captured by a semi-permeable membrane and is used for the filtration of all industrial sewage (Trishitman *et al.*, 2020). It is considered an advanced promising wastewater treatment technique.

**(b) Coagulation-flocculation**: This technique uses coagulant and flocculant agents of iron or aluminum by which pollutants are regrouped together as heavy flocs, which in turn are removed by either precipitation or filtration. This method of removing sewage pollutants especially from industrial waters is a fast, simple and an effective with solid suspension above 80% (Junior *et al.*, 2021). It reduces the use of chemicals and their harmful effects on organisms. However, there is high sludge production and increased acidity and conductivity of the treated effluent.

**(c) Electrocoagulation**: An alternative to chemical treatment, electrocoagulation is used for the removal of pollutants from industrial sewage by using soluble anodes and the pollutants are removed electrochemically (Jing *et al.*, 2021).

The anodes of aluminum or iron, which generate metal cations ($Fe^{3+}$ or $Al^{3+}$) by an electric current between these anodes are generally used. The metal cations destabilize the suspended particles by coagulation, which are then precipitated by the formation of flocs. Though the methodology is highly advantageous, it consumes high energy in removing the pollutants.

**(d) Oxidation**: Oxidation involves electrochemical removal of pollutants by electron exchange using oxidizing agents. In this technique, refractory pollutants are modified into insoluble forms for removal, or are converted into nontoxic soluble forms. The oxidizing agents commonly used include oxygen, hydrogen peroxide, ferric chloride, chlorine, potassium permanganate and ozone. Oxidation technique strongly removes bad odors, which are produced naturally or by anaerobic conditions. Further, chemical oxygen demand in a polluted water system is removed by strong oxidative free radicals produced by ozone combination with ultraviolet rays (UV) or hydrogen peroxide (Kalra *et al.*, 2011). Among free radicals, $H_2O_2$ has been reported as the main removal agent for sulfides. About 1.3 to 4.0 mg/L $H_2O_2$ is used for the removal of 1 mg/L of sulfides (Snyder, 2016). The oxidation process of sewage treatment is currently an advanced form of oxidation and is developed by using strong chemical oxidants such as $H_2O_2/Fe^{2+}$ and $H_2O_2/O_3$),photo-catalysis (by UV), or electrochemical oxidation. This is an innovative technique for the sewage treatment and is based on the principle of highly reactive and unstable hydroxyl radical production (Boczkaj and Fernandes,

2017). The oxidation process removes the pollutants at very high rates and with reduced sludge production depending on the pollutant nature and concentration.

**(e) Adsorption**: Industrial sewage is also treated by adsorption with high purification. In this technique, adsorption materials, which are organic (vegetables, fruits, bacteria, wood etc.) or inorganic (rocks, ashes, etc.), are used as a fixed bed or as a suspension with the sewage. The adsorption material can be used in natural form or in activated form. The activation is done physically by pyrolysis, carbonization and calcination, or by chemically using acids and bases, or the both. The chemicals such as sodium hydroxide, potassium hydroxide, zinc chloride, sodium bicarbonate, sulfuric and phosphoric acids are commonly used in adsorption (Malwade *et al.*, 2016). The chemical agents are cost effective for adsorption of sewage pollutants compared to the powerful adsorbent carbon. Attempts are being made to use novel adsorbents like fly ash, organic waste sawdust and olive pomace for adsorption treatment of sewage discharge (Koçer and Acemioğlu, 2016).

The adsorption can be physical or chemical. No electron exchange between the adsorbent and the adsorbate occurs in physical adsorption and involves Van der Waals non-specific physical forces forming multilayers of the pollutants. The physical adsorption is a non-specific and irreversible process, which requires low heat for operation. In chemical adsorption, exchange of electrons occurs between adsorbent and adsorbate and these are held by strong covalent bonds between the ions of the two. In this process, a single layer is formed but requires high energy.

## 2. Biological treatments

Biological treatments are considered as promising for the removal of sewage pollutants. In this treatment of sewage, non-specific microorganisms are used for activated sludge, selected microorganisms for bio-augmentation and algae or plants are removed by phytoremediation. The microorganism used biological treatments include fungi, yeasts, bacteria etc. The microorganisms used in biological treatment use pollutants from the sewage for nutrition and energy for growth and development. Biological treatment controls large scale quantities of sewage and is economical and effectively removes the nitrogen, phosphorous and biodegradable organic matter. Biological treatments of sewage can be aerobic or anaerobic.

**Aerobic treatments:** It is used for the removal of activated sludge and uses aerobic microorganisms to remove sewage pollutants. The microorganisms involved in aerobic treatments are autotrophic or heterotrophic, and include Gram-positive and Gram-negative bacteria, yeast, fungi, metazoan and protozoa. The activated sludge contains micro-organisms and organic or inorganic materials. Organic waste in sewage is degraded into carbon dioxide, new cells and non-toxic by-products. However, in this technique, nitrogen and phosphorus are not removed at the same time.

The treatment by sludge in a Sequencing batch reactor removes both nitrogen and phosphorus simultaneously under aerobic conditions. The liquid/solid separation is done in the same reactor due to large quantity of filamentous bacteria. This technique is used to treat sewages agri-food, pulp/paper, pharmaceutical, textile, chemical, petrochemical effluents, etc. The sequencing batch reactor involves effluent is put through a reactor, which is then aerated or treated in presence of oxygen by activated sludge and microorganism degrade the effluent. This is followed by settling phase, which separates the treated effluent from the activated sludge, which in turn is followed by withdrawal phase i.e., withdrawal of the treated sewage. The efficacy of sequencing batch reactor depends on the amount of organic load, treatment cycles, dissolved oxygen concentration, sewage sludge, and settling phase duration. The removal rates of pollutants are high in this technique in addition to the removal of nitrogen and phosphorus at the same time. But this technique is cost effective.

Bioremediation is one of the important aerobic treatments of sewage pollutants including heavy metals, and other toxic substance. In bioremediation, bio-augmentation and biosorption are involved for the treatment and use strong microorganisms that are resistant to extreme conditions. A Bioaugmentation process uses only living microorganism for the removal of sewage pollutants, biosorption uses both living and non-living microorganisms. Microorganisms used in biosorption process replace the activated carbon and depends on polysaccharides including amino, phosphate, carboxyl, and sulfate groups, which are components of cell wall. Heavy metals and bio-nanocomposite is removed by biosorption (Samuel *et al.*, 2015). The mechanism on which biosorption works includes adsorption, exchange of ions, chelation, and complexation. Bioaugmentation is based on detoxification of sewage pollutants by microorganisms into nontoxic or less toxic byproducts. The microorganisms involved in bioremediation resist the toxic substance of sewage by many mechanisms including biosorption, enzymatic reduction, bioaccumulation, enzymatic DNA repair system, etc. (Ramírez-Dírz *et al.*, 2008). In bioaugmentation, microorganisms can be isolated form the sewage being treated (autochthonous bioaugmentation-native or indigenous microorganism), isolated from other source (allochthonous bioaugmentation- endogenous microorganism) or are genetically modified (gene bioaugmentation). Bioremediation is considered as an efficient, eco-friendly and inexpensive technology for removal of recalcitrant pollutants such as heavy metals. Bioremediation has gained high momentum and can replace the current highly expensive physicochemical technologies.

**Anaerobic treatments**: Anaerobic treatments constitute degradation of sewage pollutants by anaerobic microorganisms. Anaerobic treatment removes mill olive, agri-food, domestic, and tannery effluents from sewage (Mekonnen *et al.*, 2016). In anaerobic treatment, biogas is produced from organic sewage pollutants. Biogas production involves four stages including intervention of fermentative bacteria, acidogenic bacteria, acetogenic bacteria followed by methanogenic bacteria. The biogas contains 50-70% methane, 30-40% carbon dioxide and other components in traces (Abbasi *et al.*, 2012). The production of renewable energy in anaerobic treatment has made it highly important.

### 3. Coupled treatments

The coupled treatment is generally for the removal of inorganic pollutants from industrial sewage. In this treatment, various processes are coupled for the removal of pollutants from sewage. For example, ozone oxidation is coupled membrane bioreactor process to treat tannery effluents (Di Iaconi *et al.*, 2002). Similarly, ferric chloride is used in chemical coagulation coupled with oxidation process (Photo-Fenton) to treat industrial sewage.

## Sewage Sludge

Sewage sludge is defined as the residue formed during domestic sewage treatment and is generally used as an organic soil conditioner and partial fertilizer. Its application in agricultural land (pastures and cropland), unfertilized land including mined lands, construction sites, etc., recreational areas (parks, golf courses, etc.), plant nurseries, forest covers, and other unexplored infertile lands. Though sewage sludge contains beneficial plant nutrients and soil improving properties; it also contains disease causing microbes including pathogenic bacteria, viruses, protozoa, parasites, and other microorganisms. Application of untreated sewage sludge on land poses risk of direct and indirect human exposure to the pathogens. The regulation of sewage sludge is important to protect public health from disease causing microbes and pollutants present in sewage sludge. The treated sewage sludge is used for beneficial use in land fertilization and other aspects.

## Conservation of water bodies from sewage

The waterbodies such as ponds, lakes, wetlands, and green architecture, which includes water sources such as storm water drains, bore wells, trenches, etc., should be identified for proper maintenance in terms of sewage discharge. An awareness campaign regarding the sewage should be conducted to avoid domestic sewage discharge into these waterbodies. The waterbody inlet outlet should be properly maintained for any flow of sewage. The sewage entry should be avoided especially the untreated effluents from industries. Only treated effluents which are under the permissible limits of State Pollution Control Board should be allowed for disposition into waterbodies. Frequent cleaning of waterbodies to avoid any sewage discharge and/or accumulation should be carried out. The cleaning of water bodies can be done by de-weeding, de-silting, aeration, nutrient reduction, removing free floating plants. The empty land around waterbodies should be notified as eco-sensitive area and dumping of solid waste, agricultural waste and sewage should be avoided in these areas. The important thing is that the water quality of these waterbodies should be monitored on regular basis. Also, waterbodies, catchment areas, command areas, feeder channels of waterbodies are to be protected from all the polluting factors including sewage deposition from domestic sources, industries, agricultural runoff etc. The pollution of waterbodies due to sewage is highly challenging and needs to be prevented from the point source by intercepting, diverting and treating the sewage and other pollution components before entering into the water body. The sewage and other pollutants from the entire catchment area should be intercepted and diverted by cleaning using de-weeding, de-silting, bioremediation, aeration, nutrient reduction, bio-manipulation, withdrawal of anoxic hypolimnion etc.

## Future Prospects

Growing population, urbanization, industrialization, and agricultural chemical runoffs discharge huge loads of domestic sewage and industrial effluents, such as heavy metals, toxins into water bodies, especially lakes. This has put the lakes' flora and fauna in danger and has made the water a source of many human health problems. Further, the water pollution by eutrophication with high levels of nutrients, which promote algal bloom cause further problems in the lake ecosystem. The pollution of lakes due to the sewage is increasing at high rate, a cross functional strategy is required to minimize the flow of sewage from domestic and industrial sources into the lakes. The proper treatment of sewage before discharging into the rivers is critical.

Water quality modeling based on key parameters such as pH, BOD, total suspended solids and Fecal Coliform is to be established. To reduce the discharge of sewage into lakes, an integrated river basin approach is very important and a more holistic approach for assessment and regulation of sewage load by facilitating the water protection and to solve the territorial management difficulties as required locally. The water body, its uses and its composition including biotic and physiological parameters should be considered for its management. There is a need to set up more stringent limits for discharge of sewage into a lake especially for the highly sensitive areas. The reuse of wastewater should be considered for irrigation purpose. Further, to prevent pollution of lakes, a risk management strategy which is comprehensive, and its assessment ishighly important for maintaining water quality and public health. It is fundamental to use "reduce, reuse, recycle" approach for the containment of water pollution in lakes.

## Bibliography

Abbasi, T., Tauseef, S. M., & Abbasi, S. A. (2012). Anaerobic digestion for global warming control and energy generation-an overview. *Renew. Sust. Energ. Rev.*, 16, 3228-3242. DOI: 10.1016/j. rser.2012.02.046

Ade, P. A., Aghanim, N., Alves, M. I. R., Armitage-Caplan, C., Arnaud, M., Ashdown, M.... &Helou, G. (2014). Planck 2013 results. I. Overview of products and scientific results. *Astronomy & Astrophysics*, 571, A1.

Ajonina, C., Buzie, C., Rubiandini, R. H., &Otterpohl, R. (2015). Microbial pathogens in wastewater treatment plants (WWTP) in Hamburg. *Journal of Toxicology and Environmental Health, Part A*, 78(6), 381-387.

Akpor, O. B., &Muchie, B. (2011). Environmental and public health implications of wastewater quality. *African Journal of Biotechnology*, 10(13), 2379-2387.

Al-Tohamy, R., Ali, S. S., Li, F., Okasha, K. M., Mahmoud, Y. A. G., Elsamahy, T., ... & Sun, J. (2022). A critical review on the treatment of dye-containing wastewater: Ecotoxicological and health concerns of textile dyes and possible remediation approaches for environmental safety. *Ecotoxicology and Environmental Safety*, 231, 113160.

Basaran, G., Kavak, D., Dizge, N., Asci, Y., Solener, M., &Ozbey, B. (2016). Comparative study of the removal of nickel (II) and chromium (VI) heavy metals from metal plating wastewater by two nanofiltration membranes. *Desalination and Water Treatment*, 57(46), 21870-21880.

Bhat, R. A., Shafiq-ur-Rehman, M. M., Dervash, M. A., Mushtaq, N., Bhat, J. I. A., & Dar, G. H. (2017). Current status of nutrient load in Dal Lake of Kashmir Himalaya. *J PharmacognPhytochem*, 6(6), 165-169.

Bill and Melinda Gates Foundation, 2020. Water, sanitation & hygiene strategy overview. Available from https://www.epa.gov/national-aquatic-resource-surveys/national -lakes-assessment-2012-key-findings. Accessed on 9 June 2020

Boczkaj, G., & Fernandes, A. (2017). Wastewater treatment by means of advanced oxidation processes at basic pH conditions: a review. *Chemical Engineering Journal*, 320, 608-633.

Chambers, P. A., & Mill, T. A. (1996). Dissolved oxygen conditions and fish requirements in the Athabasca, Peace and Slave Rivers: Assessment of present conditions and future trends. In: Northern River Basins Study, Edmonton, Alberta.

Connolly, N. M., Crossland, M. R., & Pearson, R. G. (2004). Effect of low dissolved oxygen on survival, emergence, and drift of tropical stream macro invertebrates. Journal of the North American Benthological Society, 23(2), 251-270

Das, J., & Acharya, B. C. (2003). Hydrology and assessment of lotic water quality in Cuttack City, India. *Water, Air, and Soil Pollution*, 150(1), 163-175.

Demirak, A., Yilmaz, F., Tuna, A. L., &Ozdemir, N. (2006). Heavy metals in water, sediment and tissues of Leuciscus cephalus from a stream in southwestern Turkey. *Chemosphere*, 63(9), 1451-1458.

Di Iaconi, C., Lopez, A., Ramadori, R., Di Pinto, A. C., &Passino, R. (2002). Combined chemical and biological degradation of tannery wastewater by a periodic submerged filter (SBBR), Water Res., 36, 2205Ŕ14. https://doi. org/10.1016/S0043-1354(01)00445-6

Edokpayi, J. N., Odiyo, J. O., Popoola, O. E., &Msagati, T. A. (2016). Determination and distribution of polycyclic aromatic hydrocarbons in rivers, sediments and wastewater effluents in Vhembe District, South Africa. *International journal of environmental research and public health*, 13(4), 387.

Frodge, J. D., Thamas, G. L., & Pauley, G. B. (1990). Effects of canopy formation by floating and submergent aquatic macrophytes on the water quality of two shallow Pacific Northwest lakes. *Aquatic Botany*, 38, 231-248.

Grandclément, C., Seyssiecq, I., Piram, A., Wong-Wah-Chung, P., Vanot, G., Tiliacos, N., ... &Doumenq, P. (2017). From the conventional biological wastewater treatment to hybrid processes, the evaluation of organic micropollutant removal: a review. *Water research*, 111, 297-317.

Gupta, P., Vishwakarma, M., &Rawtani, P. M. (2009). Assessment of water quality parameters of Kerwa Dam for drinking suitability. *International Journal of Theoretical and Applied Sciences*, 1(2), 53-55.

Hampel, M., Blasco, J., &Segner, H. (2015). Molecular and cellular effects of contamination in aquatic ecosystems. *Environmental science and pollution research*, 22(22), 17261-17266.

Horner, R. R., Skupien, J. J., Livingstone, E. H., and Shaver, H. E. (1994). Fundamentals of Urban Runoff Management: Technical and Institutional Issues.

Igbinosa, E. O., &Okoh, A. I. (2009). Impact of discharge wastewater effluents on the physico-chemical qualities of a receiving watershed in a typical rural community. *International Journal of Environmental Science & Technology*, 6(2), 175-182.

Jing, G., Ren, S., Pooley, S., Sun, W., Kowalczuk, P. B., & Gao, Z. (2021). Electrocoagulation for industrial wastewater treatment: an updated review. *Environmental Science: Water Research & Technology*, 7.

Junior, I. V., de Almeida, R., &Cammarota, M. C. (2021). A review of sludge pretreatment methods and co-digestion to boost biogas production and energy self-sufficiency in wastewater treatment plants. *Journal of Water Process Engineering*, 40, 101857.

Kalra, S. S., Mohan, S., Sinha, A., & Singh, G. (2011, January). Advanced oxidation processes for treatment of textile and dye wastewater: a review. In *2nd International conference on environmental science and development*, 4, 271-5). IACSIT Press Singapore.

Kirchmann, H., & Pettersson, S. (1994). Human urine-chemical composition and fertilizer use efficiency. *Fertilizer research*, 40(2), 149-154.

Kiulia, N. M., Hofstra, N., Vermeulen, L. C., Obara, M. A., Medema, G., & Rose, J. B. (2015). Global occurrence and emission of rotaviruses to surface waters. *Pathogens*, 4(2), 229-255.

Koçer, O., &Acemioğlu, B. (2016). Adsorption of Basic green 4 from aqueous solution by olive pomace and commercial activated carbon: process design, isotherm, kinetic and thermodynamic studies. *Desalination and Water Treatment*, 57(35), 16653-16669.

Lambert, M. R., Giller, G. S., Barber, L. B., Fitzgerald, K. C., & Skelly, D. K. (2015). Suburbanization, estrogen contamination, and sex ratio in wild amphibian populations. *Proceedings of the National Academy of Sciences*, 112(38), 11881-11886.

Lecerf, A., Risnoveanu, G., Popescu, C., Gessner, M. O., & Chauvet, E. (2007). Decomposition of diverse litter mixtures in streams. *Ecology*, 88(1), 219-227

Lehtiniemi, M., Engström-Öst, J., &Viitasalo, M. (2005). Turbidity decreases anti-predator behaviour in pike larvae, Esox lucius. *Environmental Biology of Fishes*, 73(1), 1-8.

Malwade, K., Lataye, D., Mhaisalkar, V., Kurwadkar, S., & Ramirez, D. (2016). Adsorption of hexavalent chromium onto activated carbon derived from *Leucaena leucocephala* waste sawdust: kinetics, equilibrium and thermodynamics. *International journal of environmental science and technology*, 13(9), 2107-2116.

Mateo-Sagasta, J., Zadeh, S. M., Turral, H., & Burke, J. (2017). Water pollution from agriculture: a global review. Executive summary.

Mekonnen, A., Leta, S., &Njau, K. N. (2016). Co-Digestion of tannery wastewater and Phragmites Karka using a laboratory scale anaerobic sequencing batch reactor (ASBR). New York Sci. J., 9(1).

Momba, M. N., Malakate, V. K., & Theron, J. (2006). Abundance of pathogenic Escherichia coli, Salmonella typhimurium and Vibrio cholerae in Nkonkobe drinking water sources. *Journal of water and health*, 4(3), 289-296.

Petrovic, M., Solé, M., López De Alda, M. J., & Barceló, D. (2002). Endocrine disruptors in sewage treatment plants, receiving river waters, and sediments: integration of chemical analysis and biological effects on feral carp. *Environmental Toxicology and Chemistry: An International Journal*, 21(10), 2146-2156.

Prüss-Ustün, A., Wolf, J., Bartram, J., Clasen, T., Cumming, O., Freeman, M. C., ... & Johnston, R. (2019). Burden of disease from inadequate water, sanitation and hygiene for selected adverse health outcomes: an updated analysis with a focus on low-and middle-income countries. *International journal of hygiene and environmental health*, 222(5), 765-777.

Ramírez-Dírz, M. I., Díaz-Pérez, C., Vargas, E., Riveros-Rosas, H., Campos-Garcéa, J., & Cervantes, C. (2008). Mechanisms of bacterial resistance to chromium compounds. Biometals, 21, 321e332.

Raschid-Sally, L., & Jayakody, P. (2009). *Drivers and characteristics of wastewater agriculture in developing countries: Results from a global assessment* (Vol. 127). IWMI.

Ravikumar, L., Kalaivani, S., Vidhyadevi, T., Murugasen, A., Kirupha, S. D., &Sivanesan, S. (2014). Synthesis, characterization and metal ion adsorption studies on novel aromatic poly (azomethine amide) s containing thiourea groups. *Open Journal of Polymer Chemistry*.

Samuel, M. S., & Chidambaram, R. (2015). Hexavalent chromium biosorption studies using Penicillium griseofulvum MSR1 a novel isolate from tannery effluent site: Box–Behnken optimization, equilibrium, kinetics and thermodynamic studies. *Journal of the Taiwan Institute of Chemical Engineers*, 49, 156-164.

Schindler, D. W. (1974). Eutrophication and recovery in experimental lakes: implications for lake management. *Science*, 184(4139), 897-899.

Snyder, E. G. (2016). Elimination of odor at six major wastewater treatment plants. *Water Environ. Federation*, 57(10), 1027-1032.

Trishitman, D., Cassano, A., Basile, A., & Rastogi, N. K. (2020). Reverse osmosis for industrial wastewater treatment. In *Current trends and future developments on (bio-) membranes* (pp. 207-228). Elsevier.

Tukura, C. S. (2009). Effect of Overhead Projector on the learning of Social Studies in Secondary Schools in Bosso Local Government area of Niger State. *Nigerian Journal of Technological Research*, 4(2), 71-79.

Vargas-González, H. H., Arreola-Lizárraga, J. A., Mendoza-Salgado, R. A., Méndez-Rodríguez, L. C., Lechuga-Deveze, C. H., Padilla-Arredondo, G., & Cordoba-Matson, M. (2014). Effects of sewage discharge on trophic state and water quality in a coastal ecosystem of the Gulf of California. *The Scientific World Journal*, 2014.

Wear, S.L., Thurber, R.V., 2015. Sewage pollution: mitigation is key for coral reef stewardship. Ann. N. Y. Acad. Sci. 1355.

Whitehead, P. G., & Hornberger, G. M. (1984). Modelling algal behaviour in the river Thames. *Water Research*, 18(8), 945-953

Wu, R. S., Zhou, B. S., Randall, D. J., Woo, N. Y., & Lam, P. K. (2003). Aquatic hypoxia is an endocrine disruptor and impairs fish reproduction. *Environmental science & technology*, 37(6), 1137-1141.

# Chapter - 8

# Microbial Pollution in Freshwater Lakes, Sources, Impact and Remediation

**Seeta Dewali[1], Uzma Siddiqui[1*], Sheetal Sharma[1], Harish Chandra Singh Bisht[1], Nityanand Pandey[2] and Satpal Singh Bisht[1]**

[1]Department of Zoology, DSB Campus, Kumaun University, Nainital-263001
[2]Directorate of Coldwater Fisheries Research, Bhimtal - 263136, Nainital, Uttarakhand
*Corresponding Author: uzma.labiba@gmail.com

## Abstract

Freshwater is a vital natural resource that we use for a variety of daily activities, including drinking and numerous developmental goals. When dangerous substances pollute a stream, river, lake, ocean, aquifer and otherwaterbodies, its quality deteriorates andbecomes poisonous to both people and the environment. Point sources and non-point sources are the two main causes of water contamination to which the city sewage and industrial waste discharge mostly contribute. Additionally, microorganisms can lower the oxygen content of the water, making it harder for aquatic plants and animals to survive. Microorganisms have been used for decades as water quality indicators. In this chapter, the literature has concurrently reported on the evaluation of the water quality of various freshwater bodies. Since the current evaluation is specific to the bacteriological examination of water, a full explanation is beyond the scope of this investigation. Thus, this article seeks to concentrate on the microbiological contamination, sources of fresh water and various health risks that have arisen as a result of the contamination of water, as well as their remediation.

**Keywords:** Freshwater; Pollution; Remediation; Microbial; Health.

## Introduction

Water is essential for life. Water supply has always been a fundamental component of society for a variety of uses, including drinking, agriculture, industry, home, etc. The lack of access to quality drinking water is a major cause of health issues in poor nations (Bylund *et al.*, 2017). Due to the potential for contamination by pathogenic bacteria, protozoa, or viruses, microbial pollution in aquatic system is one of the critical challengesrelated to the hygienic status of water bodies utilized for drinking water supply, recreational activities, and seafood harvesting. The most valuable resource available to humans is fresh water. Water makes up around 75% of the surface of Earth, getting it the nickname "blue planet" howeverthe bulk of this water is salty, accounting for less than 5% of the total and is found in ice caps, glaciers, and groundwater (Jung *et al.*, 2014).

Water pollution caused by microorganisms is regarded as one of the major threats to the aquatic environment worldwide. The number of bacteria in a water body is increased by faecal matter effluent, hospitals, industries, and Agricultureeffluent. Coliform bacterial groups have long been used as an indication of microbial water pollution and have traditionally contributed to the sense of public health security (Palit *et al.*, 2022). An estimated 80% of illnesses and over one-third of mortalities in impoverished nations are attributed to drinking water Pollution. According to the World Health Organization (WHO), there are 600 million instances of diarrhoea and dysentery cases each year and there are 46,000 new born deaths as a result of contaminated water and inadequate sanitation.

There is a risk of hazardous germs spreading whenever surface water comes into contact with local sewage or human waste (Denchak, 2018). As a result, consuming microbially tainted water for drinking, swimming, bathing, and other activities becomes extremely dangerous. Monitoring of the water quality has been done to determine whether a water supply is fit for specific use (Okoye *et al.*, 2022). The suitability of water is assessed in relation to the maximum acceptable amounts of specific water quality parameters that are specified by guidelines, standards, and other criteria (García Ávila *et al.*, 2022). Thepresence of faecal coliform bacteria in the water body shows that the water has been polluted by the faeces of warm-blooded animals. Pathogen leaks might be caused by poorly managed livestock operations.

Faecal bacteria identification is therefore essential for developing parameter-based methods for the assessment of drinking water quality as well as for basic and practical research in aquatic microbial ecology. Faecal contamination of water is also indicated by the presence of faecal *streptococci*, *Clostridium perfringens*, and *Klebsiella*.

Apart from that, *Shigella* spp., *Proteus* spp., *Salmonella enteritidis*, *Salmonella* typhi, and *Salmonella typhimurium* are the most common bacteria found in sewage samples (Korajkic *et al.*, 2018). Additionally, *Salmonella*, *Schistosoma*, *Norovirus* and parasitic worms, are occasionally found in contaminated surface waters and have been linked to human health problems.

On-site sanitation systems like septic tanks and pit latrines can generate large amounts of germs, as can inadequately treat sewage discharges. Older cities with deteriorating infrastructure may have sanitary sewer overflows as a result of leaking sewage collection systems (pipes, pumps, and valves). A number of communities also have combined sewers, which might let off untreated sewage during rainstorms. Silt (sediment) from untreated wastewater also contaminates aquatic bodies (Harrison, 2001; Kong *et al.*, 2009; Reder *et al.*, 2017; Xue *et al.*, 2020).

**Water pollution caused by Microorganisms: Sources, Fate, and Behaviour**

**Sources:**

Point sources and non-point sources are the two primary contributors of water contamination. Factories, wastewater treatment facilities, septic systems, and other sources that are obviously releasing contaminants into water sources are considered point sources. Because they cannot be linked to a specific location, non-point sources are more challenging to locate. Runoff from farms, fields, construction sites and mines can include sediment, fertiliser, pesticides, and animal faeces. If contaminants seep into nearby water sources from a landfill, they can also be considered a non-point source of pollution.

Forty percent of rivers and estuaries failing to meet ambient water quality standards fail because of pathogens, usually measured by the faecal coliform test (James & Joyce, 2004). The microbial contamination of water is often of faecal nature related to humans (water sewage treatment plants, combined Sewage Overflow (CSO), non-collective sewage systems), domesticated animals (manure spreading, pit stock overflow), or wildlife. The main origins of microbial contamination of natural aquatic resources are discharges of water treatment plants, decontamination stations, hospitals, industries considered as point sources, etc. On the other hand, diffuse sources (slurry, manure, sludge application) may also be considered. The abundance and importance of pathogens in water depend on factors such as the contamination level, pathogens' persistence in water bodies, biological reservoirs (including aquatic plants and sediments) and the ability of pathogens to be transported (Dechesne *et al.*,2006). The land use management practices and the size of the watershed also influence the survival of microorganisms (Ferguson *et al.*, 2008; Harmel *et al.*, 2010). According to George *et al.* (2004) streams flowing through areas partly or fully covered with pastures are more contaminated than those running through forests and cultivated areas.

Water microbiological contamination frequently arises from human activity (sewage treatment facilities, combined sewage overflows (CSO), non-collective sewage systems), domesticated animal activity (manure spreading, pit stock overflow), or wildlife activity. Water treatment plant outputs, decontamination facilities, hospitals, point-source industries etc., are the principal causes of microbial contamination of natural aquatic resources. Urban activities and disease concentrations are well-known to be correlated (Bates *et al.,* 2008).On the other hand, dispersed sources (such as sludge application, manure, and slurry) may also be taken into account. Pathogen concentrations, persistence in water bodies, biological reservoirs (including aquatic plants and sediments), and pathogen transportability are only a few of the variables that affect the quantity and significance of pathogens in water (Dechesne *et al.,* 2006).The size of the watershed and the methods used to control land use have animpact on microorganism survival as well (Harmel *et al.,* 2010).

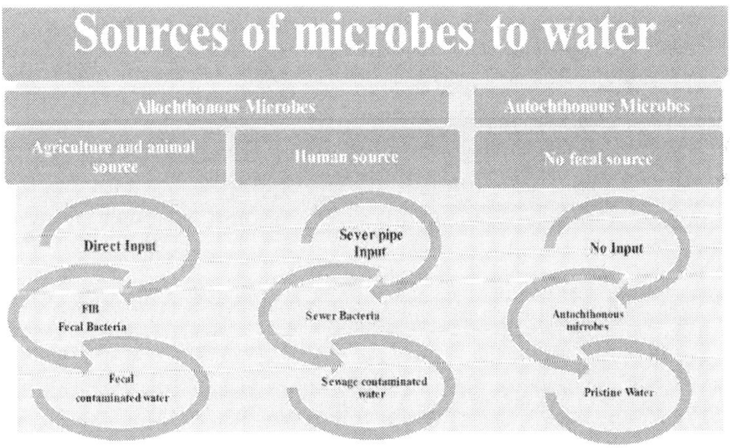


**Fig.1: Autochthonous and Allochthonous Sources of microorganisms that pollute freshwater system**

Figure 1 suggests a schematic synthesis of the origin and final destination of microbes in water. With the provision of drinking water and widespread use of recreational waterways, the human strain on resources and usage rises. The biological reservoirs are subjected to physicochemical and biological reactions on the one hand, and to hydro chemical and hydrological variables in water bodies on the other, throughout the process of sedimentation and resuspension. Additionally, depending on climate change, external runoffs from anthropogenic and natural sources are regarded as sporadic and dispersed contributions. As a result; there is a growing need for more efficient and dependable monitoring techniques, such as quick molecular methods, even if they do not offer real-time control as optical techniques do (Edge *et al.*, 2010). Sources of bacteria in water even in the absence of faecal input, autochthonous microorganisms are numerous and diverse in natural aquatic systems. Through an addition to the autochthonous bacteria, animal and human sources of allochthonous faecal germs are transported in contaminated streams.

## Microorganism's Fate in Sediments and Submerged Aquatic Vegetation

Sediments and submerged aquatic vegetation (SAV) are major microorganism reservoirs. Badgley *et al.* (2010) discovered that SAV had much greater mean densities of *Enterococci* than sediments, which had higher densities than water. The high *Enterococci* populations reported in SAV are mostly attributable to increased food availability and UV radiation protection. Sediments can contain ten times the number of viruses found in water (Dechesne *et al.*, 2006).

Garzio-Hadzick *et al.* (2010) investigated microorganism viability in different sediment types with varying particle size and organic carbon concentrations. Their findings showed that sediments with tiny particle size and high organic carbon content might help viruses survive longer. According to Chandran *et al.* (2011) sediments might operate as a reservoir of harmful microorganisms, posing significant health risks from resuspension and subsequent consumption during recreational activities. The importance of particle granulometry in microbe survival is also critical for their transit during high-flow events.

The behaviour of microorganisms is also impacted by hydro meteorological changes, such as heavy rainfall, which is expected to increase with climate change. As a result, microbes associated with sediments or SAV are placed back into suspension, potentially contaminating water (Dechesne *et al.*, 2006). Cho *et al.* (2010) investigated *E. coli* behavior in three streams with varying bottom sediment textures during or immediately after rainstorm episodes. However, more study was needed to determine which sediment features impact the parameters of streambed *E. coli* discharged into the water column.

## Transport of microorganisms is influenced by interactions between microorganisms and sediments

The interactions between microorganisms and sediments affect both their chances of survival and how they move through the environment (Jamieson *et al.*, 2005). Furthermore, the mechanisms that control sediment transport are intimately linked to the fate and movement of faecal bacteria. Particularly affected by adhesion to mineral surfaces and inactivation are the movements of viruses associated with particles (Muirhead *et al.*, 2006). Other environmental factors, such as pH changes, might cause this adsorption to reverse (Loveland *et al.*, 1996). According to Soupir & Mostaghimithe (2011) microbiological partitioning between the freely floating and particle-attached phases during transmission along overland flow paths was investigated.

The bulk of *E. coli* and *Enterococci* are carried from the fresh manure source in an unattached condition, with just 4.8% of *E. coli* and 13% of *Enterococci* connected with particles, according to rainfall simulations performed on large-scale field plots. With thinly vegetated cover, the typical percentage of connected *E. coli* or *Enterococci* is less than 3%. *E. coli* appeared to be attached mostly to tiny particles (2 m) in a laboratory-scale model system designed to study the transport mechanisms of *E. coli* in overland flow through saturated soils (Soupir & Mostaghimi, 2011). As a result, *E. coli* appeared to stay unattenuated during transit. Additionally, microbes might be thought of as biological particles carried by advection in the hydric resource.

To account for many physiochemical and biological processes, reaction kinetics, and transport mechanisms of the coupled system, they created a conceptual model (contaminant–colloids–bacteria) (Bekhit *et al.*, 2009). For naturally occurring aquatic media, such as lakes and reservoirs, various outcomes have been attained. A review by Brookes *et al.* (2004) in particular clarified the variables influencing pathogen movement and dissemination, including the relative importance of dilution, dispersion, and horizontal and vertical transport in such medium. Pathogen particle settling uses more intricate hydrodynamic dynamics than lab-scale models do.

According to Gutierrez & Nguyen (2012) findings, all of the microbial indicators put to the test—*Escherichia coli*, *Enterococci*, *Clostridium perfringens*, aerobic spores, somatic coliphages, *Cryptosporidium* spp., and *Giardia* spp.- were connected to larger-sized particles (>63.3 m), with the exception of *C. perfringens* spores, which were connected to particles in the 4. The preservation of microorganisms by settling and transfer into sediments is significantly impacted by the granulometric distribution of the related particles. The prolonged survival of protozoan parasites in various climatic conditions and disinfection procedures, as well as their capacity to interact with other organic or nonorganic particles, are all closely related to the environmental impact of these parasites. Their transfer, retention (by settling), and/or release are governed by the latter phenomena (Abudalo *et al.*, 2010).

**Effects of Hydro Meteorological Conditions on the Mobility of Microorganism.**

The hydro meteorological environment has a significant impact on the mobility of microorganisms. In fact, combined sewage overflows (CSOs) or run-off from agricultural land cause an increase in microbial contamination of surface water following heavy rains (Delpla et al., 2011). Small streams become more contaminated by faeces when it rains, with the coliforms mostly adsorbing on particles and increasing the suspended matter content of the streams. Rainfall events can also increase the amount of human enteric viruses in the natural waterways that are touched by CSOs and stormwater, which increases the risk of gastroenteritis while engaging in recreational activities or eating shellfish (Hata *et al.*, 2013). In swimming-associated gastroenteritis, norovirus predominates according to quantitative microbiological risk assessment (McBride *et al.*, 2013). Heavy rains can also cause flooding in catchment areas and a significant rise in water turbidity, the latter of which is thought to be an unmistakable sign of possible pathogen contamination. An endemic level of gastrointestinal diseases may be caused by groundwater impacted by surface water, according to a number of prospective epidemiological investigations conducted in France, notably during rainstorm events in caustic environments (Beaudeau *et al.*, 2008).

The fluctuation in water temperature affects how microorganisms are distributed hydrodynamically. Lakes in temperate regions typically stratify in the summer (with warmer water at the surface). Heavy downpours or storms cause destratification, which alters the convection of microorganism-containing particles and drives them to the surface (Brookes *et al.*, 2005). Wyer *et al.* (2011) recently examined how faecal markers changed in four streams under a moderate rainfall. They showed that the transmission time ranges from minutes to hours depending on the hydrological conditions and site features owing to microbial source tracing (MST).

**Impacts of microbial pollution**

Although water plays an essential role in supporting human life and biodiversity, it also has a great potential for transmitting diseases when contaminated (Yakasai *et al.*, 2004). Communities situated downstream or near municipal sewage outfalls or contaminated water sources are at the highest risk of illness due to increased microbial pathogens and deteriorating physicochemical parameters (Wakelin *et al.*, 2008). These pathogens can cause communicable diseases through direct or indirect body contact or ingestion of contaminated water or shellfish. Fish and other organisms in the water can also be negatively impacted.

## Effect of microbial pollution on Human Health

The discharge of very turbid wastewater, combined with thick algae blooms, frequently results in poor visibility inside these aquatic bodies, posing a threat to recreational users. Furthermore, water bodies utilised for full-contact recreational activities may be a source of numerous infectious illnesses, which can be caught either by consumption of contaminated water or through contact with contaminated water. Many waterborne illnesses are caused by biological micropollutants such as enteric bacteria, mycoplasma, viruses, and protozoa, which are major causes of death globally (Theron & Cloete, 2002). Furthermore, they have the potential to endanger human health by causing disease resistance, endocrine disruption, and chronic toxicity (Rosal *et al.*, 2010).

### Bacterial diseases

**Acute gastroenteritis**: *Campylobacter* species that are thermophilic, notably *C. jejuni* and *C. coli*, are now recognized as frequent causes of acute gastroenteritis in most regions of the world (Allos & Blaser, 1995). They are commonly spread by contaminated chicken items, unpasteurized milk, and water. *Campylobacter* cells have a low infective dosage; it is believed that 500 *C. jejuni* cells can cause human sickness (Black *et al.*, 1998). Several investigations have been undertaken to investigate the ability of campylobacters to survive in the environment (Koenraad *et al.*, 1997; Buswell *et al.*, 1998 and Höller *et al.*, 1998). Accordingto these tests, the survival potential was just a few hours at high temperatures (37°C), but when temperatures dropped, it climbed and eventually reached many weeks at 4°C. Other bacteria, especially those found in biofilms, increased the likelihood of survival (Buswell *et al.*, 1998).

Several epidemics of *campylobacter* enteritis have been linked to contaminated drinking water (Tauxe, 1992; Jones & Roworth, 1996). The primary dangers for *Campylobacter* spp. contamination are the use of unchlorinated surface water and secondary contamination of drinking water in storage reservoirs. *Campylobacters* have been detected in high concentrations (10 to 105 cfu/100 ml) in raw sewage (Höller, 1998; Stelzer & Jacob, 1991), and they are found in fecally polluted surface waterways, with high concentrations being associated with contamination by wild birds (Kapperud & Rosef, 1983).

**Helicobacter pylori infection:** *H. pylori* infection is common worldwide (Goodwin *et al.*, 1997), and it is significantly linked to gastroduodenal illness, including chronic active gastritis, peptic and duodenal ulcer disease, and gastric cancer (Graham, 1991). Several laboratory experiments have shown that *H. pylori* can live in freshwater for 10 days and sterile, distilled water (7°C) for 14 days, respectively (Bode *et al.*, 1993; Shahamat *et al.*, 1993; Fan *et al.*, 1998).

**Gastrointestinal infections:** *Yersinia enterocolitica* strains cause gastrointestinal illnesses in people. These strains are known to exhibit virulence factors (Cornelis, 1994; Bottone, 1999). There haven't been many epidemics of digestive illnesses linked to *Y. enterocolitis* in the environment, sewage, surface water with faecal influent (Fukushima *et al.*, 1984; Ziegert &Diesterweg, 1990), or drinking water (Helberg & Löfgren, 1998; Schieman, 1990).

**Pulmonary disease:** The genus *Mycobacterium* includes both strictly pathogenic species (*M. leprae*) and *Mycobacteria* that may be pathogenic (e.g., *M. avium* and *M. intracellulare*) and are most typically linked with pulmonary illness (Jenkins, 1991). The prevalence of *M. avium* complex infection in HIV-infected people has risen rapidly, fuelling discussion regarding the likely involvement of water in the spread of this agent (Singh & Yu, 1994).

### Parasitic Protozoan diseases

**Giardiasis and Cryptosporidiosis:** Waterborne outbreaks of giardiasis (Kramer *et al.*, 1998; Furness *et al.*, 2000) and cryptosporidiosis (Sorvillo *et al.*, 1992; Futardo *et al.*, 1998) have frequently been linked to contaminated recreational water (e.g., lakes, rivers, or swimming pools). *Giardia lamblia* and *Cryptosporidium parvum* produce these infections, with the infectious stages being a cyst and an oocyst, respectively. These pathogenic protozoa induce subclinical infections and self-limiting diarrhoea in healthy people. *C. parvum*, in particular, can cause severe, even deadly diarrhoea in new-borns, immunocompromised people, or individuals with underlying diseases (Current & Garcia, 1991). Between 2 and 7700 cysts or oocysts/100 litres of water have been identified in connected drinking water sources (Smith *et al.*, 1995). Cysts and oocysts have been discovered in a significant proportion of surface waters (LeChevalier *et al.*, 1991a,b), according to surveys of raw water sources (Futardo *et al.*, 1998; Rose, 1998; Rose *et al.*, 1991). The presence and quantity of organisms in surface waters are expected to be higher in developing nations where water pollution by human and animal waste is more prevalent (Smith *et al.*, 1995).

**Cyclosporiasis:** *Cyclospora cayetanensis* is a novel protozoal parasite. It has been linked to the aetiology of chronic watery diarrhoea, tiredness, and anorexia in humans. The transmission mechanism is thought to involve intake of contaminated water (Ooi *et al.*, 1995). Several investigations have recently discovered that *Cyclospora cayetanensis* is related with the use of polluted water (Marshall *et al.*, 1997; Steiner *et al.*, 1997; Meng and Doyle, 1997; Rabold *et al.*, 1994).

### Viral diseases

**Viral gastroenteritis:** Several enteric viruses (including the Norwalk virus, rotaviruses, adenoviruses, and hepatitis E virus) are the major cause of pandemic viral gastroenteritis in older children and adults in Africa, Asia, and the United States. The source of these epidemics has been traced back to fecally polluted drinking and recreational water (Moe, 1996; Naik *et al.*, 1992). Enteric viral infections, in addition to producing gastroenteritis, can also cause meningitis, pulmonary illness, and encephalitis (Hedberg & Osterholm, 1993).

### The effects of microbial contamination on disease processes in aquatic animals

When sensitive fishes are exposed to virulent microorganisms under particular environmental stress conditions, infectious diseases start to develop. If the existence of stress corresponds with the presence of harmful microorganisms, it is reasonable to expect that disease outbreaks are more likely to occur (Snieszko, 1974). Fish diseases, which have been reported to be associated with microbial contamination of the aquatic environment, include epidermal papilloma (Premdas & Metcalfe, 1994), fin and tail rot (Vethaak, 1992; Vethaak *et al.*, 1996), gill disease/hyperplasia (Kirk & Lewis 1993), liver disease (Malins *et al.*, 1988; Peters *et al.*, 1987; Vethaak *et al.*, 1996), neoplasia (Bucke & Feist, 1993; Depledge, 1996; Vethaak & Jol, 1996), parasitic diseases (Das & Shrivastava, 1984; Khan, 1987; Moller, 1987; Khan & Thulin, 1991), skin disease/ulceration (Vethaak, 1992; Vethaak & Jol, 1996) and viral diseases, e.g., lymphocystis (Vethaak & Jol, 1996; Vethaak *et al.*, 1996). Contaminated diets (Landsberg, 1995), heavy metals (Roadsaether *et al.*, 1977), hydrocarbons (Khan, 1987), nitrogenous compounds, i.e., ammonia (Kirk & Lewis, 1993) and nitrites (Hanson & Grizzle, 1985), pesticides (Voigt, 1994), untreated sewage (Austin & Stobie, 1992), and unspecified pollutants have all been blamed as causes (Vethaak & Jol, 1996). These pollutants disrupt the normal state of the aquatic medium, causing behavioural changes as well as morphological and biochemical imbalances in aquatic creatures (Yadav *et al.*, 2005) and may move down the food chain (Diana, 2007). Pollutant levels in river water systems increase biological oxygen demand (BOD), chemical oxygen demand (COD), total dissolved solids (TDS), total suspended solids

(TSS), toxic metals such as Cd, Cr, Ni, and Pb, and faecal coliform, rendering the water unfit for drinking, irrigation, and aquatic life (Kanu & Achi, 2011).

## Remediations

The freshwater environment provides important advantages to humans by providing food and ecological services. Furthermore, freshwater bodies such as lakes, rivers, and wetland ecosystems provide habitat and food for a variety of creatures such as fish, algae, phytoplankton, zooplankton, cyanobacteria, and birds. These enormous advantages must be preserved by eliminating human activities that have the potential to release hazardous substances and environmental stressors into aquatic bodies in order to prevent freshwater quality degradation (Amoatey & Baawain, 2019). Figure 2 depicts there are several approaches for preventing water contamination.

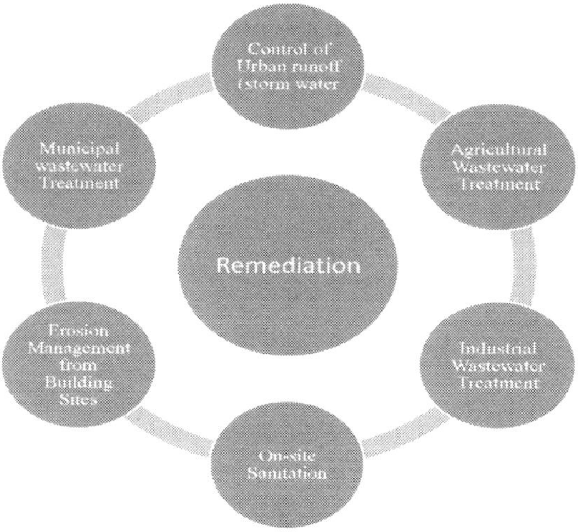


**Fig.2: Various water microbiological contamination remediation techniques**

## Remediation of Polluted Freshwater Ecosystem

Because of the high degree of pollution, remediation of polluted water is a pressing issue in many developing nations. Physical, chemical, biological, ecological, and engineering approaches are among those that may be used for polluted water remediation (Gao *et al.*, 2018; Bai *et al.*, 2020). Although a single method is not always efficient for the purification of significantly contaminated water. As a result, hybrid procedures that mix two or more single methods are more often suggested for their efficacy.

## Engineering and Physical Processes

### Aeration

The correct treatment of polluted freshwater and wastewater depends on aeration, which raises the oxygen saturation of water (Zhang, 2016). Aeration techniques can be expensive to implement even if they are easy to use, durable, and broadly applicable. The microbial populations that break down organic molecules in river water and wastewater are more diverse and robust when there is aeration. The correct treatment of dirty freshwater and wastewater depends on aeration, which raises the oxygen saturation of water (Zhang, 2016).

### Water Diversion and Water Transfer

The widespread application of water transfer engineering worldwide has the potential to change environmental conditions, agricultural practices, industrial growth, and human lifestyles (Davies *et al.*, 1992). According to Xie *et al.* (2009) and Hu *et al.* (2010), this engineering may be used for power production, nutrient reduction, navigation, flood control, irrigation, water supply, and phytoplankton reduction, as well as for irrigation, water quality improvement (Li *et al.*, 2011). Clean water is added to river water to remove the colour and odour while also diluting the pollutants in the water. By using water diversion, the watershed's capacity to cleanse itself is enhanced (Wang *et al.*, 2012; Ensink *et al.*, 2010). River water in Fuzhou, Zhongshan, and other Chinese cities was purified using this technique (Wang *et al.*, 2012). Due to its high cost and labour requirements, the task of diverting river water may at times become too onerous. In order to determine the optimum way to clean unclean river water while utilising the least amount of pure water, Martinez *et al.* (2010) developed an engineering model.

### Hydraulic Structures

In several countries, particularly India, river water contaminated by untreated urban sewage dumping has been used for irrigation and agricultural crop production. Water quality may be improved and river water made appropriate for agriculture by installing hydraulic structures on the river, such as irrigation weirs or irrigation infrastructure (Ensink *et al.*, 2010). Construction of irrigation weirs on a river reduces water flow while lengthening hydraulic retention time, allowing for the occurrence of anaerobic reactions, sedimentation, aeration, sunlight irradiation, and other processes that enhance water quality (Radspinner *et al.*, 2010).

## Ecological Engineering-Based Processes

Freshwater ecosystems may be effectively restored and treated using riparian wetlands, sedimentation and absorption of nitrogen (N) and phosphorus (P), aeration, and ecological floating beds, among other techniques (Peilin *et al.*, 2019). Artificial floating wetlands, ecological floating islands, and other bioremediation technologies are some of the most popular treatment options.

### Phytoremediation Process:

River water, wastewater effluents, and contaminated water may all be greatly improved by aquatic vegetation, particularly algae (Anawar *et al.*, 2008). Plants that can tolerate toxicants can remove them from river water by absorption, adsorption, accumulation, and breakdown. These plants showed a potent capacity for nutrient extraction from water bodies, including total N and P (Tong *et al.*, 2003). Reed (*Phragmites australis*), water hyacinth (*Eichhornia crassipes*), alligator weed (*Alternantheraphiloxeroides*), water lettuce (*Pistia stratiotes*), and whorl-leaf watermilfoil (*Myriophyllum verticillatum*) are the aquatic plants most frequently employed for wastewater treatment (Allam *et al.*, 2016). Regardless of locale, the strategy is affordable and is well-liked by the local population. In addition, plant species differ in their ability to remove pollutants and their susceptibility to contamination loading rates. Therefore, choosing the right plant species is essential for plant-based technology to work as effectively as possible (Zhao *et al.*, 2012).

## Ecological Floating Beds

Natural floating beds use natural processes and can be used to treat in-stream water in a decentralized manner (Jamwal, 2018). Among their pollution removal techniques are phytoremediation (the uptake of heavy metals by plants), microbial biodegradation of organic compounds, and removal of N and P by absorption and sedimentation processes. This method is becoming more and more popular for treating river water because of its low cost, high removal efficiency, and increased plant accommodation. These systems may be placed anywhere along the river since they are portable. According to Zheng and Wang (2017), they can provide habitat for fish and birds, but they impede phytoplankton formation. These techniques enhance the ecological diversity of freshwater ecosystems in general (Knowles *et al.*, 2011).

## Remediation by Aquatic Animals

Aquatic animals can be used to clean up contaminated water (Ateia *et al.*, 2016). The freshwater gastropod, for instance, have been employed as a bio-indicator to calculate the contamination levels of various heavy metals in river water and sediments (Li *et al.*, 2018). Some aquatic organisms, including clams, snails, silver and common carp, and other filter-feeding fish, can lower nitrogen and algal concentrations in river water, resulting in an improvement in the water quality (Xiao *et al.*, 2010). However, some studies suggested that the toxic effects of the bio toxins released by algae and the presence of inorganic or organic contaminants in contaminated water limit the ability of silver carp to control algal biomass, which ultimately reduces the effectiveness of this biological treatment system (Ateia *et al.*, 2016).

Choosing the best treatment is therefore essential for the restoration of freshwater habitats. Aeration is a common, effective, and sustainable procedure used in physical engineering that increases microbial activity and breaks down organic contaminants. However, they are expensive and harmful to river ecosystem. Other engineering solutions (water diversion, mechanical algae removal, and hydraulic structures) are also useful. Due to their high economic, environmental, and ecological benefits, ease of maintenance, and lack of secondary pollutants, ecological engineering-based approaches are preferred.

## Microbial Water Quality

Monitoring and modelingmicrobial water quality has an impact on the health of people, animals, and the environment, hence, go hand in hand. Multi-disciplinary research is focused on the management, monitoring, and prediction of microbiological water quality. Interdisciplinary research is necessary in order to address the important public health issue, which is made more challenging by the complexity of microbial populations and their habitats, the multiplicity and heterogeneity of transport pathways, the multiplicity of controls, and the multiscale nature of microbial water quality (Fujioka *et al.*, 2015; Oliver *et al.*, 2016; Bentzon-Tilia *et al.*, 2016; Lothrop *et al.*, 2018).

## Observational Design

Currently, the most of monitoring design schemes are based on a phased approach to monitoring microbiological water quality (NRC, 2004). The first stage involves screening, which identifies the potential problem; while as the second stage involves more in-depth research to confirm a public health risk and support management decisions; and the final stage involves research to identify the sources of microbial contamination so that the health risk can be decreased through a variety of engineering and policy solutions.

## Molecular Markers

Our knowledge of the formation of microbiological water quality at particular sites may be greatly improved by using a number of indicators in addition to data on faecal indicator organisms. Depending on the hydrology, land use, and likelihood of subterranean transit, the environment can affect how genetic markers employed in microbial source tracing are carried. Another helpful example of how to employ molecular markers to understand diseases' environmental pathways in agro-ecosystems is provided by Rieke *et al.* (2018). Nevers *et al.* (2018) investigated for sources of microbiological pollution and assessed a short-term management approach to decrease beach closures by deterring birds. For molecular markers and associated techniques to be used more effectively, standardization and cross-validation are crucial. Currently, little is known about the prevalence of different pathogen serotypes, serogroups, and strains in natural waterways. This information might be useful in developing new molecular markers. Marker concentrations reflect both sources and microbial transit conditions (Garca-Aljaro *et al.*, 2017). Investigating the fate and transit of molecular markers is still essential in the field of microbiological water quality.

## Antibiotic Resistance

A rapidly growing body of research has been conducted on the sources of antibiotic resistance genes (ARGs) and spatiotemporal variability in waters that may have an impact on human health, despite the fact that antibiotic resistance genes (ARGs) are typically not mentioned in microbial water quality standards but that there is general agreement regarding the potential risks of their presence (WHO, 2014). Whole genome analysis may be used to study ARG in surface waters in great depth. The quantity and prevalence of various antibiotic resistance genes can be used as indicators of the microbiological waterquality however; it is still unknown how these indicators relate to both animal and human health. Metrics relating to ARG are currently absent from assessments of microbial water quality.Whether or not ARG in ecosystems leads to the emergence of new diseases is still up for discussion. The corpus of evidence now available on ARG will elucidate the environmental factors that inspire resistance and eventually support the development of conceptual models for how resistance emerges and spreads (Bengtsson-Palme *et al.*, 2017).

## Spatiotemporal Variability of Microbial Water Quality

If the spatiotemporal variability of the microbiological water quality is recognized and measured, monitoring and management activities will be effective. High regional heterogeneity is typically seen in microbial water quality assessments. Variations in the spatiotemporal results of monitoring the microbiology of water may be an expression of observation size. In microbiological water quality study, the geographical scales are often characterized by the stream's order, the area from which runoff and/ or drainage occurs, or the hierarchical sequence of the plot-field-watershed. The amount of time that passes between water samples or the amount of time it takes to obtain a water sample determines the temporal scales. Numerous studies have demonstrated that indicator organism patterns are temporally consistent across freshwater sources (Pachepsky *et al.,* 2018).

## Conclusion

Improving water quality requires setting up a management strategy that contains realistic approaches to address water quality. Similar to this, in order to maintain the health of the global environment, it is essential to raise public awareness of the problems associated with faecal contamination of water supplies. These elemental analyses are anticipated to be paired as a well-established complementary package for future aspect in order to address the cause and source of water pollutions with enhanced specificity, efficiency, and comprehensiveness. Environmental experts are now creating a system that is more complete, rapid, and concentrated to augment the already in place aquatic monitoring technique in order to address the specific cause and source. Microbial monitoring and modelling will be essentialto establish and link different microbial populations and ecosystems through environmental transfers.

## Bibliography

Abudalo, R. A., Ryan, J. N., Harvey, R. W., Metge, D. W., & Landkamer, L. (2010). Influence of organic matter on the transport of Cryptosporidium parvum oocysts in a ferric oxyhydroxide-coated quartz sand saturated porous medium. *Water research, 44*(4), 1104-1113.

Allam, A., Tawfik, A., El-Saadi, A., & Negm, A. (2016). Potentials of using duckweed (*Lemna gibba*) for treatment of drainage water for reuse in irrigation purposes. *Desalination and Water Treatment*, *57*(1), 459-467.

Allos, B. M., & Blaser, M. J. (1995). *Campylobacter jejuni* and the expanding spectrum of related infections. *Clinical Infectious Diseases*, 1092-1099.

Amoatey, P., & Baawain, M. S. (2019). Effects of pollution on freshwater aquatic organisms. *Water Environment Research*, *91*(10), 1272-1287.

Anawar, H. M., Garcia-Sanchez, A., Kul Alam, M. T., & Rahman, M. M. (2008). Phytofiltration of water polluted with arsenic and heavy metals. *International Journal of Environment and Pollution*, *33*(2-3), 292-312.

Ateia, M., Yoshimura, C., & Nasr, M. (2016). In-situ biological water treatment technologies for environmental remediation: a Review. *Journal of Bioremediation* and *Biodegradation*, *7*(348), 2.

Austin, B., & Stobie, M. (1992). Recovery of *Serratia plymuthica* and presumptive *Pseudomonas pseudoalcaligenes* from skin lesions in rainbow trout, *Oncorhynchus mykiss* (Walbaum), otherwise infected with enteric redmouth. *Journal of Fish Diseases*, *15*(6), 541-543.

Badgley, B. D., Nayak, B. S., & Harwood, V. J. (2010). The importance of sediment and submerged aquatic vegetation as potential habitats for persistent strains of enterococci in a subtropical watershed. *Water research*, *44*(20), 5857-5866.

Bai, X., Zhu, X., Jiang, H., Wang, Z., He, C., Sheng, L., & Zhuang, J. (2020). Purification effect of sequential constructed wetland for the polluted water in urban river. *Water*, *12*(4), 1054.

Bates, B., Kundzewicz, Z., & Wu, S. (2008). *Climate change and water*. Technical Paper of the Intergovernmental Panel on Climate Change. IPCC Secretariat: Geneva, Switzerland, 2008; pp. 1–210.

Beaudeau, P., de Valk, H., Vaillant, V., Mannschott, C., Tillier, C., Mouly, D., & Ledrans, M. (2008). Lessons learned from ten investigations of waterborne gastroenteritis outbreaks, France, 1998–2006. *Journal of Water and Health*, *6*(4), 491-503.

Bekhit, H. M., El-Kordy, M. A., & Hassan, A. E. (2009). Contaminant transport in groundwater in the presence of colloids and bacteria: model development and verification. *Journal of Contaminant Hydrology*, *108*(3-4), 152-167.

Bengtsson-Palme, J., Larsson, D. J., & Kristiansson, E. (2017). Using metagenomics to investigate human and environmental resistomes. *Journal of Antimicrobial Chemotherapy*, *72*(10), 2690-2703.

Bentzon-Tilia, M., Sonnenschein, E. C., & Gram, L. (2016). Monitoring and managing microbes in aquaculture-Towards a sustainable industry. *Microbial biotechnology*, *9*(5), 576-584.

Black, R. E., Levine, M. M., Clements, M. L., Hughes, T. P., & Blaser, M. J. (1988). Experimental Campylobacter jejuni infection in humans. *Journal of infectious diseases*, *157*(3), 472-479.

Bode, G., Mauch, F., & Malfertheiner, P. (1993). The coccoid forms of Helicobacter pylori. Criteria for their viability. *Epidemiology & Infection*, *111*(3), 483-490.

Bottone, E. J. (1999). Yersinia enterocolitica: overview and epidemiologic correlates. *Microbes and infection*, *1*(4), 323-333.

Brookes, J. D., Antenucci, J., Hipsey, M., Burch, M. D., Ashbolt, N. J., & Ferguson, C. (2004). Fate and transport of pathogens in lakes and reservoirs. *Environment International*, *30*(5), 741-759.

Brookes, J. D., Hipsey, M. R., Burch, M. D., Regel, R. H., Linden, L. G., Ferguson, C. M., & Antenucci, J. P. (2005). Relative value of surrogate indicators for detecting pathogens in lakes and reservoirs. *Environmental science & technology*, *39*(22), 8614-8621.

Bucke, D., & Feist, S. W. (1993). Histopathological changes in the livers of dab, *Limanda limanda* (L.). *Journal of Fish Diseases*, *16*(4), 281-296.

Buswell, C. M., Herlihy, Y. M., Lawrence, L. M., McGuiggan, J. T., Marsh, P. D., Keevil, C. W., & Leach, S. A. (1998). Extended survival and persistence of Campylobacter spp. in water and aquatic biofilms and their detection by immunofluorescent-antibody and-rRNA staining. *Applied and environmental microbiology*, *64*(2), 733-741.

Bylund, J., Toljander, J., Lysén, M., Rasti, N., Engqvist, J., & Simonsson, M. (2017). Measuring sporadic gastrointestinal illness associated with drinking water–an overview of methodologies. *Journal of water and health*, *15*(3), 321-340.

Chandran, A., Varghese, S., Kandeler, E., Thomas, A., Hatha, M., & Mazumder, A. (2011). An assessment of potential public health risk associated with the extended survival of indicator and pathogenic bacteria in freshwater lake sediments. *International journal of hygiene and environmental health*, *214*(3), 258-264.

Cho, K. H., Pachepsky, Y. A., Kim, J. H., Guber, A. K., Shelton, D. R., & Rowland, R. (2010). Release of Escherichia coli from the bottom sediment in a first-order creek: Experiment and reach-specific modeling. *Journal of Hydrology*, *391*(3-4), 322-332.

Cornelis, G. R. (1994). Yersinia pathogenicity factors. *Current Topics in Microbiology and Immunology*, 192, 243–263.

Current, W. L., & Garcia, L. S. (1991). Cryptosporidiosis. *Clinical microbiology reviews*, *4*(3), 325-358.

Das, M. C., & Shrivastava, A. K. (1984). Fish mortality in Naini Tal Lake (India) due to pollution and parasitism. *Hydrobiological Journal*, *20*(4), 60-64.

Davies, B. R., Thoms, M., & Meador, M. (1992). An assessment of the ecological impacts of inter-basin water transfers, and their threats to river basin integrity and conservation. *Aquatic conservation: Marine and freshwater ecosystems*, *2*(4), 325-349.

Dechesne, M., Soyeux, E., Loret, J. F., Westrell, T., Stenström, T. A., Gornik, V., & Medema, G. J. (2006). Pathogens in source water. *Microbiological Risk Assessment: A Scientific basis for managing drinking water safety from source to tap*, 1-42.

Delpla, I., Baurès, E., Jung, A. V., & Thomas, O. (2011). Impacts of rainfall events on runoff water quality in an agricultural environment in temperate areas. *Science of the total environment*, *409*(9), 1683-1688.

Denchak, M. (2018). Flint Water Crisis: Everything You Need to Know. *The Natural Resources Defence Council (NRDC)*. Retrieved from https://www.nrdc.org/stories/flint-water-crisis-everything-you-need-know.

Depledge, M. H. (1996). Genetic ecotoxicology: an overview. *Journal of Experimental Marine Biology and Ecology*, *200*(1-2), 57-66.

Diana, T. C. (2007). *Studies on histological and histopathological aspects of the marine teleost Nemipterus japonicus Bloch with special reference to diseases* (Doctoral dissertation, Ph. D Thesis submitted to Andhra University, Visakhapatnam, India).

Edge, T. A., Hill, S., Seto, P., & Marsalek, J. (2010). Library-dependent and library-independent microbial source tracking to identify spatial variation in faecal contamination sources along a Lake Ontario beach (Ontario, Canada). *Water Science and Technology*, *62*(3), 719-727.

Ensink, J. H., Scott, C. A., Brooker, S., & Cairncross, S. (2010). Sewage disposal in the Musi-River, India: water quality remediation through irrigation infrastructure. *Irrigation and Drainage Systems*, *24*(1), 65-77.

Fan, X.G., Chua, A., Li, T.G., & Zeng, Q.S. (1998). Survival of Helicobacter pylori in milk and tap water. *Journal of Gastroenterology and Hepatology*, *13*(11), 1096-1098.

Ferguson, C. M., Charles, K., & Deere, D. A. (2008). Quantification of microbial sources in drinking-water catchments. *Critical Reviews in Environmental Science and Technology*, *39*(1), 1-40.

Fujioka, R. S., Solo-Gabriele, H. M., Byappanahalli, M. N., & Kirs, M. (2015). US recreational water quality criteria: a vision for the future, 7752-7776.

Fukushima, H., Saito, K., Tsubokura, M., & Otsuki, K. (1984). Yersinia spp. in surface water in Matsue, Japan. *Zentralblatt fur Bakteriologie, Mikrobiologie und Hygiene. 1. Abt. Originale B, Hygiene*, *179*(3), 236-247.

Furness, B. W., Beach, M. J., & Roberts, J. M. (2000). Giardiasis surveillance–United States, 1992–1997. *MMWR CDC Surveill Summ*, *49*(7), 1-13.

Furtado, C., Adak, G. K., Stuart, J. M., Wall, P. G., Evans, H. S., & Casemore, D. P. (1998). Outbreaks of waterborne infectious intestinal disease in England and Wales, 1992–5. *Epidemiology & Infection*, *121*(1), 109-119.

Gao, H., Xie, Y., Hashim, S., Akhtar Khan, A., Wang, X., & Xu, H. (2018). Application of microbial technology used in bioremediation of urban polluted river: a case study of Chengnan River, China. *Water*, *10*(5), 643.

García-Aljaro, C., Ballesté, E., Muniesa, M., & Jofre, J. (2017). Determination of crAssphage in water samples and applicability for tracking human faecal pollution. *Microbial Biotechnology*, *10*(6),1775–1780.

García-Ávila, F., Zhindón-Arévalo, C., Valdiviezo-Gonzales, L., Cadme-Galabay, M., Gutiérrez-Ortega, H., & del Pino, L. F. (2022). A comparative study of water quality using two quality indices and a risk index in a drinking water distribution network. *Environmental Technology Reviews*, *11*(1), 49-61.

Garzio-Hadzick, A., Shelton, D. R., Hill, R. L., Pachepsky, Y. A., Guber, A. K., & Rowland, R. (2010). Survival of manure-borne E. coli in streambed sediment: effects of temperature and sediment properties. *Water research*, *44*(9), 2753-2762.

George, I., Anzil, A., & Servais, P. (2004). Quantification of fecal coliform inputs to aquatic systems through soil leaching. *Water research*, *38*(3), 611-618.

Goodwin, C. S., Mendall, M. M., & Northfield, T. C. (1997). Helicobacter pylori infection. *The Lancet*, *349*(9047), 265-269.

Graham, D. Y. (1991). I. *Helicobacter pylori*: Its epidemiology and its role in duodenal ulcer disease. *Journal of gastroenterology and hepatology*, *6*(2), 105-113.

Gutierrez, L., & Nguyen, T. H. (2012). Interactions between rotavirus and Suwannee River organic matter: aggregation, deposition, and adhesion force measurement. *Environmental science & technology*, *46*(16), 8705-8713.

Hanson, L. A., & Grizzle, J. M. (1985). Nitrite-induced predisposition of channel catfish to bacterial diseases. *The Progressive Fish-Culturist*, *47*(2), 98-101.

Harmel, R. D., Karthikeyan, R., Gentry, T., & Srinivasan, R. (2010). Effects of agricultural management, land use, and watershed scale on *E. coli* concentrations in runoff and streamflow. *Transactions of the ASABE*, *53*(6), 1833-1841.

Harrison, R. M. (Ed.). (2001). *Pollution: causes, effects and control.* Royal society of chemistry.

Hata, A., Katayama, H., Kojima, K., Sano, S., Kasuga, I., Kitajima, M., & Furumai, H. (2014). Effects of rainfall events on the occurrence and detection efficiency of viruses in river water impacted by combined sewer overflows. *Science of the Total Environment*, *468*, 757-763.

Hedberg, C. W., & Osterholm, M. T. (1993). Outbreaks of food-borne and waterborne viral gastroenteritis. *Clinical microbiology reviews*, *6*(3), 199-210.

Hellberg, B., & Löfgren, S. (1988). Gastrointestinal disturbances due to an established bacterial contamination of a water distribution system. *Vatten (Sweden)*, 44, 277.

Höller, C., Witthuhn, D., & Janzen-Blunck, B. (1998). Effect of low temperatures on growth, structure, and metabolism of Campylobacter coli SP10. *Applied and environmental microbiology*, *64*(2), 581-587.

Hu, L., Hu, W., Zhai, S., & Wu, H. (2010). Effects on water quality following water transfer in Lake Taihu, China. *Ecological Engineering*, *36*(4), 471-481.

James, E., & Joyce, M. (2004). Assessment and management of watershed microbial contaminants. *Critical reviews in environmental science and technology*, *34*(2), 109-139.

Jamieson, R., Joy, D. M., Lee, H., Kostaschuk, R., & Gordon, R. (2005). Transport and deposition of sediment-associated Escherichia coli in natural streams. *Water Research*, *39*(12), 2665-2675.

Jamwal, P. (2018). Remediation of contaminated urban streams: A decentralized ecological wastewater treatment approach. In *Water Remediation*, Springer, Singapore, 29-41.

Jenkins, P. A. (1991). Mycobacteria in the environment. *Journal of applied bacteriology*, *70*, 137S-141S.

Jones, I. G., & Roworth, M. (1996). An outbreak of *Escherichia coli* 0157 and campylobacteriosis associated with contamination of a drinking water supply. *Public health, 110*(5), 277-282.

Jung, A. V., Le Cann, P., Roig, B., Thomas, O., Baurès, E., & Thomas, M. F. (2014). Microbial contamination detection in water resources: interest of current optical methods, trends and needs in the context of climate change. *International Journal of Environmental Research and Public Health, 11*(4), 4292-4310.

Kanu, I., & Achi, O. K. (2011). Industrial effluents and their impact on water quality of receiving rivers in Nigeria. *Journal of applied technology in environmental sanitation, 1*(1), 75-86.

Kapperud, G. A., & Rosef, O. (1983). Avian wildlife reservoir of Campylobacter fetus subsp. jejuni, Yersinia spp., and Salmonella spp. in Norway. *Applied and Environmental Microbiology, 45*(2), 375-380.

Khan, R. A. (1987). Crude oil and parasites of fish. *Parasitology Today, 3*(4), 99-100.

Khan, R. A., & Thulin, J. (1991). Influence of pollution on parasites of aquatic animals. *Advances in parasitology, 30*, 201-238.

Kirk, R. S., & Lewis, J. W. (1993). An evaluation of pollutant induced changes in the gills of rainbow trout using scanning electron microscopy. *Environmental Technology, 14*(6), 577-585.

Knowles, P., Dotro, G., Nivala, J., & García, J. (2011). Clogging in subsurface-flow treatment wetlands: occurrence and contributing factors. *Ecological Engineering, 37*(2), 99-112.

Koenraad, P. M. F. J., Rombouts, F. M., & Notermans, S. H. W. (1997). Epidemiological aspects of thermophilic Campylobacter in water-related environments: a review. *Water environment research, 69*(1), 52-63.

Kong, Y., Yuan, J., Wang, Z., Yao, S., & Chen, Z. (2009). Application of expanded graphite/attapulgite composite materials as electrode for treatment of textile wastewater. *Applied Clay Science, 46*(4), 358-362.

Korajkic, A., McMinn, B. R., & Harwood, V. J. (2018). Relationships between microbial indicators and pathogens in recreational water settings. *International Journal of Environmental Research and Public Health, 15*(12), 2842.

Kramer, M. H., Sorhage, F. E., Goldstein, S. T., Dalley, E., Wahlquist, S. P., & Herwaldt, B. L. (1998). First reported outbreak in the United States of cryptosporidiosis associated with a recreational lake. *Clinical Infectious Diseases, 26*(1), 27-33.

Landsberg, J. H. (1995). Tropical reef-fish disease outbreaks and mass mortalities in Florida, USA: what is the role of dietary biological toxins. *Diseases of Aquatic Organisms, 22*(2), 83-100.

LeChevallier, M. W., Norton, W. D., & Lee, R. G. (1991a). Occurrence of Giardia and Cryptosporidium spp. in surface water supplies. *Applied and Environmental Microbiology, 57*(9), 2610-2616.

LeChevallier, M.W., Norton, W.D., & Lee, R.G. (1991b). *Giardia* and *Cryptosporidium* spp. in filtered drinking water supplies. *Applied and Environmental Microbiology*, **57**, 2617-2621.

Li, D., Pi, J., Zhang, T., Tan, X., & Fraser, D. J. (2018). Evaluating a 5-year metal contamination remediation and the biomonitoring potential of a freshwater gastropod along the Xiangjiang River, China. *Environmental Science and Pollution Research*, *25*(21), 21127-21137.

Li, Y., Acharya, K., & Yu, Z. (2011). Modeling impacts of Yangtze River water transfer on water ages in Lake Taihu, China. *Ecological Engineering*, *37*(2), 325-334.

Lothrop, N., Bright, K. R., Sexton, J., Pearce-Walker, J., Reynolds, K. A., & Verhougstraete, M. P. (2018). Optimal strategies for monitoring irrigation water quality. *Agricultural Water Management*, *199*, 86-92.

Loveland, J. P., Ryan, J. N., Amy, G. L., & Harvey, R. W. (1996). The reversibility of virus attachment to mineral surfaces. *Colloids and Surfaces A: Physicochemical and Engineering Aspects*, *107*, 205-221.

Malins, D. C., McCain, B. B., Landahl, J. T., Myers, M. S., Krahn, M. M., Brown, D. W., & Roubal, W. T. (1988). Neoplastic and other diseases in fish in relation to toxic chemicals: an overview. *Aquatic Toxicology*, *11*(1-2), 43-67.

Marshall, M. M., Naumovitz, D., Ortega, Y., & Sterling, C.R. (1997). Waterborne protozoan pathogens. *Clinical Microbiology Reviews*, *10*, 67-85.

Martínez, A., Alvarez-Vázquez, L. J., Vázquez-Méndez, M. E., & Vilar, M. A. (2010). Optimal control for river pollution remediation. In *Numerical Mathematics and Advanced Applications 2009*, Springer, Berlin, Heidelberg, 627-635.

McBride, G. B., Stott, R., Miller, W., Bambic, D., & Wuertz, S. (2013). Discharge-based QMRA for estimation of public health risks from exposure to stormwater-borne pathogens in recreational waters in the United States. *Water research*, *47*(14), 5282-5297.

Meng, J., & Doyle, M. P. (1997). Emerging issues in microbiological food safety. *Annual review of nutrition*, *17*(1), 255-275.

Moe, C. L. (1996). Waterborne transmission of infectious agents In: Manual of Environmental Microbiology, Hurst, C.J., Knudsen, G.R., McInerney, M.J., Stetzenbach, L.D., and Walter, M.V., Eds., ASM Press, Washington, D.C, 136.

Möller, H. (1987). Pollution and parasitism in the aquatic environment. *International Journal for Parasitology*, *17*(2), 353-361.

Muirhead, R. W., Collins, R. P., & Bremer, P. J. (2006). Interaction of Escherichia coli and soil particles in runoff. *Applied and Environmental Microbiology*, *72*(5), 3406-3411.

Naik, S. R., Aggarwal, R., Salunke, P. N., & Mehrotra, N. N. (1992). A large waterborne viral hepatitis E epidemic in Kanpur, India. *Bulletin of the World Health Organization*, *70*(5), 597.

Nevers, M. B., Byappanahalli, M. N., Morris, C. C., Shively, D., Przybyla-Kelly, K., Spoljaric, A. M., & Roseman, E. F. (2018). Environmental DNA (eDNA): A tool for quantifying the abundant but elusive round goby (*Neogobius melanostomus*). *PloS one*, *13*(1), e0191720.

Okoye, C. O., Nyaruaba, R., Ita, R. E., Okon, S. U., Addey, C. I., Ebido, C. C., & Chukwudozie, K. I. (2022). Antibiotic resistance in the aquatic environment: analytical techniques and interactive impact of emerging contaminants. *Environmental Toxicology and Pharmacology*, 103995.

Oliver, D. M., Porter, K. D., Pachepsky, Y. A., Muirhead, R. W., Reaney, S. M., Coffey, R., & Quilliam, R. S. (2016). Predicting microbial water quality with models: over-arching questions for managing risk in agricultural catchments. *Science of the Total Environment*, *544*, 39-47.

Ooi, W. W., Zimmerman, S. K., & Needham, C. A. (1995). *Cyclospora* species as a gastrointestinal pathogen in immunocompetent hosts. *Journal of clinical microbiology*, *33*(5), 1267-1269.

Pachepsky, Y. A., Allende, A., Boithias, L., Cho, K., Jamieson, R., Hofstra, N., & Molina, M. (2018). Microbial water quality: monitoring and modeling. *Journal of Environmental Quality*, *47*(5), 931-938.

Palit, K., Rath, S., Chatterjee, S., & Das, S. (2022). Microbial diversity and ecological interactions of microorganisms in the mangrove ecosystem: Threats, vulnerability, and adaptations. *Environmental Science and Pollution Research*, 1-46.

Peilin, G., Meng, C., Lichao, Z., Yuejun, S., Minghao, M., & Lingyun, W. (2019). Study on water ecological restoration technology of river. In *IOP Conference Series: Earth and Environmental Science, 371*, 032025, IOP Publishing.

Peters, N., Köhler, A., & Kranz, H. (1987). Liver pathology in fishes from the lower Elbe as a consequence of pollution. *Diseases of Aquatic Organisms 2 (2)*, 87-97.

Premdas, P. D., & Metcalfe, C. D. (1994). Regression, proliferation and development of lip papillomas in wild white suckers, *Catostomus commersoni*, held in the laboratory. *Environmental biology of fishes*, *40*(3), 263-269.

Rabold, J. G., Hoge, C., Shlim, D., Kefford, C., Rajah, R., & Echeverria, P. (1994). *Cyclospora* outbreak associated with chlorinated drinking water. *The Lancet*, *344*(8933), 1360-1361.

Radspinner, R. R., Diplas, P., Lightbody, A. F., & Sotiropoulos, F. (2010). River training and ecological enhancement potential using in-stream structures. *Journal of Hydraulic Engineering*, *136*(12), 967-980.

Reder, K., Alcamo, J., & Flörke, M. (2017). A sensitivity and uncertainty analysis of a continental-scale water quality model of pathogen pollution in African rivers. *Ecological Modelling*, *351*, 129-139.

Rieke, J., Eitel, F., Weygandt, M., Haynes, J. D., & Ritter, K. (2018). Visualizing convolutional networks for MRI-based diagnosis of Alzheimer's disease. In *Understanding and*

*Interpreting Machine Learning in Medical Image Computing Applications* (pp. 24-31). Springer, Cham.

Rodsaether, M. C., Olafsen, J., Raa, J., Myhre, K., & Steen, J. B. (1977). Copper as an initiating factor of vibriosis (*Vibrio anguillarum*) in eel (*Anguilla anguilla*). *Journal of Fish Biology*, *10*(1), 17-21.

Rosal, R., Rodríguez, A., Perdigón-Melón, J. A., Petre, A., García-Calvo, E., Gómez, M. J., & Fernández-Alba, A. R. (2010). Occurrence of emerging pollutants in urban wastewater and their removal through biological treatment followed by ozonation. *Water research*, *44*(2), 578-588.

Rose, J. B. (1988). Occurrence and significance of Cryptosporidium in water. *Journal-American Water Works Association*, *80*(2), 53-58.

Rose, J. B., Gerba, C. P., & Jakubowski, W. (1991). Survey of potable water supplies for Cryptosporidium and Giardia. *Environmental Science & Technology*, *25*(8), 1393-1400.

Schiemann, D. A. (1990). Yersinia enterocolitica in drinking water. In *Drinking water microbiology* (pp. 322-339). Springer, New York, NY.

Shahamat, M., Mai, U., Paszko-Kolva, C., Kessel, M., & Colwell, R. R. (1993). Use of autoradiography to assess viability of Helicobacter pylori in water. *Applied and environmental microbiology*, *59*(4), 1231-1235.

Singh, N., & Yu, V. (1994). Potable water and *Mycobacterium avium* complex in HIV patients: Is prevention possible? *The Lancet*, *343*(8906), 1110-1111.

Smith, H. V., Robertson, I.J., & Ongerth, J.E. (1995). Crytosporidiosis and Giardiasis: The Impacts of Waterborne Transmission. *Journal* of *water* supply research and technology-*Aqua*, *44*(6), 258-274.

Snieszko, S. F. (1974). The effects of environmental stress on outbreaks of infectious diseases of fishes. *Journal of Fish Biology*, *6*(2), 197 208.

Sorvillo, F. J., Fujioka, K., Nahlen, B., Tormey, M. P., Kebabjian, R., & Mascola, L. (1992). Swimming-associated cryptosporidiosis. *American Journal of Public Health*, *82*(5), 742-744.

Soupir, M. L., & Mostaghimi, S. (2011). Escherichia coli and enterococci attachment to particles in runoff from highly and sparsely vegetated grassland. *Water, Air, & Soil Pollution*, *216*(1), 167-178.

Steiner, T. S., Thielman, N. M., & Guerrant, R. L. (1997). Protozoal agents: what are the dangers for the public water supply? *Annual review of medicine*, *48*(1), 329-340.

Stelzer, W., & Jacob, J. (1991). A study of Campylobacter in sewage, sewage sludge and in river water. *Water Science and Technology*, *24*(2), 117-120.

Tauxe, R. V. (1992). Epidemiology of Campylobacter jejuni infections in the United States and other industrialized nations. *Campylobacter jejuni: current status and future trends.*

Theron, J., & Cloete, T. E. (2002). Emerging waterborne infections: contributing factors, agents, and detection tools. *Critical Reviews in Microbiology*, *28*(1), 1-26.

Tong, C. H., Yang, X. E., & Pu, P. M. (2003). Effects and mechanism of hydrophytes on control of release of nutrient salts in lake sediment. *Journal of Agro-Environment Science*, *22*(6), 673-676.

Vethaak, A. D. (1992). Diseases of flounder (*Platichthys flesus* L.) in the Dutch Wadden Sea, and their relation to stress factors. *Netherlands Journal of Sea Research*, *29*(1-3), 257-271.

Vethaak, A. D., & Jol, J. G. (1996). Diseases of flounder *Platichthys flesus* in Dutch coastal and estuarine waters, with particular reference to environmental stress factors. I. Epizootiology of gross lesions. *Diseases of Aquatic Organisms*, *26*(2), 81-97.

Vethaak, A. D., Jol, J. G., Meijboom, A., Eggens, M. L., Rheinallt, T. A., Wester, P. W., & Marquenie, J. M. (1996). Skin and liver diseases induced in flounder (*Platichthys flesus*) after long-term exposure to contaminated sediments in large-scale mesocosms. *Environmental Health Perspectives*, *104*(11), 1218-1229.

Voigt, H. R. (1994). Fish surveys in the Väike Väin Strait between the Islands of Saaremaa and Muhu, western Estonia. *Proceedings of the Estonian Academy of Science and Ecology*, *4*, 128-135.

Wakelin, S. A., Colloff, M. J., & Kookana, R. S. (2008). Effect of wastewater treatment plant effluent on microbial function and community structure in the sediment of a freshwater stream with variable seasonal flow. *Applied and environmental microbiology*, *74*(9), 2659-2668.

Wang, J., Liu, X. D., & Lu, J. (2012). Urban river pollution control and remediation. *Procedia Environmental Sciences*, *13*, 1856-1862.

Wyer, M. D., Kay, D., Watkins, J., Davies, C., Kay, C., Thomas, R., & Moore, H. (2010). Evaluating short-term changes in recreational water quality during a hydrograph event using a combination of microbial tracers, environmental microbiology, microbial source tracking and hydrological techniques: a case study in Southwest Wales, UK. *Water research*, *44*(16), 4783-4795.

Xiao, L., Ouyang, H., Li, H., Chen, M., Lin, Q., & Han, B. P. (2010). Enclosure study on phytoplankton response to stocking of silver carp (*Hypophthalmichthys molitrix*) in a eutrophic tropical reservoir in South China. *International review of hydrobiology*, *95*(4-5), 428-439.

Xie, X., Qian, X., Zhang, Y., Qian, Y., & Tian, F. (2009). Effect on Chaohu Lake water environment of water transfer from Yangtze River to Chaohu Lake. *Research of Environmental Sciences*, *22*(8), 897-901.

Xue, N., Wang, L., Li, W., Wang, S., Pan, X., & Zhang, D. (2020). Increased inheritance of structure and function of bacterial communities and pathogen propagation in plastisphere along a river with increasing antibiotics pollution gradient. *Environmental Pollution*, *265*, 114641.

Yadav, A., Neraliya, S., & Singh, R. (2005). Effect of fertilizer industrial effluent on the behaviour and morphology of fresh water catfish, *Heteropeneustes fossilis* (Bloch). *The Proceedings of the National Academy* of *Sciences*, *India, 75*(111), 191-195.

Yakasai, I. A., Salawu, F., & Musa, H. (2004). Concentrations of Lead in Ahmadu Bello University Dam Raw, Treated (Tap) and ABUCONS Pure Waters. *Chemclass Journal, 1*, 86-90.

Zhang, L. (2016). A Study of Polluted River Remediation by Aeration. In *Proceedings of the 6th International Asia Conference on Industrial Engineering and Management Innovation*, Atlantis Press, Paris, 451-461.

Zhao, Y., Yang, Z., Xia, X., & Wang, F. (2012). A shallow lake remediation regime with *Phragmites australis*: Incorporating nutrient removal and water evapotranspiration. *Water research*, *46*(17), 5635-5644.

Zheng, L., & Wang, H. (2017). Effect of combined ecological floating bed for eutrophic lake remediation. In *AIP Conference Proceedings*, AIP Publishing LLC, *1839*, 020058.

Ziegert, E., & Diesterweg, I. (1990). The occurrence of *Yersinia enterocolitica* in sewage. *Zentralblatt fur Mikrobiologie*, *145*(5), 367-375.

# Chapter - 9

# Pesticide Removal Techniques in Lakes, Factors Governing Efficient Lake Decontamination

**[1*]Sameena Khan, [1]Imtiyaz Qayoom, [1]Adnan Abubakr, [1]Taranum Mushtaq, [2]Farheen Saba and [1]Mehjabeen Akhtar**

[1]Division of Aquatic Environmental Management,
Faculty of Fisheries, SKUAST-K, Rangil-190006, Ganderbal, J&K, India
[2]Department of Zoology, School of Sciences, MANUU Hyderabad - 500032, India
[*]Corresponding email: sameena.k003@gmail.com

## Abstract

Pesticides continue to contaminate waterbodies and their level is increasing day by day. Severe problems associated with pesticide residues are well established and need to be addressed. There are many ways to get rid of pesticide residues in waterbodies, classified as physical, chemical and biological treatments. This chapter reviews in detail all the methods of lake decontamination and in comparison, their efficiency to sequester pesticides from various aquatic compartments.

**Keywords**: Lakes, Pesticides, Contamination, Removal techniques, Decontamination

## Abbreviations

(GAC) = Granular Activated Carbon

(PAC) = Powdered Activated Carbon

(CC) = Carbon Cloth

(CF) = Carbon Fibers

(BC) = Black Carbon

(ACC) = Activated Carbon Composites
(CAC) = Cobra Advanced Composites
(AN) = acrylonitrile
(DVB) = divinylbenzene
(MAN) = methacrylonitrile
(HDPy) = hexadecylpyridinium cation
(CNTs) = carbon nanotubes
(SWCNTs) = single-walled carbon nanotubes
(MWCNTs) = multi-walled carbon nanotubes
(PAHs) = polyaromatic hydrocarbons
(IESFs) = Iron-enhanced sand filters
(CSF) = coagulation–sedimentation-filters
(AOPs) = Advanced oxidation processes
(eAOP) = electrochemical advanced oxidation processes
(cAOP) = catalytic advanced oxidation processes
(pAOP) = photo advanced oxidation processes
(DOC) = dissolved organic carbon
(ZVI) = zero-valent iron activating
(MBR) = Membrane bioreactor
(AnMBR) = Anaerobic membrane bioreactor
(BPS) = Bio-purification systems

## Introduction

The natural environment is loaded by contaminants due to anthropogenic practices and industrial outputs leading to public health concern (Arihilam and Arihilam, 2019). Pharmaceuticals (hormones, antibiotics, and others), cosmetics, synthetic dyes, and pesticides are among the emerging contaminants, forming major concern in many areas around the world (Gomes *et al.*, 2018). Worldwide production of these micro-contaminants has reached around 500 million tons/year (Thomaidis *et al.*, 2012). Pesticides, for instance, have been proven in many studies as toxic substances to humans and the environment (Rasheed *et al.*, 2019). Besides, many pesticides can be transported in water and air to contaminate areas far from their point source

(Lofrano *et al.*, 2020). The erratic use of pesticides has affected ecosystems, which threaten birds, wildlife, domestic animals, fish, and livestock (Coppock and Dziwenka, 2020). The unique chemical structure of pesticides or their patterns of use by the society and their interaction with the environment are ways of their characterization. Pesticide structures forms substitute for specific molecules in targeted biological reactions by performing the function of imitation. In view of two criteria, they can be classified mainly by: chemical classes and target organism.

The general population is exposed to the residues of pesticides, including physical and biological degradation products in air, water, and food. The primary risk factor is genotoxic potential, giving rise to enduring effects such as carcinogenic and reproductive toxicology. The majority of pesticides has been tested in a wide variety of mutagenicity assays covering gene mutation, chromosomal alteration, and DNA damage, therefore, have been considered as potential chemical mutagens (Bolognesi, 2003). The use of un-prescribed pesticides in inappropriate doses is not only disturbing the soil conditions but also destroying the healthy pool of bio-control agents that normally co-exist with the vegetation.

The progress of the synthetic organic chemical industry was seen around the time of World War II that marked the modern era of chemical pest control. Organochlorine compounds being the first synthetic organic pesticides include dichlorodiphenyltrichloroethane (DDT), whose commercial production began in 1943. The production and use of synthetic organic pesticides rapidly increased during the mid-1940s. It has been reported that today, more than 500 different formulations of pesticides are being used in the environment, and agriculture holds the largest single share of pesticides use (Gavrilescu, 2005).In the modern agricultural era, the levels of pesticides in water have increased due to their excessive use, and hence suitable water treatment method for pesticide removal has become necessary. Subsequently, choosing a particular method of removal of pesticide depends on the type of pesticide and the efficiency of the treatment process.

There are various treatment techniques for the removal of pesticides from water, which are thoroughly discussed in this chapter. The methods that have been utilized to instigate and elucidate the problem of the removal of pesticides from water include the current treatment techniques encompassing a combination of physical, chemical, and biological. However, each treatment technique is having advantages and limitations in various terms viz., efficiency, operability, reliability, environmental impact, pre-treatment requirements, and the production of sludge and toxic byproducts. Furthermore, capital and operational costs are additional factors that govern the boundaries of treatment techniques.

## Physical Treatment Techniques

### Adsorption

A surface phenomenon that depends on the number of sites available, porosity and specific surface area of adsorbent as well as various types of interactions can be termed as adsorption. By and large, carbonaceous materials have a distinctive place among the key adsorbents, as they are known, for a long time to be capable of adsorbing various organic compounds. Due to the high surface area and porosity, activated carbon is very effective in removing different varieties of pesticides from water and wastewater. It is a versatile material that can be applied in many technological processes as shown in table 1.

**Table 1: Carbonaceous AdsorbentsUsed for Pesticide Removal in Lakes**

| S. No. | Type of Activated Carbon Materials | Reference |
|---|---|---|
| 1. | Granular Activated Carbon (GAC) | Domingues *et al.*, 2005; Kim *et al.*, 2008; Sotelo *et al.*,2002; Kitous *et al.*, 2009 |
| 2. | Powdered Activated Carbon (PAC | Kouras *et al.*,1998; Hu*et al.*,1998; Matsui*et al.*, 2002; Fernndez-Prez *et al.*, 2005; Sarkar *et al.*, 2007; Ohnoa *et al.*, 2008; Qiu *et al.*, 2009; Castro *et al.*, 2009; Hamadi*et al.*, 2004. |
| 3. | Carbon Cloth (CC) | Ayranci and Hoda, 2004; Ayranci and Hoda, 2005; Lopez-ramon *et al.*, 2007. |
| 4 | Carbon Fibers (CF) | Lopez-ramon *et al.*,2007; Faur*et al.*, 2005; Martian-Gulloanand Fent, 2001. |
| 5. | Black Carbon (BC) | Qiu *et al.*, 2009. |
| 6. | Activated Carbon Composites (ACC) | Castro *et al.*, 2009 |
| 7. | Commercial Activated Carbon (CAC) | Ayranci and Hoda,2004; Ayranci and Hoda, 2005; Lopez-ramon *et al.*, 2007; Hamadi *et al.*, 2004 |

According to Kyriakopoulos and Doulia (2006), the forms GAC and PAC are reflectedas very capable and effective materials for the adsorption of a variety of pesticides and are the most used. Therefore, many researchers have tried to produce activated carbons for removal of various pollutants using renewable and cheaper precursors, which were mainly industrial and agricultural byproducts as shown in table2.

**Table 2: Adsorbents Used to Remove Pesticides**

| S.No. | Compound | Absorbents | Reference |
|---|---|---|---|
| 1. | 2, 4-D pesticide | date stones | Hameed *et al.*,2009 |
| 2 | phenoxyacetic acid pesticides | coals and coconut shells | Ignatowicz, 2009 |
| 3. | Endosulfan | Sal wood charcoal | Mishra and Patel, 2008 |
| 4. | Pesticide of the uracil group | PAC | Al Mardini and Legube, 2009 |

## Agricultural Wastes Adsorbents

Waste materials and byproducts from the agriculture and other industries form the sources of low-cost adsorbentsdue to their abundance in nature. In recent years, a new class of adsorbents called as lignocellulosic materialshas been investigated for the same purposes due to their attractiveness resulting from their availability, low cost, biodegradability and their ability to quantitatively accumulate heavy metalsand various organic compounds such as dyes and pesticides (Danish *et al.*, 2010; Ahmad *et al.*, 2007; Ahmad *et al.*, 2009a; Rafatullah *et al.*, 2009; Ibrahim *etal.*,2010; Sulaiman *et al.*, 2010; Ahmad *et al.*, 2009b; Nanseu-Njiki*et al.*, 2010; Akhtar *et al.*, 2007; Memon*et al.*, 2009b; Dulman and Cucu-Man, 2009; Febrianto *et al.*, 2009). Interactions with the hydroxyl and carboxyl groups particularly abundant in polysaccharides (cellulose and hemicelluloses) and lignin, forms the basis through which accumulation of these pesticides on agricultural adsorbents is generally achieved and both of whichconstitute about 90% of dry lignocellulosic materials (Ofomaja, 2008). Several recent publications reported the useof low-cost and locally available adsorbents are given below in table 3 with certain case studies as shown in table 4:

**Table 3: Various low-cost adsorbents used for pesticide removal**

| S.No. | Low-Cost Adsorbents | Reference |
|---|---|---|
| 1 | Ayous Sawdust | Nanseu-Njiki*et al.*, 2010. |
| 2 | Rice Husk | Akhtar *et al.*, 2007; Hsu and Pan, 2007 |
| 3 | Date Stones | El Bakouri *et al.*, 2009 |
| 4 | Watermelon Peels | Memon *et al.*, 2008 |
| 5 | Rice Bran | Akhtar *et al.*, 2007 |
| 6 | Pine Sawdust | Rodriguez-Cruz *et al.*, 2008 |

| 7 | Oak Sawdust | Rodriguez-Cruz *et al.*, 2008 |
|---|---|---|
| 8 | Tea Leaves | Islam *et al.*, 2009 |
| 9 | Wood Sawdust | Boussahel *et al.*, 2009 |
| 10 | Chestnut Shells | Memon *et al.*, 2007 |
| 11 | Bamboo Canes | El Bakouri *et al.*, 2008 |
| 12 | Straw | El Bakouri *et al.*, 2008; De Wilde *et al.*, 2008 |
| 13 | Mango Kernel | Memon *et al.*, 2009b |
| 14 | Peanut Shells | El Bakouri *et al.*, 2008 |
| 15 | And Peach Nut Shells | Memon *et al.*, 2009a |

**Table 4: Various pollutants adsorbed by some natural adsorbents**

| S. No. | Adsorbent used for removal of pesticides from water | Compound | Results | Reference |
|---|---|---|---|---|
| 1. | Ayous sawdust | Paraquat | Adsorbs the pollutant at ahigher rate | Nanseu-Njiki *et al.*, 2010 |
| 2 | Chemically and thermally treated watermelon peels | Methyl Parathion | Effective adsorbents | Memon *et al.*, 2009 |
| 3 | Different kinds of woods | Linuron and Metalaxyl | Higher by pine, having higher lignin content, than by oak | Rodriguez-Cruz *et al.*, 2008 |
| 4 | Rice Bran and Rice Husk | Triazophos | Effectively removal | Akhtar *et al.*, 2009 |

**Polymeric Adsorbents**

Alternative sorbents such as polymeric resins have been amalgamatedas potential alternatives to activated carbons due to high cost of regeneration of the mechanically fragile activated carbons. Macroporous (macroreticular) (co-) polymers of non-ionic polymeric resins are frequently used in water treatment (Tsyurupa *et al.*, 1995). Due to the large surface area and special functional groups tagged to a matrix of polymeric chains, hypercrosslinked polymeric adsorbents have been established that possess extensive applicability. Hypercross-linked resins are widely used in industries as well as analytical scales (Tsyurupa and Davankov, 2006). Several reports revealed that hypercross-linked polymers have high potential for the adsorption of organic and inorganic compounds, storage of hydrogen, etc. (Streat and Sweetland, 1998; Budd*et al.*, 2007) interpreting that they were successfully applied in the removal of both organic and inorganic compounds with their unique physical structures to remove organic compounds from aqueous solutions, to purify process streams, and to recover valuable compounds from aqueous solutions has been examined as thepractice of polymeric adsorbents (Garcia-Delgado *et al.*, 1992; Maity *et al.*, 1992; Casillas *et al.*, 1993; Kislenko *et al.*, 1996). So far, research work on adsorption on polymeric resins for the separation of pesticides from aqueous systems is very limited (Kyriakopoulos *et al.*, 2003; Biziuk *et al.*, 1996; Millet *et al.*, 1996; Rehana *et al.*, 1996; Doulia *et al.*, 1997; Waite *et al.*, 1999; Kyriakopoulos *et al.*, 2005; Qiua *et al.*, 2005; Aouada *et al.*, 2009). Amberlite adsorbent forms a category of polymeric resins that has been widely investigated for the removal of organic compounds, pharmaceuticals, and pesticides from aqueous solutions. These resins are either styrene-divinyl benzene copolymers or polymers of acrylic esters (such as XAD-2, XAD-4, and XAD-7 resins) as shown in table 5.

**Table 5: Various polymers used in pesticide removal**

| S. No. | Name of Polymer | Pesticide Removed | Results | Reference |
|---|---|---|---|---|
| 1 | polymers Macronet MN-150 | methomyl | Good removal | Chang *et al.*, 2008 |
| 2. | MN-500 | methomyl | Less reomval | Chang *et al.*, 2008 |
| 3. | XAD-4 (polystyrene–divinylbenzene copolymer) and the functionalized more hydrophilic XAD-7 (nonionic aliphatic acrylic polymer) | herbicides alachlor, amitrole, trifluralin, and prometryn | It was found that XAD-4 resin is more effective for hydrophobic herbicides, such as alachlor, prometryn, and trifluralin, while XAD-7 resin is preferable for more hydrophilicmolecules (such as amitrole molecule). | Kyriakopoulos *et al.*, 2005 |
| 4. | poly(acrylamide) and methylcellulose (PAAm-MC) hydrogels | paraquat | potentially viable absorbents | Aouada *et al.*, 2009 |

| 5. | six porous copolymers of acrylonitrile (AN)/ divinylbenzene (DVB) and methacrylonitrile (MAN)/DVB | | sorption of atrazine, imazapyr, and 2, 4-dichlorophenoxyacetic acid on AN and MAN polymers is largely dependent on the hydrophilicity of these compounds. The effect of polar, surface nitrile groups is more pronounced in AN series of polymers, and this effect decreases with an increase in the hydrophilic character of sorbates | Trochimczuk *et al.*,2003 |
|---|---|---|---|---|

## Industrial Wastes as Adsorbents

Industrial wastes such as sludge, fly ash, and carbon slurry are classified as low-cost materials because of their low cost, local availability and can be used as adsorbents for pesticides removal. The fly ash, a low-cost adsorbentis a solid waste from lignite coal-fired thermal power stations, and has shown significant adsorption capacity for organic pollutants (Wang, 2006). Few reports have highlighted fly ash use in the removal of pesticides from water and wastewater samples due to its sorption potential (Majumdar and Singh, 2007; Singh, 2009) and have recommended it (Sharma*et al.,* 2008). Coal fly ash has significantly high retention capacity for metribuzin, metolachlor, and atrazine (Singh, 2009), the maximum was Atrazine followed by metolachlor and metribuzin. The initial concentration of herbicide in the solutiondecides the herbicide sorption efficiency of fly ash and maximum removal of herbicide was observed at lower concentrations, which is generally encountered in the water samples. The study recommended that fly ash be subjugated as a low-cost adsorbent for chosen pesticides from wastewater and runoff water from agricultural soils.

## Inorganic Adsorbents

From the earliest days of civilization, natural clay minerals are well recognized and accustomed to mankind. Clay materials are strong adsorbents due to their low cost, abundance in most continents of the world, high sorption properties, and potential for ion exchange. They possess layered structure and are considered as host materials for the adsorbates and counter ions. In recent years, clay minerals such as cloisite, clinoptilolite, eluthrilithe, kerolite, faujasite, montmorillonite, and palygorskite are being used for their capacity to adsorb not only inorganic ions but also organic molecules. More recently, an alternative to activated carbonsuch as organoclay complex adsorbentshave been investigated as low-cost adsorbentswhich are often used in industrial and technological processes, have been proposed as adsorbents for the immobilization of industrial organic contaminants (Sanchez-Martin, 2006; Suciua and Capri, 2009), for the removal of pesticides from water (Cruz-Guzman *et al.*, 2005), for the development of slow-release pesticide formulations (Carrizosa *et al.*, 2000), or for the examination of bioavailability potential (Singh *et al.*, 2003) and photo stabilization of sorbed pesticides (Margulis *et al.*, 1992).

The effect of the modification of natural clay minerals (montmorillonite, illite, kaolinite, muscovite, sepiolite, and atapulgite) with hexadecylpyridinium cation (HDPY) on the adsorption-desorption of fungicides such as penconazole and metalaxyl was studied by Andrades *et al.*, 2004. The results obtainedpoints out the potential value of clay minerals modified with HDPY forimmobilizing pesticides with different degrees of hydrophobicity.

**Miscellaneous Adsorbents**

There are numerous other materials that have also been put to use for preparing alternative adsorbents. Carbon nanotubes (CNTs) discoveredby Iijima in 1991, have attracted great interest for their unique chemical structure and intriguing physical properties. They have been examined for use as adsorbents due to their high surface area and unique structure. It has been found that CNTs are more effective for the removal of dioxin than activated carbon because of strong interaction between dioxin and carbon nanotubes (Long and Yang, 2001). Carbon nanotubes (CNTs) have been referred as single-walled (SWCNTs) or multi-walled carbon nanotubes (MWCNTs) subject to the layers involved,

CNTs have been considered beneficial in pollution prevention strategies and are known to have widespread applications as environmental adsorbents and high-flux membranes (Mauter and Elimelech, 2008) and are also potentially important for in situ environmental remediation due to their uniqueproperties and high reactivity (Fugetsu *et al.*, 2004). Inquiries dealing with the sorption of organic contaminants, such as dioxin (Long and Yang, 2001), 1,2-dichlorobenzene (Peng *et al.*, 2006), polyaromatic hydrocarbons (PAHs) (Yang *et al.*, 2006), and *o*-xylene, *p*-xylene (Chin *et al.*, 2007), and reactive dyes (Wu, 2007) on CNTs proposesthat they may also be suitable materials for the pre-concentration and solidification of pollutants from large volumes of wastewater (Andrades *et al.*, 2004).

**Bio adsorbents**

Biosorption is a practice that is used to indicate a number of metabolism-independent processes (physical and chemical adsorption, electrostatic interaction, ion exchange, complexation, chelation, and micro precipitation taking place fundamentally in the cell wall rather than oxidation through anaerobic or aerobic metabolism (biodegradation). High selectivity and efficiency, cost effectiveness, and good removal performance forms the main attractions of biosorption. Abundant (sea weeds) or wastes from other industrial operations (fermentation wastes, activated sludge process wastes), forms raw materials and can be used as biosorbents, presenting performances often comparable with those of ion exchange resins. Both living and dead (heat killed, dried, acid, and/or otherwise chemically treated) biomass can be used to remove pesticides. However, upholding a viable biomass during adsorption is challengingdue to the requirement of a constant supply of nutrients and avoidance of organic toxicity to the microorganisms.

The profitable practice of using dead microbial cells in biosorption should be considered for water treatment as dead organisms are not affected by toxic wastes and do not require a continuous supply of nutrients, thereby, can be regenerated and reused for many cycles. Dead microbial cells may be stored or used for prolonged periods at room temperature without putrefaction. Their operation is easy and their regeneration is simple. Moreover, dead cells have been shown to accumulate pollutants to the same or greater extent of growing or resting cells.

The chemical nature of pollutant (species, size, and ionic charge), type of biomass, its preparation and its specific surface properties, and environmental conditions (pH, temperature, and ionic strength, existence of competing organic or inorganic ligands in solution) affects the mechanism of binding via inactivated biomass (Aksu, 2005). Due to the environmental persistence and toxicity of these chemicals, the capacity of microorganisms to metabolize some pesticides has received much thoughtfulness. Yet, in some cases, microbial metabolism of contaminants may produce toxic metabolites, a variety of microorganisms (many aerobic bacteria and fungi) such as *Pseudomonas pickettii*, *Alcalilgenes eutrophus*, *Desulfomonile tiedjei*, *Phanerochaete chrysosporium*, and so on are known to utilize organic pesticides as the sole carbon or energy source. However, due to the low biodegradability and toxicity or inhibition of organic pesticides to microorganisms, conventional activated sludge systems often fail to achieve high efficiency in removing pesticides from wastewater (Sotelo *et al.*, 2002, Gupta *et al.*, 2002, Aksu and Kabasakal, 2004).

## Chemical Treatment Techniques

There are varieties of chemical reactions that help us to convert the contaminants into safer chemicals, by hydrolyzing them. Salimi *et al.*, 2017 stated that, coagulation and advanced oxidation process (AOPs), including ozonation and Fenton treatmentare taken as the leading methods among the chemical methods. Photo-catalysis and/ or membrane techniques are the two methods that are usually used in combination with chemical treatments (Meghwal *et al.*, 2019).

### Iron-enhanced sand filters (IESFs)

Various researchers like Fairbairn *et al.* (2018) and (Westerhoff *et al.*, 2018) worked on stormwater samples considering the urban runoff or stormwater, a major source of contaminants. They found that upon filtration of the stormwater samples with IESFs, which consist of conventional sand filters with around 5% iron filings, the removal rate of the contaminants was between 26% and 100%. However, some of the moderately hydrophobic types had a removal rate of only 10 to 30%, when pesticides are mentioned. The large-scale study implies that IESFs is not a promising method when it comes to pesticides. IESFs method was used by Erickson *et al.* (2012) for dissolved phosphorus removal from stormwater since most practices of stormwater treatment do not capture phosphate, it was seen that average of 88% of phosphate was captured for at least 200 m of treated depth which is considered as more efficient than standard sand filtration technique.

### Chlorination

A chemical water treatment technique that is helps to put restrictions on bacterial growth in wastewater, helping in regulating odor and taste can be termed as chlorination. It has the advantage of being cheap and easy to implement. A study investigating chlorination as a pretreatment method of pesticides showed a negative impact. As the oxidation of micro-contaminants prior to pre-chlorination can cause the formation of intermediate byproducts that could be more toxic and less removable than the parent compounds (Ahmed *et al.*, 2017). When chlorination techniquewas used by Li *et al.* 2016 prior to filtration using powdered activated carbon with coagulation–sedimentation-filters (PAC-CSF), the result showed pre-chlorination of OPPs oxidized them into more toxic oxons that are hard to remove by PAC-CSF. Therefore, water contaminated with OPPs should not be treated by chlorination in order to avoid removal complications. Some case studies are given in table 6.

**Table 6: Chlorination Techniques**

| S.No. | Contaminant | Concentration | Conclusion | Reference |
|---|---|---|---|---|
| 1. | chlorpyrifos, diazinon, malathion, and tolclofos-methyl | 10 μM. | Did not remove, changed into compound that are hard to remove. | Li *et al*,. 2016 |
| 2. | Nine different organophosphate pesticides from water including methyl parathion | 100 ng/L | Formation of oxons and other oxidation products that are stable in water and require further toxicity tests. | Kamel *et al.*, 2009 |
| 3. | Isoproturon, diuron, and parathion methyl | 500 ng/L | 100% removed | Ormad *et al.*, 2008. |
| 4. | Atrazine | 551 ng/L | only 20% | Ormad *et al.*, 2008 |

## Advanced oxidation processes (AOPs)

The chemical practices that can degrade organic contaminants into harmless products. This technique neither transfer contaminants from one phase to another and nor produce massive amounts of sludge. The processes (AOPs) consist of the usage of oxidizing agents to oxidize contaminants (Wang and Zhuan, 2019). Hydroxyl radical (•OH) is considered one among the strongest oxidizing radicals used oxidizing contaminants. The Sulfate radical (SO4•-) has also been widely studied for the removal of organic contaminants from water (Tsydenova *et al.*, 2015).

Furthermore, having various advantages, including rapid reaction rates that leads to less retention time compared to other conventional treatment techniques, it does not require a large area for processing the needed flow rate for the system. However, it's high operating, and maintenance costs forms its biggest limitation. The AOPs are divided, in general, into ozone-based treatment, ultraviolet (UV)-based treatment, electrochemical advanced oxidation processes (eAOP), catalytic advanced oxidation processes (cAOP), and photo advanced oxidation processes (pAOP). Some case studies are given in table 7.

## Table 7: Case Studies of Advanced Oxidation Processes (Aops)

| S. No | Method | Contaminant | Contaminant Conc. | Process | Result | Reference |
|---|---|---|---|---|---|---|
| 1. | Free radical (sodium persulfate ($Na_2S_2O_8$) as oxidizing species) | 17 different pesticides (pymetrozine, flonicamid, imidacloprid, acetamiprid, cymoxanil, thiacloprid, spinosad, chlorantraniliprole, triadimenol, tebuconazole, fluopyram, difenoconazole, cyflufenamid, hexythiazox, spiromesifen, folpet and acrinathrin) | between 0.02 and 1.17 mg/L | Sunlight was used. Persulfate photolysis was used to produce ($SO_4^{\bullet -}$) free radicals that have the ability to oxidize pesticides into less harmful substances, which will ultimately lead to the mineralization of the parent pesticide into $CO_2$ and water. | The initial dissolved organic carbon (DOC) was reduced by 87%. | Vela *et al.* 2019 |
| 2. | iron-catalyzed photo-activation of the persulfate system | Atrazine | 4 mg/L | | Removed 90% of the pesticide atrazine | Popova *et al.*, 2019 |
| 3. | oxidizing agent; magnesium dioxide ($MnO_2$) by bisulfite ($HSO_3^-$) activation | methyl parathion and methyl paraoxon | | | Removal of 72% for methyl parathion and 86% for methyl paraoxon | Wang *et al.*, 2019 |
| 4. | Peroxymonosulfate (PMS) | 2,4-D | 20 mg/L | Activating PMS using photocatalytic activation in the presence of $TiO_2$ anchored on copper ferrite ($TiO_2$@$CuFe_2O_4$) | 97.2%. 2,4-D removal | Golshan *et al.*,2018. |
| 5. | advanced oxidation with zero-valent iron activating persulfate (ZVI/PS) of alachlor | DBPs | | (ZVI/PS) of alachlor process followed by chlorination | decreasing the amounts of DBPs | Wang *et al.*, 2020 |

## Photo-chemical Degradation

Ultraviolet (UV) light is electromagnetic radiation that is usually used for photodegradation of contaminants in wastewater. Based on its wavelength, UV-A (315–400 nm), UV-B (280–315 nm), UV-C (180–280 nm), and vacuum UV light (10–180 nm) are four categories, based on which UV light can be categorized. It is a known fact that lower the wavelength; the stronger its effect. Various UV lamps have been used in wastewater treatment plants. However, UV light is usually not enough on its own for contaminant degradation so it is used in combination with other techniques among which photochemical degradation has been widely studied (Goodwin *et al.*, 2017). Some case studies are given below in table 8:

## Table 8: Photo-Chemical Degradation of pesticides

| S. No. | Method | Contaminant | Contaminant Conc. | Process | Result | Reference |
|---|---|---|---|---|---|---|
| 1. | The Photodegradation of triazophos (10 mg/L) in the presence of titanium dioxide (TiO2) | triazophos | (10 mg/L) | | Promising results | Aungpradit *et al.*, 2007 |
| 2. | Fe2O3(0.5)/TiO2 | 2,4-dichlorophenoxyacetic acid | | Exposure of the photocatalyst to UV light leads to the excitement of the photogenerated electron from the valence band (VB) to the conduction band (CB) of the TiO2 and the photogenerated holes are left in the VB. These holes oxidize 2,4-dichlorophenoxyacetic acid directly to 2,4-dichlorophenol (2,4-DCP) before mineralizing it. The superoxide radicals are also formed due to the reduction of oxygen by the photogenerated holes.<br>Good charge transfer was promoted by the formation of heterojunction between Fe2O3 and TiO2 nanoparticles, which also suppressed the electron–hole recombination | Good results | Lee *et al.* (2017) |
| 3. | Photocatalytic degradation in the presence of TiO2 or zinc oxide (ZnO) as photocatalysts | Carbofuran | from 50 to 250 mg/L | | Removal in the presence of TiO2 was more effective than in the presence of ZnO | Mahalakshmi *et al.*, 2007 |
| 4. | ZnO/SnO2 nanocomposite as a photocatalyst for the degradation of triclopyr. | Triclopyr | | Heterojunction formation between ZnO and SnO2 which causes the recombination of charge carriers to decrease that in turn increase the efficiency of photodegradation activity. | | Yadav*et al.* 2019 |
| 5. | Reducing agent (Co3O4/MCM-41) to be used as a green nanocomposite photocatalyst | Acephate | | | Complete dgedation of pesticide | AbuKhadra *et al.* 2020 |

| 6. | Highly efficient and recyclable nanocomposite | Dicofol | | Highly efficient and recyclable nanocomposite photocatalyst containing MoS2/ZnS nanoparticles embedded in N/S doped graphite carbon throughhydrothermal and post-calcination method for the degradation | 85% degradation of dicofol was achieved | Ahamad *et al.* 2020 |
|---|---|---|---|---|---|---|

One of the studies has focused on the result of UV light combined with chlorination on the degradation of the pesticides, UV/chlorine is a common method that has gained attention in the last few years, and however, there is an apprehension of the formation of the DBPs. The study focused on a molluscicide, known as metaldehyde, and two herbicides (carbetamide and mecoprop). It was seen that the UV/chlorine combination is an effective and viable option, especially when the pH of the treatment was adjusted to lower DBPs. At high pH, the total removal of mecoprop was attained (Carra *et al.*, 2020). Additionally, photodegradation is not anappropriate approach for all types of organic contaminants, and the lab analysis should be conducted carefully before adopting any method on a large scale.

## Ozonation

The practice of using ozone for the removal of contaminant either by direct effects of the ozone molecules on the contaminants or indirectly by the oxidation effect of the free radicals resulting from the decomposition of ozone in water. The free radicals Produced are highly reactive and less selective than chemical oxidants. Ozone should be generated on-site considering the fact that ozone has a very short lifetime, which increases the cost of the treatment. Some case studies are listed below in table 9.

**Table 9: Ozonation of pesticides**

| S . No. | Method | Concentration | Result | Reference |
|---|---|---|---|---|
| 1. | 23 pesticides using this method dimethoate, chlortoluron, diuron, isoproturon, metoxuron, and vinclozolin including dikegulak. | | showed efficient removal for six pesticides, namely dimethoate, chlortoluron, diuron, isoproturon, metoxuron, and vinclozolin | Meijers *et al.* 1995 |
| | ozonation + H2O2 oxidation | | Good removal rate for all the tested pesticides except for dikegulak. | |

| | | | | |
|---|---|---|---|---|
| 2. | The removal of 40 different pesticides using ozonation | | The results showed 70% removal of isoproturon, 75% of diuron, and parathion methyl while only 50% of the atrazine level were removed. | Ormad *et al.* 2008. |
| | Ozonation + activated carbon and 20 mg Al/L | | Atrazine removal rate reached 90%, while 100% of the remaining three pesticides were removed. | |
| 3. | Effect of ozone on five types of pesticides in water (alachlor, atrazine, chlorfenvinphos, diuron, and isoproturon) using a pilot-scale reactor | Initial concentration of the five pesticides was 16.6 mg/L of atrazine; 20.1 mg/L of alachlor; 16.9 mg/L of diuron; 18.1 mg/L of isoproturon; and 18.1 mg/L of chlorfenvinphos. | All pesticides have completely disappeared from the water after a treatment period between 30 min (for isoproturon) and 270 min (for alachlor). However, the total organic carbon (TOC) levels showed only a 26% reduction after 1000 min of exposure to ozone. | Maldonado *et al.* 2006 |

Maldonado *et al.*, 2006 worked on five pesticides and found that ozonation alone is not an appropriate method for the total mineralization of the five tested pesticides, yet it can serve as a primary treatment tobreak down these organic molecules into more biodegradable ones or to facilitate photodegradation or oxidation by AOPs. Therefore, ozonation alone is not an appropriate method for the total decomposition. However, in another study, the ozonation technique was used with nano-ZnO (nZnO) to treat atrazine from contaminated water; the removal efficiency increased compared to ozonation.

## Fenton technology

Fenton technology is one of the most operative oxidation procedures for the elimination of organic contaminants. Fenton's reagents are a combination of hydrogen peroxide solution ($H_2O_2$) with ferrous iron (typically iron (II) sulfate, $FeSO_4$). Zhang *et al.*, 2019 gave the reactions (1 to 6) that illustrate the redox reactions taking place in a solution during Fenton's treatment

$$H_2O_2 + Fe^{2+} \rightarrow HO^{\cdot} + HO- + Fe^{3+} (1)$$

$$HO^{\cdot} + Fe^{2+} \rightarrow HO^{-} + Fe^{3+} (2)$$

$$HO^{\cdot} + H_2O_2 \rightarrow H_2O + HOO^{\cdot} (3)$$

$$HOO^{\cdot} + H_2O_2 \rightarrow H_2O + HO^{\cdot} + O_2 \ (4)$$

$$HO\bullet + \text{Polysaccharide-H} \rightarrow H_2O + \text{Polysaccharide}\bullet \ (5)$$

$$\text{Polysaccharide}\bullet + \text{Monomer} \rightarrow \text{Polysaccharide-g-polymer} \ (6)$$

98.5% to 100% efficacy has been shown by Fenton technology in removing organophosphate insecticide known as fenitrothion from pesticides contaminated wastewater. In addition to 97.1% to 100% removal rate of chlorfenvinphos (another organophosphorus pesticide) and 90% removal rate of the organochlorine pesticides (Barbusiński and Filipek, 2001).

The best AOPs were evaluated by Saritha *et al.* (2007) for the removal of 4-chloro-2-nitrophenol (4C-2-NP) as a model contaminant from the pesticide industry. The initial concentration of 4C-2-NP was 100 mg/L. The final target was always the total mineralization of the chemical to avoid more toxic intermediate products. Correspondingly, the reduction of chemical oxygen demand (COD) was monitored together with the 4C-2-NP concentration. It was seen that not all used AOPs were effective, UV light alone and $H_2O_2$ alone were not capable of total COD removal. However, 50% reduction was seen in COD when processes combined. The Better results were shown in the presence of $TiO_2$ as a catalyst with UV (80% reduction). 90% COD reduction was seen with the photo-Fenton treatment (UV/Fenton) considering the best results of this study. UV/Fenton techniques are the most affordable in view of the cost analysis among effective techniques.

There is the development of a new novel nitrogen-doped graphene-iron based electro-catalyst recently and was used as a heterogeneous electro-Fenton (EF) catalyst in a three-dimensional Fenton system in research carried by Ghanbarlou *et al.*, 2020. 2,6-dichlorobenzamide (BAM), 2-methyl-4-chlorophenoxy acetic acid (MCPA), and mecoprop; the three commonly used herbicides were the target pesticides of the study, the initial concentration of the three pesticides together was 50 mg/L. The new technology showed high removal rates when compared to other Fenton techniques with MCPA and mecoprop removal reaching 94%

## Biological Treatment Techniques

There are two types in which biological treatment can be divided viz., aerobic and anaerobic. Aerobic treatments are known to digest the dichlorinated pesticides. The method includes the oxidation and cleavage of the ether bond and the hydroxylation of the chlorophenol to form chloro-catechol, and once the aromatic ring is open, the compound can be easily digested by regular bacterial metabolism into water and carbon dioxide (Javaid *et al.*, 2016). Reductive dehalogenation is another process by which Di-chlorinated pesticides can also be absorbed anaerobically. The compound will be degraded into methane and carbon dioxide when the amount of chlorine atoms is low enough (Maltseva *et al.*, 1996). Since Biological digestion of pesticides is difficult but once established, the biodegradation system can be easily maintained. There is a necessity of pre-treatment in biological treatment techniques sometimes, for example, the breakdown of certain pesticides by photochemical degradation or enzymatic reactions might facilitate their biological digestion (Huang *et al.*, 2018).

Aerobic biological treatment can be adopted as a secondarytreatment as done by Becerra *et al.*, (2020), the results of the study showed that the biological treatment chosen had a high removal efficacy when Aerobic biological treatment was carried out after photocatalytic treatment of water contaminated with chlorpyrifos (insecticide).

## Treatment of fungicides with a bio-mixture augmented with ligninolytic fungi

Ligninolytic enzymes are the enzymes that are secreted by Ligninolytic fungi. These enzymes depolymerize and mineralize lignin. The addition of ligninolytic fungi to the bio-mixture that targets pesticide degradation is a reasonable option because some pesticides are incorporated into lignin-based formulations. Humic-rich substances, soil, microorganisms, and lignocellulosic substrate form the matrix of such bio-mixture for the colonization of ligninolytic fungi (Castillo Mdel *et al.*, 2008).

Research was carried out on the effect of BPS (biopurification system) on seven fungicides (tebuconazole, carbendazim, metalaxyl, triadimenol, Edifenphos, Epoxiconazole, and Fenbuconazoleafter bioaugmentation with the ligninolytic fungi Trametes versicolor) by Murillo-Zamora *et al* (2017). They found that Fungicides belong to different groups of agrochemicals among which triazole is the largest in number and the most persistent with a half-life in soil of 100 days. It was seen that there was not any significant improvement of the fungicide removal upon the addition of Trametes versicolor to the bio-mixture.

## Treatment using bio-augmented activated sludge

Quan *et al.* (2004) haveworked upon (2,4-DCP) and developed conventional activated sludge augmented by 2,4-dichlorophenol (2,4-DCP) degrading bacteria isolated from soil polluted by 2,4-DCP. This 2,4-DCP is a chlorinated derivative of phenol (Cl2C6H3OH) known as the precursor of the herbicide 2,4-D. When the initial concentration was between 10 and 100 mg/L, the removal efficiency of the sludge reach to 88% to 89%, showing the process has improved.

Carboneras *et al.* (2018) worked upon activated sludge developed from wastewater of an oil refinery was acclimated and used to digest fluoxil from the water. Fluoxil is herbicide withan organochlorinated active compound (oxyfluorfen). The initial concentrations tested were ranged between 85 and 500 mg/L. Although oxyfluorfen was not totally removed, the removal rate of the herbicide reached 80% after 70 h of exposure. The phosphorus herbicide known as glyphosate was also treated from water using activated sludge. Results showed that at a low initial concentration (0.1, 0.5, and 1 g/L), the herbicide was totally removed in 4, 13, and 18 h, respectively. However, at higher herbicide concentrations (2 and 5 g/L), the cell growth was inhibited in the sludge (Tazdait *et al.*, 2018).

## *In-Vitro* Treatment using Microbial Consortium

HCH belongs to the organochlorine pesticides. Hexachlorocyclohexane (HCH) is an insecticide made of a mixture of eight or more stereoisomers having environmentalhalf-life of 8 to 10 years making it difficult to be eliminated from water.

Murthy and Manonmani conducted lab-based researchin India in 2007 where the study focused on developing a bacterial consortium capable of digesting HCH in liquid culture. The bacteria which include*Pseudomonas, Burkholderia, Flavobacterium, and Vibrio* strains forms the most active consortium and were isolated upon enrichment of HCH contaminated sewage and soil.

Baghapour *et al.* conducted a study in 2013 on the biological treatment of atrazine (0.01–10 mg/L). it was seen that submerged biological aerated filter (SBAF) containing a consortium of adapting bacteria can remove atrazine from wastewater in a pilot-scale trial up to 97.9%.

PHs is broad-spectrum herbicides that inhibit photosynthesis, PHs is nowadays known as contaminants of coastal water, groundwater, and soil. Toxicity analysis showed that some of the PHs is carcinogenic. A study was conducted by Zhang *et al.*, in 2018 where bacterial consortium with Diaphorobacter sp. strain LR2014-1 and Achromobacter sp. strain ANB-1 were isolated from linuron mineralization consortium and was used to mineralize a combination of phenylurea herbicides (PHs). The PHs included in this water purification method had varied initial concentrations ranging from 4 to 56.4 μM, and they included: linuron, diuron, chlortoluron, fluometuron, and siduron. The degradation percentages after 72 h were 99.9%, 94.3%, 96.4%, 53.5%, and 88.3%, respectively. Moreover, it was prominent that the consortium successfully degraded most of the herbicides byproducts. Furthermore, Sorensen *et al.* (2008) combined Arthrobacter globiformis strain D47 as diuron-degrading organism and Variovorax sp. strain SRS16 as linuron-mineralizing organism to construct a linuron and diuron-mineralizing two-member consortium. The results showed that a consortium consisting of both strains can remediate diuron and linuron in the contaminated soil and water.

**Pressurized activated sludge**

In the pressurized activated sludge, the amount of soluble oxygen increases by enhancing total air Pressurized. Activated sludge is a highly efficient treatment technique that does not demand large land area. It is less costly when compared to AOPs, and more environmentally friendly when compared to chlorination. However, it requires a sludge disposal area as well as skilled labor for maintenance and operation (Luo *et al.*, 2014). The difficulty of treating water with high COD in large volumes due to the limited amount of dissolved oxygen under atmospheric pressure becomes the limitations of activated sludge as the aeration process increases the technology cost.

According to the Jin *et al.* (2010), the results showed more COD removal with the increase of the operating pressure, aeration time, and sludge concentration to reach an optimum level of 85.0% to 92.5% COD removal. Whereas Pan *et al.*, 2004 illustrated that the COD removal increased in such treatment with the increased air pressure, aeration time, and sludge concentration. At 0.2 MPa pressure and 6 h aeration time, the COD removal of initial concentration between 2500 and 5000 mg/L has reached to 92.5%.

**Anaerobic–aerobic biological treatment**

The effectiveness of bioremediation was checked through a laboratory-scale study that aimed to purify triadimenol from water. Triadimenol is a fungicide that blocks fungal ergosterol biosynthesis, which jeopardizes membrane integrity. To reach 96% triadimenol removal rate water has to be in an aerobic bioreactor at 22 ◦C for 172 days. While the same removal rate required 230 days in the anaerobic bioreactor at 30 ◦C. The biomass of active bacteria in the bioreactors increased when the alternation of anaerobic and aerobic biological processes took place and the efficacy of the treatmentwas improvedthereby lowering the retention time. The results showed that 12 h of anaerobic treatment followed by 24 h of aerobic treatment were the optimum combination for the treatment of simulated wastewater containing triadimenol in the laboratory (Shawaqfeh, 2010).

## Membrane bioreactor (MBR)

MBR is one of the latest technologies which combine membrane filtration techniques with biological treatment to treat in wastewater. Despite its high efficiency, it should be noted that this technique requires high energy, in addition to the cost of anti-fouling techniques needed to keep it functioning that might become its one of the limitations. Navaratna *et al.*, 2016 worked on a broad-spectrum herbicide known as ametryn (S-triazine), where it was seen that within 15 h a removal rate of 65% for a herbicide known as ametryn (S-triazine) of initial concentration from 1 to 4 mg/L was seen in an anoxic MBR. Increasing theretention time to 1.5, 2.5, and 7.5 days causes the removal rate to reach 83%, 92%, and 99%, respectively. Another study conducted by Monsalvo *et al.*, 2014 on organic contaminant removal by an anaerobic membrane bioreactor (AnMBR) showed that pesticides were difficult to be treated anaerobically with very low removal rates of atrazine (6.8%) and linuron (10.5%)

Pesticides are not the only chemicals used for pest control in agriculture, in order to avoid bacterial spoilage of crops, antibiotics are also added. Hence, pesticides-rich wastewater is also rich with antibiotics. Considering the mode of action of antibiotics, this might limit the action of bio-remediating bacteria in a bio-mixture. Moreover, when choosing the best bioremediation approach, the presence of antimicrobial agents with the rest of the pesticides should be taken into consideration. Isolating adapted strains from soil already exposed to agricultural chemicals might guarantee the presence of antimicrobial-resistant strains that are capable of digesting pesticides and probably antibiotics as well (Castillo-Gonzalez *et al.*, 2017).

## Various bio-purification systems (BPS)

Bio-mixtures are usually made of humic-rich substances to increase pesticide retention, soil, and microorganisms. It is a process where pesticides-containing wastewater is treated using microorganisms capable of digesting pesticides. Bacteria present in the bio-mixture can be of endogenous or exogenous origins. Soil samples collected from places rich with pesticides will have adapted microorganisms that would serve better as endogenous biodegrading microorganisms. However, due to biomass limitations of some endogenous species, selected exogenous species might be used (Karanasios *et al.*, 2012). It should be noted that this essentially considered technique might not easily digest pesticides using microorganisms because of the toxicity of pesticides on various bacteria and fungi (Goodwin *et al.*, 2017).

## Bibliography

Abu Khadra, M. R., Mohamed, A. S., El-Sherbeeny, A. M. and Elmeligy, M. A. 2020. Enhanced photocatalytic degradation of acephate pesticide over MCM-41/Co3O4 nanocomposite synthesized from rice husk silica gel and Peach leaves. *J. Hazard. Mater*. **389**(5): 122129.

Ahamad, T., Naushad, M., Al-Saeedi, S., Almotairi, S. and ALshehri, S. 2020. Fabrication of $MoS_2$/ZnS embedded in N/S doped carbon for the photocatalytic degradation of pesticide. *Mater. Lett.***263**: 127271.

Ahmad, A., Rafatullah, M. and Danish, M. 2007. Removal of Zn (II) and Cd (II) ions from aqueous solutions using treated sawdust of sissoo wood as an adsorbent. *Holz als Roh-undWerkstoff.***65**: 429–436.

Ahmad, A., Rafatullah, M., Sulaiman, O., Ibrahim, M. H., Chii, Y. Y. and Siddique, B. M. 2009a. Removal of Cu(II) and Pb(II) ions from aqueous solutions by adsorption on sawdust of meranti wood. *Desalination***247**: 636–646.

Ahmad, A., Rafatullah, M., Sulaiman, O., Ibrahim, M. H. and Hashim, R. 2009b. Scavenging behaviour of meranti sawdust in the removal of methylene blue from aqueous solution. *JHazard Mater*. **170**: 357–365.

Ahmed, M., Zhou, J., Ngo, H., Guo, W., Thomaidis, N. and Xu, J. 2017. Progress in the biological and chemical treatment technologies for emerging contaminant removal from wastewater: A critical review. *J. Hard Mater*. **323**: 274–298.

Akhtar, M., Hasany, S. M., Bhanger, M. I. and Iqbal, S. 2007. Lowcost sorbents for the removal of methyl parathion pesticide from aqueous solutions. *Chemosphere***66**: 1829– 1838.

Akhtar, M., Iqbal, S., Bhanger, M. I. and Moazzam, M. 2009. Utilization of organic by-products for the removal of organophosphorous pesticide from aqueous media. *J Hazard Mater*. **162**: 703–707.

Aksu, Z. 2005. Application of biosorption for the removal of organic pollutants: A review. *Process Biochem*.**40**: 997–1026.

Aksu, Z. and Kabasakal, E. 2004, Batch adsorption of 2, 4-dichlorophenoxyacetic acid (2, 4-D) from aqueous solution by granular activated carbon. *Sep PurifTechnol*. **35**: 223–240.

Andrades, M. S., Rodr´ıguez-Cruz, M. J., S´anchez-Mart´ın, M. and S´anchez-Camazano. 2004. Effect of the modification of natural clay minerals with hexadecylpyridinium cation on the adsorption-desorption of fungicides. *Intern J Environ Anal Chem*.**84**: 133–141.

Aouada, F. A., Pan, Z., Orts. W. J. and Mattoso, L. H. C. 2009. Removal of paraquat pesticide from aqueous solutions using a novel adsorbent material based on polyacrylamide and methylcellulose hydrogels. *J App Polym Sci*. **114**: 2139–2148

Arihilam, N. and Arihilam, E. 2019. Impact and control of anthropogenic pollution on the ecosystem – A review. *J. Biosci. Biotechnol. Discov*. **4**: 54–59.

Aungpradit, T., Sutthivaiyakit, P., Martens, D., Sutthivaiyakit, S. and Kettrup, A. A. F. 2007. Photocatalytic degradation of triazophos in aqueous titanium dioxide suspension: Identification of intermediates and degradation pathways. *J Hazard Mater.* **146**: 204–213.

Ayranci, E. and Hoda, N. 2004. Adsorption of bentazon and propanil from aqueous solutionsat the high area activated carbon-cloth. *Chemosphere* **57**: 755–762.

Ayranci, E. and Hoda, N. 2005. Adsorption kinetics and isotherms of pesticides onto activated carbon-cloth. *Chemosphere* **60**: 1600–1607.

Baghapour, M.A., Nasseri, S. and Derakhshan, Z. 2013. Atrazine removal from aqueous solutions using submerged biological aerated filter. J. Environ. *Health Sci. Eng.* **11**(1): 6.

Barbusiński, K. and Filipek, K. 2001. Use of Fenton's reagent for removal of pesticides from industrial wastewater. *Polish J. Environ. Stud.* **10**(4): 207-212.

Becerra, D., Arteaga, B. L., Ochoa, Y. E., Barajas-Solano, A. F., García-Martínez, J. B. and Ramírez, L. F. 2020. Coupling of heterogeneous photocatalysis and aerobic biological process of activated sludge to treat wastewater containing chlorpyrifos. *Ing. Compet.* **22**(1).

Biziuk, M., Namiesnik, J., Czerwinski, J., Gorlo, D., Makuch, B., Janicki, W., Polkowska, Z. and Wolska, L. 1996. Occurrence and extermination of organic pollutants in tap and surface waters of the Gdansk district. *J Chromatogr A.* **733**: 171–183.

Bolognesi, C. 2003. Genotoxicity of pesticides: A review of human biomonitoring studies. *Mutat Res.* **543**: 251–272.

Boussahel, R., Irinislimane, H., Harik, D. and Moussaoui, K.M. 2009. Adsorption, kinetics, and equilibrium studies on removal of 4, 4-DDT from aqueous solutions using low-cost adsorbents. *Chem Eng Comm.* **196**: 1547–1558.

Budd, P. M., Peter, M. and Butler, A. 2007. The potential of organic polymer-based hydrogen storage materials. *Phys Chem Chem Phys.* **9**: 1802–1808.

Carboneras, M. B., Villaseñor, J., Fernández-Morales, F. J., Rodrigo, M. A. and Cañizares, P. 2018. Biological treatment of wastewater polluted with an oxyfluorfen-based commercial herbicide. *Chemosphere* **213**: 244–251.

Carra, I., Fernandez Lozano, J., Autin, O., Bolton, J.R. and Jarvis, P. 2020. Disinfection by-product formation during UV/Chlorine treatment of pesticides in a novel UV-LED reactor at 285 nm and the mitigation impact of GAC treatment. *Sci. Total Environ.* **712**: 136413

Carrizosa, M. J., Calderon, M. J., Hermosin, M. C. and Cornejo, J. 2000. Organosmectites as sorbent and carrier of the herbicide bentazone. *Sci Total Environ.* **247**: 285–293.

Casillas, J. L., Martinez, M., Addo-Yobo, F. and Aracil, J. 1993. Modeling of the adsorption of cephalosporin C on the modified resins in a stirred tank. *Chem Eng J.* **52**: B71–B75.

Castillo Mdel, P., Torstensson, L. and Stenstrom, J. 2008. Biobeds for environmental protection from pesticide use–a review. *J. Agric. Food Chem.* **56**(15): 6206–6219.

Castillo-Gonzalez, H., Perez-Villanueva, M., Masis-Mora, M., Castro-Gutierrez, V. and Rodriguez-Rodriguez, C. E. 2017. Antibiotics do not affect the degradation of fungicides and enhance the mineralization of chlorpyrifos in biomixtures. *Ecotoxicol. Environ. Saf.* **139**: 481-487.

Castro, C. S., Guerreiro, M. C., Goncalves, M., Oliveir, L. C. A. and Anast´acio, A. S. 2009. Activated carbon/iron oxide composites for the removal of atrazine from aqueous medium. *J HazardMater*. **164**: 609–614.

Chang, C. F., Chang, C. Y., Hsu, K. E., Lee, S. C. and H¨oll, W. 2008. Adsorptive removal of the pesticide methomyl using hypercrosslinked polymers. *J Hazard Mater*. **155**: 295–304.

Chin, C. J. M., Shih, L. C., Tsai, H. J. and Liu, T. K. 2007. Adsorption of *o*-xylene and *p*-xylene from water by SWCNTs. *Carbon* **45**: 1254–1260.

Coppock, R. W. and Dziwenka, M. M. 2020. Threats to wildlife by chemical and warfare agents. In: Gupta, R.C. (Ed.), Handbook of Toxicology of Chemical Warfare Agents. *Academic Press, Boston*, pp. 1077–1087.

Cruz-Guzman, M., Celis, R., Hermosin, M. C., Koskinen, W. C. and Cornejo, J. 2005. Adsorption of pesticides from water by functionalized organobentonites. *J Agric Food Chem*. **53**: 7502–7511.

Danish, M., Sulaiman, O., Rafatullah, M., Hashim, R. and Ahmad, A. 2010. Kinetics for the removal of paraquat dichloride from aqueous solution by activated date (*Phoenix dactylifera*) stone carbon. *J Disp Sci Tech*. **31**: 248–259.

De Wilde, T., Mertens, J., Spanoghe, P., Ryckeboer, J., Jaeken, P. and Springael, D. 2008. Sorption kinetics and its effects on retention and leaching. *Chemosphere*.**72**: 509–516.

Domingues, V., Alves, A., Cabral, M. and Delerue-Matos, C. 2005. Sorption behaviour of bifenthrin on cork. *J Chromatogr A*. **1069:** 127–132.

Doulia, D., Hourdakis, A., Rigas, F. and Anagnostopoulos, E. 1997. Removal of atrazine from water by use of nonionic polymeric resins. *J Environ Sci Health A*.**32**: 2635– 2656.

Dulman, V. and Cucu-Man, S. M. 2009. Sorption of some textile dyes by beach wood sawdust. *J Hazard Mater*. **162**: 1457–1464.

El Bakouri, H., Morillo, J., Usero, J. and Ouassini, A. 2008. Potential use of organic waste substances as an ecological technique to reduce pesticide ground water contamination. *J Hydrol*. **353**: 335–342.

El Bakouri, H., Usero, J., Morillo, J., Rojas, R. and Ouassini, A. 2009. Drin pesticides removal from aqueous solutions using acid-treated date stones. *Bioresour Technol*. **100**: 2676–2684.

Erickson, A., Gulliver, J. and Weiss, P. 2012. Capturing dissolved phosphorus with iron enhanced sand filtration. *Water Res*. **46**(9): 6601–6608.

Fairbairn, D. J., Elliott, S. M., Kiesling, R. L., Schoenfuss, H. L., Ferrey, M. L. and Westerhoff, B. M. 2018. Contaminants of emerging concern in urban stormwater: Spatiotemporal patterns and removal by iron-enhanced sand filters (IESFs). *Water Res.* **145**: 332–345.

Faur, C., Pignon, H. M. and Cloirec, P. L. 2005. Multicomponent adsorption of pesticides onto activated carbon fibers. *Adsorption* **11**: 479–490.

Febrianto, J., Kosasih, A. N., Sunarso, J., Ju, Y., Indraswati, N. and Ismadji, S. 2009. Equilibrium and kinetic studies in adsorption of heavy metals using biosorbent: A summary of recent studies. *J Hazard Mater.* **162**: 616–645.

Fernndez-Prez, M., Villafranca-Snchez, M., Flores-Cspedes, F., Garrido-Herrera, F. J. and Prez-Garca, S. 2005. Use of bentonite and activated carbon in controlled release formulations of carbofuran. *J Agric Food Chem.* **53**: 6697–6703.

Fugetsu, B., Satoh, S., Shiba, T., Mizutani, T., Lin, Y., Terui, N., Nodasaka, Y., Sasa, K., Shimizu, K. and Akasaka, T. 2004. Caged multiwalled carbon nanotubes as the adsorbents for affinity-based elimination of ionic dyes. *Environ Sci Technol.* **38**: 6890–6896.

Garcia-Delgado, R. A., Cotoruelo, L. M. and Rodriguez, J. J. 1992. Adsorption of anionic mixtures by polymeric resins. *Sep Sci Technol.* **27**: 1065–1076.

Gavrilescu, M. 2005. Fate of pesticides in the environment and its bioremediation. *EngLife Sci.* **5**: 497–526.

Ghanbarlou, H., Nasernejad, B., Fini, M. N., Simonsen, M. E. and Muff, J., 2020. Synthesis of an iron-graphene based particle electrode for pesticide removal in three-dimensional heterogeneous electro-Fenton water treatment system. *Chem. Eng. J.* **395**: 125025.

Golshan, M., Kakavandi, B., Ahmadi, M. and Azizi, M. 2018. Photocatalytic activation of peroxymonosulfate by $TiO_2$ anchored on cupper ferrite (TiO2@CuFe2O4) into 2, 4-D degradation: Process feasibility, mechanism and pathway. *J. Hazard. Mater.* **359**: 325–337.

Gomes, I. B., Simões, L. C. and Simões, M. 2018. The effects of emerging environmental contaminants on stenotrophomonas maltophilia isolated from drinking water in planktonic and sessile states. *Sci. Total Environ.* **643**: 1348–1356.

Goodwin, L., Carra, I., Campo-Moreno, P. and Soares, A. 2017. Treatment Options for Reclaiming Wastewater Produced by the Pesticide Industry. *International Journal of Water and Wastewater Treatment* **4**(1).

Gupta, V. K., Jain, C. K., Ali, I., Chandra, S. and Agarwal, S. 2002. Removal of lindane and malathion from wastewater using bagasse fly ash-a sugar industry waste. *Water Res.* **36**: 2483–2490.

Hamadi, N. K., Swaminathan, S. and Chen, X. D. 2004. Adsorption of paraquat dichloride from aqueous solution by activated carbon derived from used tires. *J Hazard Mater B.* **112**: 133–141.

Hameed, B. H., Salman, J. M. and Ahmad, A. L. 2009. Adsorption isotherm and kinetic modeling of 2, 4-D pesticide on activated carbon derived from date stones. *J Hazard Mater.* **163**: 121–126.

Hsu, S. and Pan, T. 2007. Adsorption of paraquat using methacrylic acid-modified rice husk. *Bioresour Technol.* **98**: 3616–3621.

Hu, J. Y., Aizawa, T., Ookubo, Y., Morita, T. and Magara, Y. 1998. Adsorptive characteristics of ionogenic aromatic pesticides in water on powered activated carbon. *Water Res.* **32**: 2593–2600.

Huang, Y., Xiao, L., Li, F., Xiao, M., Lin, D., Long, X. and Wu, Z. Microbial Degradation of Pesticide Residues and an Emphasis on the Degradation of Cypermethrin and 3-phenoxy Benzoic Acid: A Review. *Molecules* **23**(9): 2313.

Ibrahim, M. M., Ngah, W. S. W., Norliyana, M. S., Daud, W. R. W., Rafatullah, M., Sulaiman, O. and Hashim, R. 2010. A novel agricultural waste adsorbent for the removal of lead (II) ions from aqueous solutions. *J Hazard Mater.* **182**: 377–385.

Ignatowicz, K. 2009. Selection of sorbent for removing pesticides during water treatment. *J Hazard Mater.* **169**: 953–957.

Iijima, S. 1991. Helical microtubules of graphitic carbon. *Nature*354: 56–58.

Islam, M. A., Sakkas, V. and Albanis, T.A. 2009. Application of statistical design of experiment with desirability functions for the removal of organophosphorus pesticide from aqueous solution by low-cost material. *J Hazard Mater.* **170**: 230–238.

Javaid, M. K., Ashiq, M. and Tahir, M. 2016. Potential of Biological Agents in Decontamination of Agricultural Soil. *Scientifica***9**.

Jin, Z., Pan, Z., Yu, S. and Lin, C. 2010. Experimental study on pressurized activated sludge process for high concentration pesticide wastewater. *J. Environ. Sci.* **22**(9): 1342–1347.

Kamel, A., Byrne, C., Vigo, C., Ferrario, J., Stafford, C., Verdin, G., Siegelman, F., Knizner, S. and Hetrick, J. 2009. Oxidation of selected organophosphate pesticides during chlorination of simulated drinking water. *Water Res.* **43**(2): 522–534.

Karanasios, E., Tsiropoulos, N. G. and Karpouzas, D. G. 2012. On-farm biopurification systems for the depuration of pesticide wastewaters: recent biotechnological advances and future perspectives. *Biodegradation***23**(6): 787-802.

Kim, T. Y., Park, S. S., Kim, S. J. and Cho, S. Y. 2008. Separation characteristics of some phenoxy herbicides from aqueous solution. *Adsorption* **14**: 611–619.

Kislenko, V. N., Berlin, A. A., Kawaguchi, M. and Kato, T. 1996. Mathematical models of polymer adsorption at a porous adsorbent surface. *Langmuir.* **12**: 768–773.

Kitous, O., Cheikh, A., Lounici, H., Grib, H., Pauss, A., Mameri, N., 2009. Application of the electrosorption technique to remove metribuzin pesticide. *J Hazard Mater.* **161**: 1035–1039.

Kouras, A., Zouboulis, A., Samara, C. and Kouimtzis, T. 1998. Removal of pesticides from aqueous solutions by combined physicochemical processes-the behaviour of lindane. *EnvironPollut.* **103**: 193–202.

Kyriakopoulos, G. and Doulia, D., 2006. Adsorption of pesticides on carbonaceous and polymeric materials from aqueous solutions: A review. *Sep Purif Rev.* **35**: 97–191.

Kyriakopoulos, G., Hourdakis, A. and Doulia, D. 2003. Adsorption of pesticides on resins. *J Environ Sci Health B.* **38**: 157–168.

Kyriakopoulosa, G., Douliaa, D. and Anagnostopoulos, E. 2005. Adsorption of pesticides on porous polymeric adsorbents. *Chem Eng Sci.* **60**: 1177–1186.

Lee, S. C., Lintang, H. O. and Yuliati, L. 2017. High photocatalytic activity of $Fe_2O_3$/ $TiO_2$ nanocomposites prepared by photodeposition for degradation of 2, 4-dichlorophenoxyacetic acid. *Beilstein J. Nanotechnol.* **8**: 915–926.

Li, R., Duan, J., Saint, C. P. and Van Leeuwen, J., 2016. Impact of prechlorination on organophosphorus pesticides during drinking water treatment: Removal and transformation to toxic oxon byproducts. *Water Res.* **105**: 1–10.

Lofrano, G., Libralato, G., Meric, S., Vaiano, V., Sacco, O., Venditto, V., Guida, M. and Carotenuto, M. 2020. Occurrence and potential risks of emerging contaminants in water. In: Sacco, O., Vaiano, V. (Eds.), Visible Light Active Structured Photocatalysts for the Removal of Emerging Contaminants. *Elsevier*, pp. 1–25.

Long, R. Q. and Yang, R. T., 2001. Carbon nanotubes as superior sorbent for dioxin removal. *J Am Chem Soc.***123**: 2058–2059.

Lopez-ramon, M. V., Fontecha-Camara, M. A., Alvarez-Merino, M. A. and Moreno-Castilla, C. 2007. Removal of diuron and amitrol from water under static and dynamic conditions using activated carbon in form of fiber, cloth and grains. *Water Res.* **41**: 2865–2870.

Luo, Y., Guo, W., Ngo, H., Nghiem, L., Hai, F., Zhang, J., Shuang, L. and Wang, X. 2014. A review on the occurrence of micropollutants in the aquatic environment and their fate and removal during wastewater treatment. *Sci. Total Environ. J.***473**: 619–641.

Mahalakshmi, M., Arabindoo, B., Palanichamy, M. and Murugesan, V. 2007; Photocatalytic degradation of carbofuran using semiconductor oxides.*J Hazard Mater.* **143**:240–245.

Maity, N., Payne, G. F., Ernest, Jr. M.V. and Albright, R. L. 1992. Caffeine adsorption from aqueous solutions onto polymeric sorbents-the effect of surface chemistry on the adsorptive affinity and adsorption enthalpy. *React Polymb.* **17**: 273–287.

Majumdar, K. and Singh, N. 2007. Effect of soil amendments on sorption and mobility of metribuzin in soils. *Chemosphere***66**: 630–637.

Maldonado, M. I., Malato, S., Perez-Estrada, L. A., Gernjak, W., Oller, I., Domenech, X. and Peral, J. 2006. Partial degradation of five pesticides and an industrial pollutant by ozonation in a pilot-plant scale reactor.*J. Hazard. Mater.***138**(2): 363–369.

Maltseva, O., McGowan, C., Fulthorpe, R. and Oriel, P. 1996. Degradation of 2, 4-dichlorophenoxyacetic acid by haloalkaliphilic bacteria. *Microbiology* **142**(Pt 5), 1115–1122.

Margulis, L., Stern, T., Rubin, B. and Ruzo, L.O. 1992. Photostabilization of trifluralin adsorbed on a clay matrix. *J Agric Food Chem*. **40**: 152–155.

Martian-Gulloan, I. and Font, R., 2001. Dynamic pesticide removal with activated carbon fibers. *Water Res.* **35**: 516–520.

Matsui, Y., Knappe, D. R. U. and Takagi, R. 2002. Pesticide adsorption by granular activated carbon adsorbers. 2. Effects of pesticide and natural organic matter characteristics on pesticide breakthrough curves. *Environ Sci Technol*. **36**: 3432–3438.

Mauter, M. S. and Elimelech, M. 2008. Environmental applications of carbon-based nanomaterials. *Environ Sci Technol*. **42**: 5843–5859.

Meghwal, K., Agarwal, R., Kumawat, S., Jangid, N. and Ameta, C. 2019. Chemical and Biological Treatment of Dyes In.

Meijers, R. T., Oderwald-Muller, E. J., Nuhn, P. A. N. M. and Kruithof, J. C. 1995. Degradation of pesticides by ozonation and advanced oxidation. *Ozone Sci. Eng.* **17**(6):673–686.

Memon, G. Z., Bhanger, M. I. and Akhtar, M. 2007. The removal efficiency of chestnut shells for selected pesticides from aqueous solutions. *J Colloid Interface Sci*. **315**: 33–40.

Memon, G. Z., Bhanger, M. I. and Akhtar, M. 2009a. Peach-nut shells—An effective and lowcost adsorbent for the removal of endosulfan from aqueous solutions. *Pak J Anal EnvironChem*. **10**: 14–18.

Memon, G. Z., Bhanger, M. I., Memon, J. R. and Akhtar, M. 2009b. Adsorption of methyl parathionfrom aqueous solutions using mango kernels: Equilibrium, kinetic and thermodynamicstudies. *Bioremed J*. **13**: 102–106.

Memon, G. Z., Bhanger, M. I., Akhtar, M., Talpur, F. N. and Memon, J.R. 2008. Adsorption of methyl parathion pesticide from water using watermelon peels as a lowcost adsorbent. *ChemEng J*. **138**: 616–621.

Millet, M., Wortham, H., Sanusi, A. and Mirabel, P. 1996. A multi residue method for determination of trace levels of pesticides in air and water. *Arch Environ Contam Toxicol*. **31**: 543–556.

Mishra, P. C. and Patel, R. K. 2008. Removal of endosulfan by sal wood charcoal. *J Hazard. Mater*. **152**: 730–736.

Monsalvo, V. M., McDonald, J. A., Khan, S. J. and Le-Clech, P. 2014. Removal of trace organics by anaerobic membrane bioreactors. *Water Res*. **49**: 103–112.

Murillo-Zamora, S., Castro-Gutierrez, V., Masis-Mora, M., Lizano-Fallas, V. and Rodriguez-Rodriguez, C.E. 2017. Elimination of fungicides in biopurification systems: Effect of fungal bioaugmentation on removal performance and microbial community structure. *Chemosphere* **186**: 625–634.

Murthy, H. M. and Manonmani, H. K. 2007. Aerobic degradation of technical hexachlorocyclohexane by a defined microbial consortium. *J. Hazard. Mater.* **149**(1): 18–25.

Nanseu-Njiki, C. P., Dedzo, G. K. and Ngameni, E. 2010. Study of the removal of paraquat from aqueous solution by biosorption onto Ayous (Triplochiton schleroxylon) sawdust. *J. Hazard. Mater.* **179**: 63–71.

Navaratna, D., Shu, L. and Jegatheesan, V. 2016. Evaluation of herbicide (persistent pollutant) removal mechanisms through hybrid membrane bioreactors. *Bioresour. Technol.* **200**: 795–803.

Ofomaja, A. E. 2008. Kinetic study and sorption mechanism of methylene blue and methyl violet onto mansonia wood sawdust. *Chem Eng J.* **143**: 85–95.

Ohnoa, K., Minamia, T., Matsuia, Y. and Magara, Y. 2008. Effects of Chlorine on Organophosphorus Pesticides adsorbed on activated carbon: Desorption and Oxon formation. *WaterRes.* **42**: 1753–1759.

Ormad, M. P., Miguel, N., Claver, A., Matesanz, J. M. and Ovelleiro, J. L. 2008. Pesticide removal in the process of drinking water production. *Chemosphere***71**(1): 97–106.

Pan, Z., Yu, S., Huang, H., Lang, J., Yang, Y. and Lin, C. 2004. Treatment of high concentration pesticide wastewater by pressurized biochemical process. *Asian Pac. J. Chem. Eng.* **624.**

Peng, X., Li, Y., Luan, Z., Di, Z., Wang, H., Tian, B. and Jia, Z. 2003. Adsorption of 1, 2-dichlorobenzene from water to carbon nanotubes. *Chem Phys Lett.* **376**: 154–158

Popova, S., Matafonova, G. and Batoev, V. 2019. Simultaneous atrazine degradation and E. coli inactivation by UV/S2O8(2-)/ Fe(2+) process under KrCl excilamp (222 nm) irradiation. *Ecotoxicol. Environ. Saf.* **169**: 169–177.

Qiu, Y., Xiao, X., Cheng, H., Zhou, Z. and Sheng, G. D., 2009. Influence of environmental factors on pesticide adsorption by black carbon: pH and model dissolved organic matter. *EnvironSci Technol.* **43**: 4973–4978.

Qiua, Y. P., Chen, J. L., Li, A. M., Zhang, Q. X. and Huang, M. S. 2005. Adsorption of 2,4-D on modified hypercrosslinked polystyrene (NDA-99) and XAD-4 resin. *Chinese J Polym Sci.* **23**: 435–440.

Quan, X., Shi, H., Liu, H., Lv, P. and Qian, Y., 2004. Enhancement of 2, 4-dichlorophenol degradation in conventional activated sludge systems bioaugmented with mixed special culture. *Water Res.* **38**(1): 245–253.

Rafatullah, M., Sulaiman, O., Hashim, R. and Ahmad, A. 2009. Adsorption of copper (II), chromium (III), nickel (II) and lead (II) ions from aqueous solutions by meranti sawdust. *J Hazard Mater.* **170**: 969–977.

Rasheed, T., Bilal, M., Nabeel, F., Adeel, M. and Iqbal, H. M. N. 2019. Environmentally-related contaminants of high concern: Potential sources and analytical modalities for detection, quantification, and treatment. *Environ. Int.* **122**: 52–66.

Rehana, Z., Malik, A. and Ahmad, M. 1996. Genotoxicity of the Ganges water at Narora (U.P.) India. *Mutat Res.* **367**: 187–193.

Rodriguez-Cruz, M. S., Andrades, M. S., Parada, A. M. and Sanchez-Martin, M. J. 2008. Effect of different wood pretreatments on the sorption-desorption of linuron and metalaxyl by woods. *J Agric Food Chem*. **56**: 7339–7346.

Salimi, M., Esrafili, A., Gholami, M., Jafari, A. J., Kalantary, R. R., Farzadkia, M., Kermani, M. and Sobhi, H. R. 2017. Contaminants of emerging concern: a review of new approach in AOP technologies. *Environ. Monit. Assess*. **189**(8): 414.

Sanchez-Martin, M. J., Rodriguez-Cruz, M. S., Andrades, M. S. and Sanchez-Camazano, M. 2006. Efficiency of different clay minerals modified with a cationic surfactant in the adsorption of pesticides: influence of clay type and pesticide hydrophobicity. *Appl Clay Sci*. **31**: 216–228.

Saritha, P., Aparna, C., Himabindu, V., Anjaneyulu, Y., 2007. Comparison of various advanced oxidation processes for the degradation of 4-chloro-2 nitrophenol. *J. Hard Mater* **149**(3): 609–614.

Sarkar, B., Venkateswralu, N., Rao, R. N., Bhattacharjee, C. and Kale, V.2007. Treatment of pesticide contaminated surface water for production of potable water by a coagulation–adsorption nanofiltration approach. *Desalination* **212**: 129–140.

Sharma, R. K., Kumar, A. and Joseph, P. E. 2008. Removal of atrazine from water by lowcost adsorbents derived from agricultural and industrial waste. *Bull Environ Contamin Toxi*. **80**: 461–464.

Shawaqfeh, A. T. 2010. Removal of pesticides from water using anaerobic-aerobic biological treatment. *Chin. J. Chem. Eng.* **18**(4): 672–680.

Singh, N. 2009. Adsorption of herbicides on coal fly ash from aqueous solutions. *J HazardMater*. **168**: 233–237.

Singh, N., Megharaj, M., Gates, W. P., Churchman, G. J, Anderson, J., Kookana, R. S., Naidu, R., Chen, Z., Slade, P. G. and Sethunathan, N. 2003. Bioavailability of an organophosphorous pesticide, fenamiphos, sorbed on an organo clay. *J Agric Food Chem*. **51**: 2653–2658.

Sorensen, S., Albers, C. and Aamand, J. 2008. Rapid mineralization of the Phenylurea herbicide diuron by Variovorax sp. strain SRS16 in pure culture and within a two-member consortium. *Appl. Environ. Microbiol.* **74**(8): 2332–2340.

Sotelo, J. L., Ovejero, G., Delgado, J. A. and Mart´ınez, I. 2002. Adsorption of lindane from water onto GAC: Effect of carbon loading on kinetic behavior. *J Chem Eng*. **87**: 111– 120.

Streat, M. and Sweetland, L. A. 1998. Removal of pesticides from water using hypercrosslinked polymer phases. Part 1. Physical and chemical characterization of adsorbents. *Process Saf Environ Prot*. **76**: 115–126.

Suciua, N. A. and Capri, E. 2009. Adsorption of chlorpyrifos, penconazole and metalaxyl from aqueous solution by modified clays. *J Environ Sci Health B*. **44**: 525–532.

Sulaiman, O., Amini, M. H. M., Rafatullah, M., Hashim, R. and Ahmad, A. 2010. Adsorption equilibrium and thermodynamic studies of copper (II) ions from aqueous solutions by oil palm leaves. *Int J Chem React Eng*. **8**: A108.

Tazdait, D., Salah, R., Grib, H., Abdi, N. and Mameri, N. 2018. Kinetic study on biodegradation of glyphosate with unacclimated activated sludge. *Int. J. Environ. Health Res*. **28**(4): 448–459.

Thomaidis, N., Asimakopoulos, A. and Bletsou, A. 2012. Emerging Contaminants: A Tutorial Mini-Review. *Global NEST* **14**(1): 72-79.

Trochimczuk, A. W., Streat, M. and Malik, D. J. 2003. Sorption of pesticides and herbicides on highly polar polymeric sorbents. *Sep Sci Technol*. **38**: 1813–1827.

Tsydenova, O., Batoev, V. and Batoeva, A. 2015. Solar-enhanced advanced oxidation processes for water treatment: Simultaneous removal of pathogens and chemical pollutants. *Int. J. Environ. Res. Publ. Health***12**(8): 9542–9561.

Tsyurupa, M. P. and Davankov, V. A. 2006. Porous structure of hypercrosslinked polystyrene: State-of-the-art mini-review. *React Funct Polym*. **66**: 68–779.

Tsyurupa, M. P., Maslova, L. A., Andreeva, A. I., Mrachkovskaya, T. A. and Davankov, V. A. 1995. Sorption of organic compounds from aqueous media by hypercrosslinked polystyrene sorbents "Styrosorb." *React Polym*. **25**: 69–78.

Vela, N., Fenoll, J., Garrido, I., Pérez-Lucas, G., Flores, P., Hellín, P. and Navarro, S. 2019. Reclamation of agro-wastewater polluted with pesticide residues using sunlight activated persulfate for agricultural reuse. *Sci. Total Environ*. **660**: 923–930.

Waite, T. D., Cessna, J. A., Gurprasad, P. N. and Banner, J. 1999. A new sampler for collecting separate dry and wet atmospheric depositions of trace organic chemicals. *Atmosph Environ*. **33**: 1513–1523.

Wang, J., Teng, Y., Zhang, C., Liao, X., Zhai, Y. and Zuo, R. 2019. Activation of manganese dioxide with bisulfite for enhanced abiotic degradation of typical organophosphorous pesticides: Kinetics and transformation pathway. *Chemosphere***226**: 858–864.

Wang, S. and Wu, H. 2006. Environmental benign utilization of fly ash as low-cost adsorbents. *J Hazard Mater*. **136**: 482–501.

Wang, O., Shao, Y., Gao, N., Liu, S., Dong, L., Rao, P., Chu, W., Xu, B., An, N., Deng, J., 2020. Impact of zero valent iron/persulfate preoxidation on disinfection byproducts through chlorination of alachlor. *Chem. Eng. J*. **380:** 122435.

Wang, J. and Zhuan, R., 2019. Degradation of antibiotics by advanced oxidation processes: An overview. *Sci. Total Environ.***701**: 135023.

Westerhoff, B. M., Fairbairn, D. J., Ferrey, M. L., Matilla, A., Kunkel, J., Elliott, S. M., Kiesling, R. L., Woodruff, D. and Schoenfuss, H. L. 2018. Effects of urban stormwater and iron enhanced sand filtration on Daphnia magna and pimephales promelas. *Environ. Toxicol*. Chem. **37**(10): 2645–2659.

Wu, C. 2007. Adsorption of reactive dye onto carbon nanotubes: Equilibrium, kinetics and thermodynamics. *J Hazard Mater* **144**: 93–100.

Yadav, S., Kumar, N., Kumari, V., Mittal, A. and Sharma, S. 2019. Photocatalytic degradation of Triclopyt, a persistent pedticide by ZnO/SnO2 nano-composities. *Mater. Today: Proc.* **19**(2): 642–645.

Yang, K., Zhu, L., Xing, B., 2006. Adsorption of polycyclic aromatic hydrocarbons by carbon nanomaterials. *Environ Sci Technol.* **40**: 1855–1861.

Zhang, M., Dong. H., Zhao. L., Wang. D. X. and Meng. D. 2019. A review on Fenton process for organic wastewater treatment based on optimization perspective. *Sci. Total Environ.* **670**: 110–12.

Zhang, H. P., Hu, Q., Chen, X. L., Zhou, X. Y., Chen, K. and Jiang, J. D. 2018. Degradation of phenylurea herbicides by a novel bacterial consortium containing synergistically catabolic species and functionally complementary hydrolases. *J. Agric. Food Chem.* **66**(47): 12479–12489.

# Chapter - 10

## Water Pollution and Anthropocene: Trends, Effects and Remedies

**Sadiya Farooq[1*], Mansoor Ahmad Rather[1] and Anayituallah Chesti[1]**

[1]Division of Aquaculture, Faculty of Fisheries, SKUAST-K, Rangil-190006, Ganderbal, J&K, India

Correspondence: sadiyafarooq119@gmail.com

### Abstract

Water is a vital resource for continuity of life on earth. Human race is regarded as the largest driver of environmental change at the planetary scale. During the past few decades, factors such as increasing population, rapid industrialization, growing urbanization, and careless utilization of natural resources have adversely impacted water quality. In the Anthropocene epoch, water pollution has emerged as a central issue since it is directly related to the present anthropogenic activities in all its complexity. The detrimental effects of water pollution on human, animal and ecosystem have prompted strategies to develop robust, eco-friendly, easily operable and economically viable treatment methods for the removal of contaminants from the aquatic system and for safeguarding the environment.

**Keywords:** Anthropocene, Pollutants, Wastewater, Environmental deterioration, Remediation

### Introduction

Anthropogenic activities have the dominant influence on myriad earth system parameters. The influence has significantly altered ecosystems worldwide to the extent that they have heralded a new epoch in geological timeline: the Anthropocene

(Griggs *et al.*, 2013; Steffen *et al.*, 2007). Evidently, a human being has become a geological agent itself (Chakrabarty, 2009). A life link tends to connect every biotic and abiotic component of any ecosystem. Therefore, the deterioration that occurs in a part of the system affects the whole system over time. Globally, the environmental impacts of human action are numerous and tend to modify water, soil, atmosphere composition and climate.

Water is an abundant natural resource, which is unevenly distributed on the planet. All living things depend mainly on the water as a means of life, energy and growth. About 71% of the earth's surface is covered by water. Around 97% of the total water available on earth is found in oceans; however, it is too saline for drinking or irrigation. Fresh water constitutes 3% of the total water available on earth. Of this, 2.997% is locked in ice caps or glaciers hence, only 0.003% of the earth's total volume of water is easily available to as water vapor, soil moisture, groundwater, and water in rivers, streams, lakes and wetlands. The existence of life on the surface of earth would be impossible without the presence of water. It can be rightly said that the word, "water" is a true synonym for the word, "life". Undoubtedly, water is an essential resource for development and prosperity of any society. And, as a person advances, the demand for water increases. Moreover, water resources are the only source of aquatic wealth in the form of plants and fishes thriving in aquatic ecosystems, which are one of the key food sources in the world (Fadlallah, 2001).

## Water Pollution and Anthropocene

Water pollution is either directly or indirectly caused by human activities. It is only part of an interactive array of immense and boundless environmental changes that have occurred in tandem with the emergence of *Homo sapiens*, being the primary agent of global environmental change in Anthropoecenic era. Water pollution occurs when some unwanted substances/constituents enter into the water bodies and cause alteration in the water characteristics, thus making it harmful to human beings and /or the ecosystem (Alrumman *et al.*, 2016; Briggs, 2003). The abundance of the pollutant, its effect on the environment, and the usage of water affect the intensity of pollution. Water residence time is crucial to pollution problems because it affects pollution potential. It varies in accordance with the type of water body. For example, river water has a relatively short residence time (the average time that a water molecule spends in a water body), so a polluting agent typically exists there for a brief period of time before it finds its way into another water body, such as the sea or eventually an ocean, where it can cause further problems. In contrast, groundwater is usually characterized by slow flow and prolonged residence time, which makes it particularly problematic. Finally, pollution residence time is greater than the water residence time since a pollutant may be absorbed onto sediment or retained within the ecosystem for a long time (Fisher, 2017).

### Sources

The Anthropocene is primarily recognized by three factors: the acceleration in technological era after the First Industrial Revolution, the rapid growth in population and the increase in levels of production and consumption. Over the last two centuries, the interaction of these three factors in human evolution has changed the face of our planet. Almost every hominid action has the potential to contribute towards pollution. In the present geological epoch, water is considered as the most vulnerable natural resource to pollution. The classical categorization broadly divides sources of water pollution into the following:

#### a) On the basis of nature of sources:

Natural and Anthropogenic Sources: Pollution includes even a sudden rise in a substance's concentration if it occurs naturally; these sources are referred to as natural sources. One such naturally occurring occurrence that occurs in the majority of water bodies is siltation, which is formed of dirt, sand, and mineral particles (Bharathy, 2018). Human activities that cause water contamination are known as anthropogenic or manmade sources of water pollution. Anthropogenic sources include waste from homes (including sewage and waste water), businesses, and agriculture that enters rivers, lakes, streams, seas, and oceans. This category also includes certain pollutants that runoff from the land which finally carries into various water bodies.

#### b) On the basis of specificity of sources:

Specific and Unspecified sources of pollution: Specific/Point sources of pollution are the easily identifiable type of pollutant sources where all the pollutant materials enter into water bodies from a single, definite and distinguishable source. Unspecified/ Nonpoint sources of pollution result from widespread sources that are difficult to be controlled directly. These sources of pollution are advocated as the leading cause of water pollution with much greater impact than point sources.

### Trends

It is usually difficult to measure pollution trends in rivers, lakes, and other surface waters is more challenging than measuring trends in air pollution (Sharpio, 2022).

Water pollution trends around the world have been discussed in various aspects given below (Tiseo, 2022):

**a) Clean-water index:** The clean water index (on a scale of 0-100) measures pollution by quantifying excessive nutrients, pathogens, and waste in estuarine, coastal, and open ocean waters. An index score of 70-100 is an indicator of clean water. The reference point is taken as zero suggesting water is completely free from chemicals, nutrient, human pathogens and any other polluting agents. With the minimization in pollution of estuarine, coastal and open waters, the goal score is higher and vice versa. With a score of just 19 out of 100, Nauru had the lowest clean water score in the world in the year, 2021. This score was 70 points lower than the clean water index in Canada, which had the highest clean waters score worldwide. Togo, Bangladesh, and Cameroon were some nations which had high levels of pollutants in estuarine, coastal, and open ocean waters that year. Other developing countries typically recorded lower clear water scores.

**b) Global share of wastewater:** In year 2020, domestic households generated 44% of the world's waste water flow which was not safely treated owing to the regional disparities for wastewater treatment worldwide. In North America and Europe, 80% of household wastewater flow is treated with adequate treatment, but less than 30% is treated safely in Sub-Saharan Africa and Central and Southern Asia.

**c) Global oil tanker spills:** Approximately, an amount of one thousand metric tons of oil spill was reported from tankers globally in year, 2021. The largest quantity oil spill (MT Sanchi, East China Sea) incident in 24 years was reported in 2018 with a total of 116 thousand metric tons of oil leaked from oil tanker incidents. However, the average annual amount of oil spills from tankers has significantly decreased since 1970's and 1980's.

**d) Plastic waste generation:** Every year, huge quantities of improperly disposed plastic waste finds its way into aquatic environments, causing devastating pollution. This may result from unplanned waste management issues, illegal dumping, or simply littering. The estimations indicate the coastal population of the United States (117.94 million) generated as much as 1.45 million metric tons of mismanaged plastic waste. This was one of the highest in the world, just after India and Indonesia.

**e) Quantity of plastic in major marine areas:** It is estimated that North Pacific Ocean constitutes almost two trillion pieces of plastic. This is the largest quantity of plastic when compared to other significant major marine areas. Also, microplastics make their way into water sources around the world, causing harm to aquatic life.

These can come from a variety of sources, such as city dust, car tyres and synthetic textiles.

**Nature, types and effects of pollutants**

**a) Domestic and Municipal Pollutants:** Sewage constitutes trash material, garbage, soaps, detergents, waste food and human excreta. Pathogens (disease causing bacteria, viruses, protozoa and parasitic worms) enter water systems from domestic sewage and untreated human, animal wastes and/or excreta. Generally, these organisms do not pose a threat in low numbers. However, the susceptibility of contracting diseases from them increases with the increase in number of wastes in the water. Unfortunately, nearly a billion people worldwide are exposed to waterborne pathogen pollution every day and 1.5 million children, mostly in underdeveloped countries die every year from waterborne diseases. In most of developing nations, sewage is frequently discharged into local waters either fully untreated or after only rudimentary treatment. Moreover, overflows of combined sewer systems, poorly managed municipal factories, and leaky or broken sewage collection systems can result in untreated sewage discharge into water bodies. Several water-borne diseases (cholera, typhoid, dysentery bacillus, infectious hepatitis, and giardia) and water-based diseases (schistosomiasis, filariasis) are commonly caused by drinking contaminated water. Sewage water is home to insects transmitting the germs associated with water, which cause yellow fever, dengue fever, malaria and others. Also, bacteria and other organisms in water use oxygen to metabolize the sewage they accompany. While breaking down the sewage, these micro-organisms can cause hypoxic (oxygen-depleted) dead zones. These dead zones lack oxygen that fish and other native organisms need to survive. Anaerobic action may result in release of toxic compounds such as hydrogen sulphide which is known to cause death of fishes in waters receiving untreated sewage. It is also reported that heavy organic pollution can cause an increase in disease, parasitism and abnormalities among fishes (Häder, 2021).

**b) Industrial pollutants:** Hazardous organic and inorganic wastes, as well as highly toxic metals (chromium, arsenic, lead and mercury) along with other chemical compounds (e.g., acids, alkalis, cyanides, chlorides, etc.) are discharged into rivers by industries that are typically situated near rivers or streams. The majority of these pollutants are water-soluble and resistant to breakdown by microorganisms (called non-biodegradable). Higher concentration of chemicals can make the water unfit to drink, harm fish and other aquatic life, reduce crop yields and accelerate corrosion of equipment which comes in contact with this water.

**c) Heavy metals:** Heavy metals are not biodegradable and have a propensity to bioaccumulate, which is the process by which their concentration in living things gradually increases (Cabral Pinto *et al.*, 2017; Fu and Wang, 2011; Kumar *et al.*, 2020). Most of the heavy metal ions are toxic or carcinogenic. They can cause damage to eyes, lungs, kidneys, liver, prostate, oesophagus, stomach, and skin even at very low concentrations. With prolonged exposure, these substances can also cause neurodegenerative disorders and diseases, such as Alzheimer's and Parkinson's diseases (Cabral Pinto *et al.*, 2017; Cabral Pinto and Ferreira da Silva, 2019; Jaishankar *et al.*, 2014; Yuan et *al.*, 2016). Metals have adverse impact on aquatic organisms (phytoplankton, zooplankton, and fish), accumulating in several organs and causing oxidative damage, endocrine disruption, and depression of the immune system (Maurya *et al.*, 2019; Guerra *et al.*, 2012; Le *et al.*, 2019). Aquatic organisms that spend most of their life in metal-contaminated regions (e.g., lead and cadmium) have showed slower exploration tendencies with high levels of metal in their blood profile (Grunst *et al.*, 2019).

Factories manufacturing plastic, caustic soda and chemical fertilizers release mercury (a heavy metal) along with other effluents in a nearby water body. Like any other heavy metal, mercury enters the food chain through bacteria, algae, fish and finally into the human body. During 1953-1960, the toxicity of mercury became evident by the Minamata Bay tragedy in Japan where mass mortality of fish was reported. Many people who ate mercury-contaminated fish were affected by poisoning and quite a few deaths were reported (Feng *et al.*, 2022). Depression and irritability are milder symptoms of mercury poisoning but acute toxic effects can cause blindness, paralysis, insanity, birth defects and even death. The formation of soluble monomethyl mercury ion, ($CH_3$, $Hg^+$) and volatile dimethylmercury [$(CH_3)_2$ Hg] by anaerobic bacteria in sediments is the cause of the high concentration of mercury in water and fish tissues.

**b) Acids and alkalis:** Either acidic or alkaline compounds alter the pH of water. If water is contaminated with acids, this will result in corrosion of the pipes. This corrosion causes risks to human health according to the type of contaminated acid. Besides, alkali pollution is caused salts such as carbonates, bicarbonate, hydroxides and chlorides. Carbon, calcium and magnesium bicarbonate causes hardness of water. Elevated levels of hardness have been reported to cause stress and even death of fish (Paul *et al.*, 2019; Mohajan, 2018)

**e) Agricultural waste:** Run-off from agriculture land contains silt, manure, fertilizers, pesticides and it also includes untreated waste from cattle farms, slaughterhouses, and poultry farms. A large quantity of fertilizers composed of phosphates and nitrates or organic manures make a water body rich in nutrients, which lead to eutrophication and consequent depletion of dissolved oxygen. Consumption of nitrate-rich water is bad for human health especially for young children. Pesticides (Dichlorodiphenyltrichloroethane (DDT), dieldrin, aldrin, malathion, carbaryl, etc.) are used to eradicate insect and rodent pests. Toxic pesticide residues enter the human body through drinking water or through food chain (biomagnification). These substances have relatively low solubility in water but are highly soluble in fats. For instance, the concentration of DDT in river water may be very low initially but some fish tend to bioaccumulate DDT over a period of time that they become unfit for human consumption. Several studies suggest that chemical fertilizers distress normal body functioning and spatial memory in fishes (Hong and Zha, 2019).

**f) Physical pollutants**: Following are some of the examples of physical pollutants mentioned discussed below:

**a) Thermal Pollutants**: Various thermal plants and nuclear power plants require water for cooling and the resultant hot water is discharged directly into nearby rivers or lakes. Water with higher temperature decreases the solubility of dissolved oxygen, which is crucial for aquatic life. (Issakhov, 2017). A sudden change in water temperatures affects normal physiology and functioning of fish and other aquatic organism. Overall, the ecology of the water body is disturbed as a result of this type of pollution.

**b) Radioactive pollutants**: Accidental leaks of water-soluble radioactive waste from nuclear power plants, mines, research laboratories and hospitals also cause water pollution. As these radioactive isotopes move across food webs and food chains, they tend to concentrate in various tissues and vital organs. Fish can develop genetic, mutagenic, and teratogenic defects as a result of radioactive waste that enters water bodies (Sharma, 2008).

**c) Suspended matter**: Deforestation is a common cause of soil erosion. It results in insoluble particles of soil and other solids that become suspended in water. The turbid conditions interfere with the penetration of sunlight. This reduces the photosynthetic activity of aquatic plants and algae disrupting the ecological balance of the aquatic bodies. The suspended particles in streams and rivers settle to the bottom as sediments and the velocity of water decreases. Extreme turbid conditions can cause gill choking in fishes, leading to stress and mortality. Excessive sediments that settle down damages clogs and fills lakes, artificial reservoirs, etc. damaging feeding and spawning grounds of fish. For filter-feeding macrobenthos, a high concentration of suspended solids can suffocate the feeding structures and reduce feeding effectiveness, which lowers growth rates, increases stress levels, and even causes mortality (Laitano *et al.*, 2022).

**d) Plastics**: Plastic contamination is advocated as one of the biggest threats to both human development and ecosystem integrity with many referring to Anthropocene as the Plastic Age (Porta, 2021). Plastic contamination presents a major hazard to fish, seabirds and other aquatic animals. The ingestion, suffocation, and entanglement of hundreds of aquatic species are the most visible effects of plastic debris. Micro-plastics have been linked with cancer, reduced fertility, psychological illnesses, and birth defects. Numerous harmful substances, such as neurotoxins, carcinogens, and endocrine disruptors, are frequently found in plastics.

**e) Oil-based Pollutants**: The most common way that crude oil and other related fuels end up in the water is through accidental or unintentional spills from tankers, ships, pipelines, etc. In addition to the unintentional spills, oil refineries, oil exploration sites, and automobile service centers also pollute various water bodies. During an oil spill on water, oil tends to floats on the surface as it is less dense than water, and the lightest hydrocarbons evaporate, decreasing the surface area covered by the spill but polluting the atmosphere. Finally, certain bacteria begin to degrade the remaining oil by a process which can take many years. Only 15% of the original volume might remain after several months, but it is in the form of thick asphalt lumps, which is particularly harmful to birds, fish, and shellfish and detrimental to the aquatic **ecosystem (Fisher, 2017).**

**f) Human and veterinary pharmaceuticals**: These substances are used to treat or prevent a number of diseases. Pharmaceuticals for animals may also be used to improve food production efficiency. Pharmaceuticals are widely used and unavoidably released into the environment via one of these channels: indiscriminate landfill leaching, landfill disposal of household and medical waste, drainage water, and sewage. Although, pharmaceuticals have been discovered in a variety of environmental samples at concentrations between ng/l and g/l, but these substances are classified as endocrine disruptors (EDs) and pose a risk to both aquatic life and people (Archer *et al.*, 2017). Pharmaceuticals in water are known to be toxic to aquatic organisms in both acute and chronic phases (Kar & Roy, 2010; Aguirre-Martínez *et al.*, 2016). As implied by their name, EDs alter normal endocrine functions and increase the risk of cancer in people. Their effects on aquatic lives may range from endocrine system disruption through production to feminization of female aquatics to the reduction in eggs and sperm cells (Akanyeti *et al.*, 2017; Saha *et al.*, 2010).

**g) Alien species**: While sewage, toxic metals, and oil spills are the most common pollutants associated with water pollution, pollution can also be biological in the form of alien species (Occhipinti-Ambrogi, 2021). An important issue is invasive alien species. The effects on ecological quality include genetic changes in the population, altered or degraded habitats, and the spread of pathogens and parasites. The establishment of alien species happens due to the movement of organisms, rate and mechanism of transport and the characteristics of the organisms that make them successful invaders. The invasive species negatively impact native species and ecosystems due to the competition for food and nourishment, which makes it difficult to coexist in any ecological niche.

## Remedies

Either *in-situ* water treatment or pollution control at the source point can be used to remediate polluted water. In contrast to conventional treatment methods, it is necessary to detect and measure the concentration of pollutants contaminating given water before choosing an effective remediation treatment. The method/technique that is chosen for this purpose must be economical, environmentally friendly, selective, and sensitive enough to detect traces with good accuracy (Malik *et al.*, 2019).

Traditional remediation methods did not aim on detection of pollutants however, the evolution and combination of various techniques (Spectroscopic, electrochemical and optical) allow the accurate detection of pollutants and the rapid measurement of their concentrations. This has also allowed gaining knowledge about the magnitude of water pollution problems around the world.

Different methods that can be applied for remediation of polluted water are categorized into physical, chemical, biological, ecological and engineering techniques, but a single method is not sometimes effective for the purification of heavily contaminated water. Therefore, hybrid techniques, which combine two or more single methods, are more widely recommended for more efficient treatment.

### a) Membrane-based sewage treatment:

By taking into account, the removal of nutrients (such as nitrate and phosphate) and easily degradable organic compounds as well as the presence of toxic compounds and variations in pH of the wastewater, microbial treatment methods go further than traditional municipal wastewater treatment. When designing a treatment system, certain factors should be taken into account such as biomass yield, nutrient addition, oxygen or another electron donor supply, pH control/kinetics, or the rates of biological reactions (biotransformation). The first stage of treatment involves separating bulky material from the water source, and the second stage concentrates on getting rid of more toxic materials and other debris. The tertiary phase entails complete water purification, typically accomplished through chemical disinfection.

Membrane bioreactors have been used more recently and have proven to be very effective at removing contaminants. These are combinations of microbial communities and high-efficiency membranes that remove contaminants much more efficiently. Three main functions of the membrane are possible. First of all, membranes (hollow-fiber gas-permeable membranes) can be used as a surface for microbial attachment, growth, and oxygen permeation of the biofilm. The membrane is made of microporous, hydrophobic polypropylene, ensures high biomass density in the area while allowing almost 100% oxygen transfer. Membranes can also be used as selective barriers. Such membranes allow organic compounds in wastewater to permeate but do not transport ions into the bioreactor. Thus, selectivity of biodegradable organic compounds is made possible. Rubber silicone is an illustration of a substance used for such a membrane. And finally, membranes can be used to separate biomass as well.

**b) Activated sludge treatment:**

The activated sludge process is a biological treatment method. It consists of an aeration basin where aeration equipment mixes wastewater properly with oxygen to maintain a uniformly mixed liquor suspended solids. Following the liquid-solid separation, which is typically done in a clarifier by gravity, the settled biomass is then returned to the activated sludge basin. Examples of aeration basin configurations are –completely mixed basins, basins in series, plug-flow systems, etc. A bio membrane may occasionally be used in place of a clarifier after treating activated sludge. When such membranes are employed, the effluent produced has less sludge. Also, automated processing can be easily employed. The disadvantage however is the higher cost of the investment for initial start-up as well as maintenance (Mutamim *et al.*, 2012; Sharrer & Summerfelt, 2010).

**c) Bio-ecological remediation:**

Technologies for bio-ecological treatment include contact oxidation, ecological ponds, plant purification treatment, ecological floating beds, and constructed wetlands, among others, which are described in the literature (Peilin *et al.*, 2010). Polluted water exhibits odor, turbidity, high concentrations of chemical oxygen demand (COD), biological oxygen demand (BOD) and organic and inorganic contaminants. Some studies have suggested to employ the processes of microorganism, plant and aquatic animal-assisted bioremediation or biodegradation processes to destroy or decompose the organic contaminants, absorbed metals (inorganic contaminants) and also to completely remove COD, BOD, odor, turbidity and organic and inorganic contaminants from river water (Wang *et al.*, 2012; Zhao *et al.*, 2013; Cui *et al.*, 2019;

Zheng & Wang, 2017). Both *in-situ* and microbial dosing techniques were used in these procedures. By utilizing indigenous bacteria, *in-situ* microbial techniques are more environmentally and financially viable. *In-situ* microbial techniques have therefore gained more popularity in wastewater treatment because they are widely accepted and applicable.

The plant-assisted floating bed techniques and constructed wetlands are the most widely used *in-situ* microbial methods for treating river water. These techniques can naturally induce growth of bacteria, fungi and other fauna, which play important roles in biodegradation of organic contaminants in river water. Compared to physical and chemical processes, biological methods are more cost-effective, self-sustaining, and environmentally friendly. However, the processes function from several months to years for microbial growth and, sometimes, different environmental factors such as temperature and moisture affect their performance. Some of these technologies require high initial cost, skilled labor and maintenance as well.

**d) Phytoremediation process:**

Aquatic plants such as algae, demonstrate a high potential to purify water, wastewater effluents and contaminated water (Anawar *et al.*, 2008). The tolerant plants have the ability to purify the river water by absorption, adsorption, accumulation and degradation of contaminants. These plants exhibit significant capacity to remove nutrients such as total Nitrogen and Phosphorus from water bodies (Tong *et al.*, 2003). Strong and extensive rhizosphere systems are provided by plant roots, which promote the development of bacteria and other microorganisms as well as the secretion of root exudates (Steenhoudt & Vanderleyden, 2000; Zimmels, *et al.*, 2008). These microorganisms play a significant role in the removal of organic and inorganic contaminants from wastewater by biodegradation of organic chemicals and accumulation of nutrients and other metal contaminants. Phytoremediation process can also be applied along the wastewater/stormwater discharge point. The method is less expensive and more widely accepted in the community, regardless of location. The most commonly used aquatic plants for wastewater treatment are reed (*Phragmites australis*), water hyacinth (*Eichhornia crassipes*), alligator weed (*Alternanthera philoxeroides*), water lettuce (*Pistia stratiotes*), Whorl-leaf watermilfoil (*Myriophyllum verticillatum*), pondweed (*Potamogeton* spp.), cattail (*Typha latifolia*), duckweed (*Lemna gibba*) and canna (*Canna indica*) (Sato *et al.*, 2002; Zimmels *et al.*, 2008; Allam *et al.*, 2016). These aquatic plants can be used in vegetative biofiltration systems, constructed wetlands, traditional wastewater treatment processes and ecological floating bed systems. Besides this, plant species have different contaminant removal efficiencies and variable tolerances to contaminant loading rates. As a result, choosing the right plant species is crucial to ensuring that plant-based technologies perform to their full potential (Zhao *et al.*, 2012).

**e) Chemical means of remediation:**

To remove suspended solids (SS) and completely eradicate pathogens, polluted water undergoes chemical treatment using flocculation, precipitation, ion exchange, and electrochemical methods. Although, chemical processes provide a quick remediation of polluted water, but they are only temporary and may produce secondary wastes, which can pose other dangers. The development of more affordable, secure, and efficient technologies has recently attracted a lot of attention, both to decrease the amount of wastewater produced and to improve the quality of the treated effluent. Adsorption has emerged as one of the most effective treatments, and in recent years, the hunt for inexpensive adsorbents with metal-binding capabilities has intensified (Leung *et al.*, 2000).

The adsorbents may be organic or biological origin, composed of minerals, industrial byproducts, agricultural wastes, biomass, and materials made of polymers (Kurniawan *et al.*, 2005). Due to its convenient operation, membrane separation has gained momentum for the treatment of inorganic effluent. There are several types of membrane filtration, including ultrafiltration (UF), nanofiltration (NF), and reverse osmosis (RO) (Kurniawan *et al.*, 2006). Electro-treatments such as electrodialysis have aided in environmental protection (Pedersen, 2003). Photocatalytic process is an innovative and promising technique for efficient destruction of pollutants in water (Skubal *et al.*, 2002). Although many techniques can be employed for the treating inorganic effluents, the ideal treatment should be able to meet the established maximum contaminant level (MCL) standards in addition to being suitable, appropriate, and applicable to the local conditions.

## Conclusion

Since water forms a core of the existence of human and other living things, its preservation and sustainable availability cannot be overemphasized. To reduce Anthropoecenic impacts on water resources, several personal and social attributes, such as collaborative consumption, organic food production, enforcing reverse logistic laws, restricting fertilizers and pesticides use, investment in technology for alternate energy sources and pollutants removal in treatment facilities, purchasing eco-friendly products, etc. While various wastewater treatment methods are being explored by industries and various treatment plants, untreated wastewater is still being discharged into the water bodies by some industries. Thus, effective environmental protection policies compliance drive will be of immense benefit to the environment and by extension to human. Factoring these environmental protection policies into the goals and objectives of various factors involved in environmental deterioration will help policies performance. This will serve as a step forward in the direction of ameliorating water pollution.

## Bibliography

Aguirre-Martínez, G. V., Okello, C., Salamanca, M. J., Garrido, C., Del Valls, T. A., & Martín-Díaz, M. L. (2016). Is the step-wise tiered approach for ERA of pharmaceuticals useful for the assessment of cancer therapeutic drugs present in marine environment? *Environmental research*, *144*, 43-59.

Akanyeti, İ., Kraft, A., & Ferrari, M. C. (2017). Hybrid polystyrene nanoparticle-ultrafiltration system for hormone removal from water. *Journal of water process engineering*, *17*, 102-109.

Allam, A., Tawfik, A., El-Saadi, A., & Negm, A. (2016). Potentials of using duckweed (Lemna gibba) for treatment of drainage water for reuse in irrigation purposes. *Desalination and Water Treatment*, *57*(1), 459-467.

Alrumman, S. A., El-kott, A. F., & Keshk, S. M. A. S. (2016). Water pollution: Source and treatment. *American Journal of Environmental Engineering*, *6*(3), 88-98.

Anawar, H. M., Garcia-Sanchez, A., Kul Alam, M. T., & Rahman, M. M. (2008). Phytofiltration of water polluted with arsenic and heavy metals. *International Journal of Environment and Pollution*, *33*(2-3), 292-312.

Archer, E., Petrie, B., Kasprzyk-Hordern, B., & Wolfaardt, G. M. (2017). The fate of pharmaceuticals and personal care products (PPCPs), endocrine disrupting contaminants (EDCs), metabolites and illicit drugs in a WWTW and environmental waters. *Chemosphere*, *174*, 437-446.

Bharathy, N. (2018). Water pollution and water quality standards for livestock. *International Journal of Science, Environment and Technology*, *7*(6), 1905-1913.

Briggs, D. (2003). Environmental pollution and the global burden of disease. *British medical bulletin*, *68*(1), 1-24.

Cabral Pinto, M. M., & Ferreira da Silva, E. A. (2019). Heavy Metals of Santiago Island (Cape Verde) alluvial deposits: Baseline value maps and human health risk assessment. *International journal of environmental research and public health*, *16*(1), 2.

Cabral Pinto, M. M., Silva, M. M., Ferreira da Silva, E. A., & Marinho-Reis, A. P. (2017). The cancer and non-cancer risk of Santiago Island (Cape Verde) population due to potential toxic elements exposure from soils. *Geosciences*, *7*(3), 78.

Chakrabarty, D. (2009). The climate of history: Four theses. *Critical inquiry*, *35*(2), 197-222

Cui, N., Chen, G., Liu, Y., Zhou, L., Cai, M., Song, X., & Zou, G. (2019). Comparison of two different ecological floating bio-reactors for pollution control in hyper-eutrophic freshwater. *Scientific Reports*, *8*(1), 1-9.

Fadlallah, S. (2001). Environmental pollution and its impact on agricultural economic development. *Assiut Journal of Environmental Studies* 20 PP. 82-84.

Feng, X., Li, P., Fu, X., Wang, X., Zhang, H., & Lin, C. J. (2022). Mercury pollution in China: implications on the implementation of the Minamata Convention. *Environmental Science: Processes & Impacts, 25*(4), 634-648.

Fisher, M. R. (2017). 7.3 Water Pollution. *Environmental Biology*. Open Oregon Educational Resources.

Fu, F., & Wang, Q. (2011). Removal of heavy metal ions from wastewaters: a review. *Journal of environmental management*, *92*(3), 407-418.

Griggs, D., Stafford-Smith, M., Gaffney, O., Rockström, J., Öhman, M. C., Shyamsundar, P., ... & Noble, I. (2013). Sustainable development goals for people and planet. *Nature*, *495*(7441), 305-307.

Grunst, A. S., Grunst, M. L., Daem, N., Pinxten, R., Bervoets, L., & Eens, M. (2019). An important personality trait varies with blood and plumage metal concentrations in a free-living songbird. *Environmental Science & Technology*, *53*(17), 10487-10496.

Guerra, F., Trevizam, A. R., Muraoka, T., Marcante, N. C., & Canniatti-Brazaca, S. G. (2012). Heavy metals in vegetables and potential risk for human health. *Scientia Agricola*, *69*, 54-60.

Häder, D. P. (2021). Effects of Pollution on Fish. *Anthropogenic Pollution of Aquatic Ecosystems*, 39-60.

Hong, X., & Zha, J. (2019). Fish behavior: A promising model for aquatic toxicology research. *Science of the total environment*, *686*, 311-321.

Issakhov, A. (2017). Numerical study of the discharged heat water effect on the aquatic environment from thermal power plant by using two water discharged pipes. *International Journal of Nonlinear Sciences and Numerical Simulation*, *18*(6), 469-483.

Jaishankar, M., Tseten, T., Anbalagan, N., Mathew, B. B., & Beeregowda, K. N. (2014). Toxicity, mechanism and health effects of some heavy metals. *Interdisciplinary toxicology*, *7*(2), 60.

Kar, S., & Roy, K. (2010). First report on interspecies quantitative correlation of ecotoxicity of pharmaceuticals. *Chemosphere*, *81*(6), 738-747.

Kumar, A., Cabral-Pinto, M., Kumar, A., Kumar, M., & Dinis, P. A. (2020). Estimation of risk to the eco-environment and human health of using heavy metals in the Uttarakhand Himalaya, India. *Applied Sciences*, *10*(20), 7078.

Kurniawan, T. A., Chan, G. Y., Lo, W. H., & Babel, S. (2006). Comparisons of low-cost adsorbents for treating wastewaters laden with heavy metals. *Science of the total environment*, *366*(2-3), 409-426.

Kurniawan, T. A., Chan, G. Y., Lo, W. H., & Babel, S. (2005). Physico–chemical treatment techniques for wastewater laden with heavy metals. *Chemical engineering journal*, *118*(1-2), 83-98.

Laitano, M. V., Chiaradia, N. M., & Nuñez, J. D. (2022). Human Impacts over Sandy Beaches. In *Sandy Beaches as Endangered Ecosystems* (54-88). CRC Press.

Le, T. T. N., Le, V. T., Dao, M. U., Nguyen, Q. V., Vu, T. T., Nguyen, M. H., ... & Le, H. S. (2019). Preparation of magnetic graphene oxide/chitosan composite beads for effective removal of heavy metals and dyes from aqueous solutions. *Chemical Engineering Communications*, *206*(10), 1337-1352.

Leung, W. C., Wong, M. F., Chua, H., Lo, W., Yu, P. H. F., & Leung, C. K. (2000). Removal and recovery of heavy metals by bacteria isolated from activated sludge treating industrial effluents and municipal wastewater. *Water science and technology*, *41*(12), 233-240.

Malik, L. A., Bashir, A., Qureashi, A., & Pandith, A. H. (2019). Detection and removal of heavy metal ions: a review. *Environmental Chemistry Letters*, *17*(4), 1495-1521.

Maurya, P. K., Malik, D. S., Yadav, K. K., Kumar, A., Kumar, S., & Kamyab, H. (2019). Bioaccumulation and potential sources of heavy metal contamination in fish species in River Ganga basin: Possible human health risks evaluation. *Toxicology Reports*, *6*, 472-481.

Mohajan, H. (2018). Acid rain is a local environment pollution but global concern.

Mutamim, N. S. A., Noor, Z. Z., Hassan, M. A. A., & Olsson, G. (2012). Application of membrane bioreactor technology in treating high strength industrial wastewater: a performance review. *Desalination*, *305*, 1-11.

Occhipinti-Ambrogi, A. (2021). Biopollution by invasive marine non-indigenous species: a review of potential adverse ecological effects in a changing climate. *International Journal of Environmental Research and Public Health*, *18*(8), 4268.

Paul, S., Mandal, A., Bhattacharjee, P., Chakraborty, S., Paul, R., & Mukhopadhyay, B. K. (2019). Evaluation of water quality and toxicity after exposure of lead nitrate in fresh water fish, major source of water pollution. *The Egyptian Journal of Aquatic Research*, *45*(4), 345-351.

Pedersen, A. J. (2003). Characterization and electrodialytic treatment of wood combustion fly ash for the removal of cadmium. *Biomass and Bioenergy*, *25*(4), 447-458.

Peilin, G., Meng, C., Lichao, Z., Yuejun, S., Minghao, M., & Lingyun, W. (2019, December). Study on water ecological restoration technology of river. In *IOP Conference Series: Earth and Environmental Science* (Vol. 371, No. 3, p. 032025). IOP Publishing.

Porta, R. (2021). Anthropocene, the plastic age and future perspectives. *FEBS Open Bio*, *11*(4), 948-953.

Saha, B., Karounou, E., & Streat, M. (2010). Removal of 17β-oestradiol and 17α-ethinyl oestradiol from water by activated carbons and hypercrosslinked polymeric phases. *Reactive and Functional Polymers*, *70*(8), 531-544.

Sato, K., Sakui, H., Sakai, Y., & Tanaka, S. (2002). Long-term experimental study of the aquatic plant system for polluted river water. *Water science and technology*, *46*(11-12), 217-224.

Shapiro, J. S. (2022). Pollution trends and US environmental policy: Lessons from the past half century. *Review of Environmental Economics and Policy*, *16*(1), 42-61.

Sharma, P. D. (2008). Population Growth and Environmental Degradation, Partha Das Sharma's weblog on keeping world environment safer and greener.

Sharrer, M. J., & Summerfelt, S. T. (2010). Evaluation of a membrane biological reactor for reclaiming water, alkalinity, salts, phosphorus, and protein contained in a high-strength aquacultural wastewater. *Bioresource technology*, *101*(12), 4322-4330.

Skubal, L. R., Meshkov, N. K., Rajh, T., & Thurnauer, M. (2002). Cadmium removal from water using thiolactic acid-modified titanium dioxide nanoparticles. *Journal of photochemistry and photobiology A: Chemistry*, *148*(1-3), 393-397.

Steenhoudt, O., & Vanderleyden, J. (2000). Azospirillum, a free-living nitrogen-fixing bacterium closely associated with grasses: genetic, biochemical and ecological aspects. *FEMS microbiology reviews*, *24*(4), 487-506.

Steffen, W., Crutzen, P. J., & McNeill, J. R. (2007). The Anthropocene: are humans now overwhelming the great forces of nature. *AMBIO: A Journal of the Human Environment* *36*(8), 614-621.

Tiseo, 2022. https://www.statista.com/topics/4739/environmental-pollution/#topicOverview

Tong, C. H., Yang, X. E., & Pu, P. M. (2003). Effects and mechanism of hydrophytes on control of release of nutrient salts in lake sediment. *Journal of Agro-Environment Science*, *22*(6), 673-676.

Wang, J., Liu, X. D., & Lu, J. (2012). Urban river pollution control and remediation. *Procedia Environmental Sciences*, *13*, 1856-1862.

Yuan, W., Yang, N., & Li, X. (2016). Advances in understanding how heavy metal pollution triggers gastric cancer. *BioMed research international*, *2016*.

Zhao, F., Zhang, S., Ding, Z., Aziz, R., Rafiq, M. T., Li, H., He, Z., Peter, J., Stofella & Yang, X. (2012). Enhanced purification of eutrophic water by microbe-inoculated stereo floating beds. *Polish Journal of Environmental Studies*, *22*(3), 557-564.

Zhao, Y., Yang, Z., Xia, X., & Wang, F. (2013). A shallow lake remediation regime with Phragmites australis: incorporating nutrient removal and water evapotranspiration. *Water research*, *46*(17), 5635-5644.

Zheng, L., & Wang, H. (2017). Effect of combined ecological floating bed for eutrophic lake remediation. In *AIP Conference Proceedings* (Vol. 1839, No. 1, p. 020058). AIP Publishing LLC.

Zimmels, Y., Kirzhner, F., & Malkovskaja, A. (2008). Application and features of cascade aquatic plants system for sewage treatment. *Ecological engineering*, *34*(2), 147-161.

# Chapter - 11

# Effect of Microplastics on Aquatic Food Chain and Food Web Altering Phytoplankton Community Structure

**Uzma Siddiqui[1*], Rashi Miglani[1], Nagma Parveen[1], Ankit Kumar[2], Sheetal Sharma[1] and Harish Chandra Singh Bisht[1]**

[1]Department of Zoology, DSB Campus, Kumaun University, Nainital-263001
[2]Department of Pharmaceuticals Sciences, Sir J C Bose Technical Campus, Bhimtal-263136
*Corresponding Author: uzma.labiba@gmail.com

## Abstract

In aquatic ecosystems, planktonic organisms play a crucial role in ecosystem services, such as biogeochemicaland nutrient cycling. Microplastics (MPs) influence on plankton may extend through the whole aquatic food chain via., trophic transfers. MPs in the ocean are not only present in the water but are also eaten by planktons and transported to other trophic levels by them. Coastal habitats and littoral zone vegetation are particularly vulnerable to plastic pollution in aquatic systems. The abundance of microplastics in aquatic habitats forces researchers to investigate their effects on ecosystem processes and food webs.

**Keywords:** Microplastics, Contamination, Biogeochemical cycling,

## Introduction

Microplastics are frequently referred to as synthetic polymers with a 5mm maximum size limit and no lower size limit. They can be divided into primary and secondary microplastics. Primary microplastics are defined as those plastics that were initially and are most frequently found in clothing, medications and personal care items like face and body washes. Secondary microplastics are formed as a result of the breakdown of bigger materials, such as macro-plastic fragmentation (e.g., plastic bags, food packaging, ropes) or microplastic release from terrestrial transportation (e.g., abrasion of automobile tyres during use is the most major source). Plastics in the environment can disintegrate slowly into tiny pieces due to a variety of reasons such as sunshine, bacteria and mechanical wear. Due to widespread usage and poor management of plastic products, more microplastics are entering the environment and have gained government and professional attention due to their contamination and risks. Microplastics have lately been found on a number of substrates around the world (Xiang *et al.*, 2022). Microplastics were first found in saltwater but have since then spread to coastal shallows, marine sediments, freshwater lakes, wetlands, inland soils and urban atmospheres (Silva *et al.*, 2018; Su *et al.*, 2018; Lv *et al.*, 2022). The potential negative effects of microplastic exposure on aquatic organisms have given rise to concerns. Aquatic animals that consume microplastics may suffocate or starve as a result of false satiety. Microplastics may also leak hazardous monomers or additives. Additionally, heavy metals, polybrominateddiphenyl ethers (PBDEs) and polycyclic aromatic hydrocarbons (PAHs) can all be transported through microplastics. Microplastics have also been shown to move tropically, and this process could be dangerous for animals at high trophic levels (Grigorakis *et al.*, 2017; Hermabessiere *et al.*, 2017; Zhang *et al.*, 2018a). Globally, microplastic pollution poses a serious threat to ecological communities. In aquatic ecosystems, phytoplankton plays an important ecological role by contributing to food webs and performing critical ecosystem functions, such as carbon cycling. There have been very few studies done so far on how microplastics affect phytoplankton, primarily using monocultures of algae in laboratories. Gowing microplastic pollution, phytoplankton communities will be influenced in unknown ways (Hitchcock, 2022).

Microplastics have been detected in terrestrial, aquatic, and atmospheric ecosystems worldwide. Plastic waste from several sources enters the ocean. Plastics are haphazardly tossed into or beside rivers along the coast, polluting the water. The majority of marine plastics originate on land. Due to the fact that half of the world's population lives within 50 miles of the ocean, various waterways transport plastic garbage to the coast. In water, drainage procedures destroy cosmetic and air-blasting polymers. Plastics in rivers contaminate freshwater systems from source to sea. Marine plastic pollution is mostly caused by tourism, fishing and maritime operations (aquaculture and oil rigs) (Chen *et al.*, 2021; Bhuyan *et al.*, 2021). Sources of microplastic in the aquatic ecosystem are explained in Figure 1. Aquaculture settings and products must be tested for pollution and human health consequences. This chapter covers the current distributions and characteristics of microplastics in aquaculture animals and ecosystems and discusses sources and potential health risks to animals and humans, raising concerns about microplastics and aquaculture products.

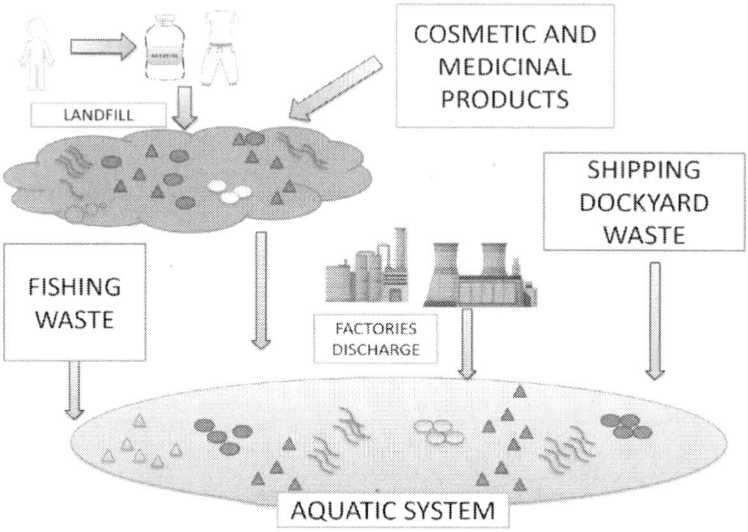


**Fig 1. Sources of microplastics in Aquatic Ecosystem**

## Distribution of microplastic at various trophic levels

The more human activity, the higher the concentration of microplastics (MPs). Aquatic animals devour MPs after exposure. Microplastic dispersion affects aquatic animals (Lyu *et al.*, 2020). MPs scattered in the aquatic system (water and sediment) produce a wide variety of toxic assaults on aquatic animals at different trophic levels, including behavior, morphology physiological function and abnormalities which may constitute a risk to human health (AnbumaniandKakkar, 2018; Rezania *et al.*, 2018).

*Chlorella* and *Scenedesmus*, two algae of the first nutrient grade, bind to charged polystyrene beads in a lab setting and prevent them from photosynthesizing (Bhattacharya *et al.*, 2010). *Skeletonema costatum* algae growth also slows downafter being exposed to MPs (Zhang *et al.*, 2017). Zooplankton absorbs MPs into the body during the second nutritional stage, dramatically reducing algal feeding and inducing oxidative stress reactions. Low-density polyethene MPs were also found in wild zooplankton using spectroscopic and digestive techniques. A number of species, including coelenterates (jellyfish), crustaceans (shrimps and crabs), mollusks (bivalves and sea cucumbers), fish, amphibians (tadpoles), reptiles (turtles) and mammals (whales), have been demonstrated in studies to consume MPs at higher trophic levels. It has been stated that MPs have been seen in human excrement at the top of the food chain (Zhang *et al.*, 2017; Sun *et al.*, 2017; Macali*et al.*, 2018; Yu *et al.*, 2018; Liu *et al.*, 2019; Lyu *et al.*, 2020). A simple distribution of microplastic in different trophic levels is explained in Fig 2.

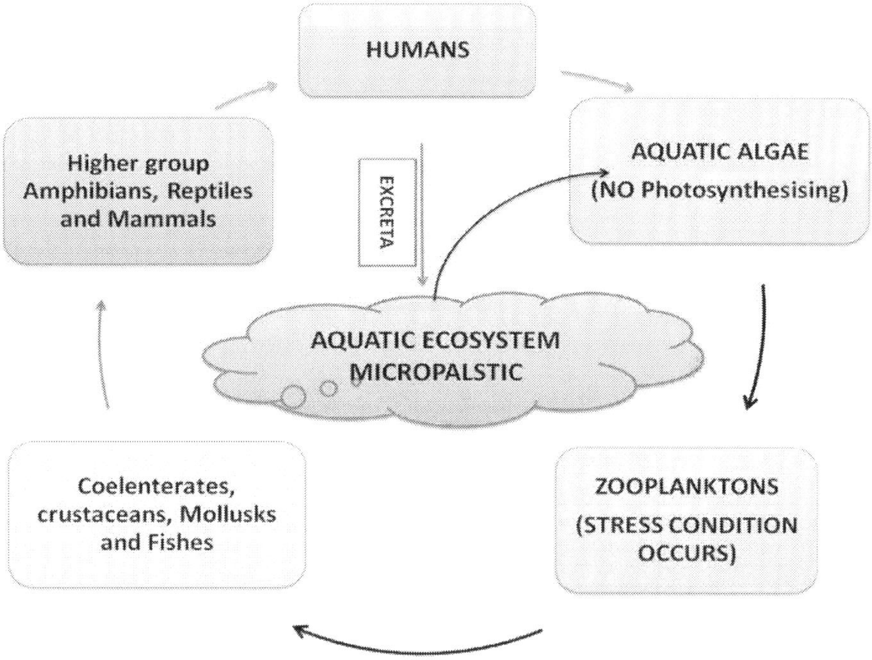


**Fig 2: Microplastic Distribution at Different trophic Levels**

## Bio-indicators of microplastic pollution in freshwater

Physical disintegration of big plastic debris is a primary entry point for MP into freshwater systems in underprivileged countries with poor waste management practices (Akindele*et al.*, 2019). Microplastics in lakes and rivers can come from a variety of sources, including tributaries, on-water activities, tourism, and inappropriate disposal of obsolete or discarded plastic wastes from terrestrial sources. Furthermore, rivers and lakes may become active secondary MP sources as a result of this process. The fragmentation and distribution of plastic waste left on riverbanks and wetlands that have grown brittle due to weathering and are readily broken by the flowing water and waves, similar to what happens at sea (Bellasi *et al.*, 2020).

Microplastic buildup in aquatic biota is often accomplished by oral intake or breathing in the aquatic environment and/or sediment (Watts *et al.*, 2014). This means that plankton, nekton, and macro zoobenthos all have a high chance of consuming MPs. Macro-zoobenthosmay be able to be used as MP bioindicators in lotic freshwater systems because most members live long enough to be able to measure environmental damage/pollution over time. Some macrozoobenthic organisms are also opportunistic feeders, which mean they eat whatever is in front of them. This makes them more likely to feed on MP (Nel *et al.*, 2018; Windsor *et al.*, 2019; Akindele *et al.*, 2019).

Akindele *et al.* (2019) investigated the accumulation of MPs in three gastropods i.e., *Melanoides tuberculate, Theodoxus fluviatilis* and *Lanistesvaricus* assess their usefulness as bioindicators of MPs. Windsor *et al.* (2019) and Akindele *et al.* (2020) assessed the presence of MPs in insects' bodies and found that the majority of insects from the families *Heptageniidae*, *Baetidae*, and *Hydropsychidae*, as well as the species *Siphlonurus*sp., *Chironomus* sp., and *Lestesviridis*, had consumed MPs.*Chironomus sp.* exhibited a much greater MP body burden per gram of wet body weight than the other species (Akindele *et al.*, 2020). In freshwater systems, invertebrates like *Chironomus sp.* have been highly suggested as MP bio-indicators. Several researchers have recently reported that deposit feeders like ephemeropterans and dipterans consume microplastics (Nel *et al.*, 2018; Windsor *et al.*, 2019).Vertebrate and invertebrate organisms are both considered to be bioindicators of MPs in freshwater. *Corbiculafluminea* is suggested as an invertebrate bioindicator of MP concentrations in contaminated aquatic environments. While vertebrates such as freshwater fishes are important bioindicators of MP pollution, they are also excellent model species for elucidating how MPs alter the biology and physiology of animals from the single organism to the community level. Some important freshwater organisms have been studied by researchers worldwide for the accumulation of MPs in their body/tissue and their potential use as bioindicators (Table 1).

**Table 1: Some important organisms accumulating MPs in their body/tissue in Freshwater Ecosystem**

| Organism | Taxonomic Position | Reference |
|---|---|---|
| *Siphlonurus* sp. | Phylum- Arthropoda<br>Class- Insecta<br>Order- Ephemeroptera<br>Family- Siphlonuridae | Akindele *et al.*, 2020 |
| *Chironomus* sp. | Phylum- Arthropoda<br>Class- Insecta<br>Order- Diptera<br>Family- Chironomidae | |
| *Lestes viridis* | Phylum- Arthropoda<br>Class- Insecta<br>Order- Odonata<br>Family- Lestidae | |
| *Melanoides tuberculata* | Phylum- Mollusca<br>Class- Gastropoda<br>Order- Caenogastropoda<br>Family-Thiaridae | Akindele *et al.*, 2019 |
| *Theodoxus fluviatilis* | Phylum- Mollusca<br>Class- Gastropoda<br>Order- Neritidae<br>Family- Lestidae | |
| *Lanistes varicus* | Phylum- Mollusca<br>Class- Gastropoda<br>Order- Architaenioglossa<br>Family- Ampullariidae | |
| *Carcinus maenas* | Phylum- Arthropoda<br>Class- Malacostraca<br>Order- Decapoda<br>Family- Portunidae | Watts *et al.*, 2014 |
| *Corbicula fluminea* | Phylum- Mollusca<br>Class- Bivalvia<br>Order- Veneroida<br>Family- Corbiculidae | Su *et al.*, 2018 |
| *Tubifex tubifex* | Phylum-Annelida<br>Class- Clitellata<br>Order- Tubificida<br>Family-Naididae | Hurley *et al.*, 2017 |

| | | |
|---|---|---|
| *Acipenser transmontanus* | Phylum- Chordata<br>Class-Actinopterygii<br>Order- Acipenseriformes<br>Family- Acipenseridae | Parker *et al.*, 2021 |
| *Ambassis dussumieri* | Phylum- Chordata<br>Class- Actinopterygii<br>Order- Perciformes<br>Family- Ambassidae | |
| *Barbodes gonionotus* | Phylum- Chordata<br>Class- Actinopterygii<br>Order- Cypriniformes<br>Family- Cyprinidae | |
| *Carassius auratus* | Phylum- Chordata<br>Class- Actinopterygii<br>Order- Cypriniformes<br>Family- Cyprinidae | |
| *Clarias gariepinus* | Phylum- Chordata<br>Class- Actinopterygii<br>Order- Siluriformes<br>Family- Clariidae | |
| *Cyprinus carpio* | Phylum- Chordata<br>Class- Actinopterygii<br>Order- Cypriniformes<br>Family- Cyprinidae | |
| *Daniorerio* | Phylum- Chordata<br>Class- Actinopterygii<br>Order- Cypriniformes<br>Family- Cyprinidae | |
| *Dicentrarchus labrax* | Phylum- Chordata<br>Class- Actinopterygii<br>Order- Perciformes<br>Family- Moronidae | |
| *Fundulus heteroclitus* | Phylum- Chordata<br>Class- Actinopterygii<br>Order- Cyprinodontiformes<br>Family- Fundulidae | |

## Transfer of microplastic in the aquatic food chain including phytoplankton

Quantification of microplastics in the field-collected organism, their natural predators and controlled feeding studies mimicking microplastic transmission through artificial food chains provide evidence for trophic transfer (Farrell and Nelson, 2013; Setälä*et al.*, 2014; Santana *et al.*, 2017; Carbery*et al.*, 2018).Marine organisms can consume microplastics (MPs) directly or indirectly via trophic transfer from polluted prey.In the food web of the marine ecosystem, zooplankton is an important link between phytoplankton and higher trophic levels (Botterell*et al.*, 2019; Costa *et al.*, 2020). The principal producer in the ocean is phytoplankton, which may use $CO_2$ absorbed from the atmosphere or the ocean to make organic materials and $O_2$ through photosynthesis.The growth and photosynthesis of phytoplankton can be impacted by microplastic. Despite their small size, phytoplankton is extremely important to marine ecosystems.

The primary production of marine ecosystems is responsible for around 80% of the world's total oxygen generation. Because of this, scientists are interested in learning more about phytoplankton because of the significant function; they play in the ocean carbon cycle (Witman, 2017).The extensive presence of microplastics in the ocean, on the other hand, has been shown to have a deleterious effect on the growth of phytoplankton, leading to a shift in the phytoplankton community and, as a result, a disruption in the equilibrium of the marine environment.The presence of a significant quantity of microplastics floating in the water has the potential to obstruct the passage of light, which in turn can reduce the phytoplankton's capacity for photosynthesis. Pollution caused by microplastics can have negative effects on phytoplankton, including their feeding, metabolism, development, and reproduction; the size of the microplastic particle also has a considerable impact on the degree to which it is hazardous (Sjollema *et al.*, 2016).When microplastics are present in water, phytoplankton can easily combine with them and aggregate, and the co-existence of microplastics is likely to inspire combination with phytoplankton as well as phytoplankton aggregation, which lowers the chance of phytoplankton encountering light and lowers the amount of primary production in marine environments (Shen *et al.*, 2020).

The first investigation to confirm trophic transmission of plastics came almost a decade after their discovery in sea lion scats, with noticeable effects on the feeding behavior of *Carassius carassius* exposed to MPs. Polystyrene nanoparticles (24 nm) were fed through a freshwater food chain from algae (*Scenedesmus sp.)* to zooplankton (*Daphnia magna*) to goldfish (*C. carassius*) (Cedervall*et al.*, 2012). *Chlorella and Scenedesmus* can produce more active oxygen, but their chlorophyll content is decreased when exposed to polystyrene nano plastics (1.6–40 mg/mL). Phytoplankton (*Dunaliella tertiolecta*) exposure to microplastics (250 mg/L) decreased the photosynthetic rate by 45% (Bhattacharya *et al.*, 2010).

Chytrids play a significant role in the biotic regulation of phytoplankton, influencing community dynamics and modulating or suppressing hazardous phytoplankton blooms. The dynamics of algal blooms, the makeup of the phytoplankton community, and overall primary output may all be affected if chytrid epidemics are reduced, for example by nano plastics exposure (Frenken *et al.*, 2017; Gerphagnon *et al.*, 2019). On the other hand, disease alters the trophic relationships that already exist in the food web and creates new trophic links (Selakovic *et al.*, 2014) (e.g., by serving as food for other organisms; Johnson *et al.*, 2010). By acting as zooplankton prey, chytrids release carbon that would otherwise be held in poorly consumable phytoplankton hosts for consumers (Agha *et al.*, 2016; Kagami*et al.*, 2007). The mechanism allows for trophic transfer in pelagic food webs by upgrading phytoplankton carbon into lipids with high nutritional value (Gerphagnon *et al.*, 2019) and enhancing herbivory in the system by undermining phytoplankton resistance to grazers (Frenken *et al.*, 2020). Finally, there is growing evidence that chytrid parasitism is a significant force in the evolution of phytoplankton. Field investigations and experimental evolution tests demonstrate that chytrid-induced selective pressure encourages genetic diversity in phytoplankton communities (Agha *et al.*, 2018; De Bruin *et al.*, 2008; Gsell *et al.*, 2013). Overall, nano plasticscan influence the scope and severity of underlying ecological and evolutionary processes via., influencing disease processes.

According to Besseling et *al.* (2014), growth of *Scenedesmus obliquus* and development were impeded after being exposed to polystyrene nano-plastics, and the amount of chlorophyll production considerably decreased. Additionally, when growth circumstances are constrained, *Chaetoceros sp.* and *Rhodopsis salina* can release mucilaginous compounds such as polysaccharides to create an algal mass and polymerize with the nearby microplastics (Underwood *et al.*, 2004). According to Long *et al.* (2017), this behavior can alter the density of the algal mass and alter how it is distributed in the seawater in addition to facilitating the passage of low-density microplastics to the sea floor (Ward and Kach, 2009). Quantification of microplastics in field-collected creatures, plastic particles in their natural predators, and controlled feeding trials mimicking the transfer of microplastics through artificial food chains provide evidence for trophic transfer. Laboratory exposure studies are popular for studying bioaccumulation and biomagnification, but the consequences for higher-order predators and humans are unclear. It is possible that the reactions to being exposed to plastic could have repercussions on the aquatic food chain's ability to remain viable. Therefore, to reduce the detrimental effect of MP's further research is required to comprehend the impact's underlying mechanisms, magnitudeand breadth.

**Factors determining the interaction between microplastics and phytoplankton**

Several factors affect the toxicity of microplastics, including their concentrations in theenvironment, polymer type and size, presence of additives, surface chemistry and charge. These factors summarize how microplastics interact with phytoplankton, resulting immeasurable changes in their growth.

**1. Polymer type and size**: The term 'plastic' includes many different polymers, but themost abundant classes of polymer detected in aquatic environments are polyethylene (PE), polypropylene (PP), polystyrene (PS), polyester (PEST), polyamide (PA), andacrylic (Nava and Leoni. 2020). Each microplastic shape may be represented bydifferent polymer type as manufacturers seek to produce plastics with specificproperties (e.g., flexibility, roughness, resistance, and durability) (Espí *et al.*, 2006). However, these polymer types are composed of different monomers, which canpotentially be hazardous for the environment. For instance, polyurethane (PU), apolymer used to produce flexible foams, is made of monomers highly toxic forhumans (Lithner *et al.*, 2011). Hence, several studies have shown that polymer typeand sizes are the factors that may contribute to phytoplankton toxicity. Lagarde *et al.* (2016) reported growth inhibition for polypropylene but not for high densitypolyethylene in *Chlamydomonas reinhardtii*. Furthermore, polymer type and sizehave a direct impact on its distribution across the water column and may lead tosedimentation (Davarpanah and Guilhermino, 2015) or buoyancy (Zhang *et al.*, 2017), modulating exposure. Regarding size, generally, smaller microplastics are more toxicto microalgae. For instance, polymer size of 0.05 μm induce a more marked decreasing cell density than 0.5 and 6 μm (Sjollema *et al.*, 2016) and polyvinyl chloride (PVC) cause negative effects on growth, chlorophyll and photosynthesis at the size of 1 μmbut not of 1 mm (Zhang *et al.*, 2017). Mao *et al.* (2018) observed cell changes in *Chlorella pyrenoidosa* after exposure topolystyrene microplastic with particle sizes ranging from 0.1 m and 1 m. The resultsshowed that cells exposed to microplastics had indistinct pyrenoids with cell wallsdetached from the plasma membrane. Thylakoid cells also appear indistinct anddistorted, indicating a disturbance in photosynthesis. Polymer and size also determinesurface chemistry, leading to variations in toxicity (Casado-Gavalda *et al.*, 2013).

**2. Charge**: Unlike many natural particles, microplastics can be charged electrostatically. Positively charged microplastics have higher interaction and toxicity to microalgae thannegatively charged microplastics (Bhattacharya *et al.*, 2010; Casado-Gavalda *et* *al.*, 2013; Nolte *et al.*, 2017). This probably results from the anionic cellulose in the cell wall, with carboxyland sulphate groups repelling negatively charged microplastics and adsorbing positivelycharged microplastics through electrostatic interaction, hydrogen bonds and hydrophobicinteraction, varying with algae morphology (Bhattacharya *et al.*, 2010).

**3. Concentration**: Exposure to microplastics with higher concentrations showed a moresevere impact on the damage to the morphology of phytoplankton cells. Wang *et al.* (2020a) observed the appearance of folds, small holes and cracks on the cell surface of the phytoplankton type *Platymonas helgoglandica* after exposure to microplastic type PS (polystyrene) up to 2000 g/L. Along with the increase in the concentration ofmicroplastics, the surface of the microalgae cells increasingly showed deformationcharacterized by tearing and withering of the cell surface. An increase in theconcentration of microplastics also causes cell wall damage and cytoplasmic vacuolation, namely the formation of vacuoles in the cell cytoplasm due to foreign material enteringthe cell or bacterial and viral infections (Aki *et al.*, 2012). Cytoplasmic vacuolation whichis irreversible indicates that the cell is under continuous stress which can eventually leadto cell death (Shubin *et al.*, 2016).Cell volume has also been reported in several studies to be affected by microplasticexposure. *Chlorella pyrenoidosa* exposed to PS microplastics at concentrations of 0-3mg/L and 10.5-30 mg/L reduced cell volume, while concentrations of 3- 10.5 mg/L wereeffective in increasing cell size (Li *et al.*, 2020). Nutrient loss and cell shrinkage due toplasmolysis, vacuolation and cytoderm are thought to be the main causes of reduced cellvolume at low exposure concentrations. When the concentration increases, hetero-aggregation occurs which protects the phytoplankton cells, but as the concentrationcontinues to increase it causes damage to chlorophyll so that it interferes with thephotosynthetic process of cells. Exposure to microplastics is also reported to reduce the content of chlorophyll A, whichwill result in decreased photosynthetic activity. A total of 62.82% reduction inchlorophyll A content occurred in *Phaeodactyl umtricornutum* exposed to 200 mg/L PVCmicroplastic for 96 hours (Lang *et al.*, 2022). Hence, concentration and duration ofexposure to microplastics has been shown to have an impact on the morphology ofphytoplankton cells.

**4. Plastic additives**: Plastic additives, such as plasticizers, flame retardants, stabilizers, antioxidants, and pigments are chemical compounds added to improve the performance and functionality of a plastic product (Hahladakis *et al.*, 2018; Hermabessiere *et al.*, 2017). These additives are not usually bound to the polymer andthey can readily leach from plastics into the water (Oliviero *et al.*, 2019). Once theplastic waste enters the aquatic ecosystem such as rivers, lakes, and oceans, leachingof chemical additives from these plastics is a critical issue (Romera-Castillo *et al.*, 2018). Previous studies have shown that the plastic leachates could cause acute toxicity to the freshwater species *Daphnia magna* (Bejgarn *et al.*, 2015).

**5. Aggregation**: Several studies show that phytoplankton cells can carry outdetoxification mechanism as a form of defense. One such mechanism is formingaggregates with fellow phytoplankton cells as well as with microplastic particles, which biochemically boosts the activity of enzymes involved in detoxification. Thehydrophobic properties of microplastics can lead to the formation of aggregations. When the phytoplankton cells form an aggregate, a process known as homo- aggregation, the phytoplankton community protects the cells from further damage. This cell aggregation reduces the surface area to cell volume ratio, thereby minimizing the number of microplastic particles that can adhere to cells (Chen *et al.*, 2012). Phytoplankton cells can also form aggregates with microplastic particles, which is known as hetero-aggregation. The formation of aggregate betweenphytoplankton cells and microplastic particles will facilitate microplastics to bedeposited to the bottom of the water, thereby reducing the concentration ofmicroplastics in the water column. Thus, the phytoplankton cells that are stillsuspended in the water column are protected from exposure to microplastics.

**6. Membrane thickening**: Increased extracellular polymeric substance (EPS) secretion canhelp phytoplankton cells to reduce the negative impact of microplastics by thickeningtheir cell walls. Microplastics can adhere to the surface of the EPS and preventmicroplastics from entering the cell plasma. This thickening can also reduce the physicaldamage that can occur from exposure to microplastics. Microplastics can adhere to EPSby forming hydrogen bonds or through electrostatic interactions (Quigg *et al.*, 2013). Theinteraction between EPS and microplastics will reduce the surface charge andhydrodynamic diameter of microplastics, thereby reducing their bioavailability to cells (Bellingeri *et al.*, 2019). However, due to the influence of several factors such as particlesize that is too small (nanoplastic) and surface charge, microplastic particles can enter thecell plasma even though the cell has carried out a detoxification mechanism by secretingEPS. Microplastic particles can enter the plasma of phytoplankton cells by endocytosisand phagocytosis mechanisms (Long *et al.*, 2017).

## Effects of microplastics on aquatic organism

Nanoplastics, like plastic beads present in specialty care products, are released into the environment (Gewert *et al.*, 2015; Bouwmeester *et al.*, 2015). Microplastics have a direct harmful influence on aquatic species at the molecular and physical levels

(Teuten *et al.*, 2009). Microplastic releases toxic leachates that impair development and survival. When microplastic moves through the food web and is eaten by plankton, invertebrates, and fish, essential species for commercial fisheries end up there. There have already been reports of plastic fibers and trash in fish and shellfish intended for human consumption (Rochman *et al.*, 2015, Li*et al.*, 2015a).

Research on the effects of microplastics (MP) on aquatic creatures is currently quite active. Numerous studies on macro-plastic ingestion in vertebrates (Yamashita *et al.*, 2011; Denuncio *et al.*, 2011; Lazar and Gracan, 2011) have reported widespread effects, such as internal and/or external abrasions and ulcers as well as blockages of the digestive tract that can cause satiation, starvation, and physical deterioration. In turn, this may result in decreased ability to reproduce, drowning, decreased ability to evade predators, decreased ability to feed, the potential transmission of harmful toxicants from seawater, and finally death (Gregory, 2009). Plastic microspheres have the potential to be prey analogues for planktivorous and may be handled and consumed in a similar way to algae because of the similar specific gravities of both (Brillant and MacDonald, 2000). There aren't many studies on the bioaccumulation of POPs and related polymers across the marine food chain. Microplastics are likely to enter the food chain because lower trophic creatures, particularly invertebrates, can consume and collect microplastic particles. Although in situ research has found microplastic ingestion in various vertebrate species (Wright *et al.*, 2013a).

Benthic deposit- and suspension-feeders are likely to consume sedimentary and sinking microplastics. The types of microplastics that are most frequently found in the marine environment are fibers. The selective consumption of microplastics by benthic holothurians was discovered, with a predilection for fibrous forms. Furthermore, according to the discovery of nylon fibres in *N. norvegicus*' stomach during a gut content investigation, benthic scavengers are vulnerable to exposure to fibrous microplastics. This means that their environment serves as a fiber sink.

### Microplastic effecting zooplankton and phytoplankton

Microplastics being small size are highly bioavailable and easy ingested by aquatic organisms like molluscs, echinoderms, corals, vertebrates' larval stage, green sea turtle, cetaceans and seabirds having adverse effects on their health (Botterell *et al.*, 2019; Caron *et al.*, 2018). Zooplankton in aquatic environments are a food source for many other aquatic organisms, connecting phytoplankton (primary producers) with carnivores such as fish that play a key role in carbon and nutrient cycling. Microplastic can hinder the food web and therefore affect element cycling according to recent research (Lin, 2016). Microplastic beads ingested by the copepod species

*C. helgolandicus* change feeding behavior which results in preference for smaller prey, and negative effect on reproduction, that is manifested by smaller eggs and lower hatching success. Microplastics block feeding appendages and curb food intake, and may hinder or bruise the alimentary canal (Cole *et al.*, 2013). It was reported that plankton, bivalves, fish and mammals could be affected by the presence of MPs (Besseling *et al.*, 2015; Van Cauwenberghe and Janssen, 2014; Desforges *et al.*, 2015; Foekema *et al.*, 2013). Microplastics have accumulated in the ocean and threaten marine and fresh life from plankton to fish and mammals when present in large concentration (Avio *et al.*, 2017).

Setälä *et al.* (2018) showed that 33% of the gooseneck barnacles (*Lepas* spp.) ingested 1 - 30 microplastic particles. Microplastics in the marine environment destined up in seafood as fish, bivalves and crustaceans (Gallo *et al.*, 2018). Microalgae being the primary producers in the aquatic ecosystem play an important role in maintaining the long-term stability of the aquatic environment. Studies mainly focus on the toxic effects of microplastics on aquatic organisms only on animals but rarely on phytoplankton, especially microalgae. Growth inhibition is the toxic effect of microplastics on algae (Yang *et al.*, 2022; Wang *et al.*, 2020b; Zhu *et al.*, 2020).

Decreasing size of microplastics increases the growth inhibition (Yang *et al.*, 2020; Zhao *et al.*, 2020). Along with the inhibition effects of microplastics on microalgae, the metabolic responses of algal cells were also affected obviously (González *et al.*, 2019; Zheng *et al.*, 2021). Microplastic have different shape and sharp edges, due to which it causes mechanical damage to the cells as cell wall damage and cracked cell, causing algal death (Zhu *et al.*, 2018; Yi *et al.*, 2019; Wang *et al.*, 2020c; Song *et al.*, 2020). As there is an increase in microplastic concentration mechanical damage becomes more severe, as this damage is also because of nanoparticles of plastic like Nano-TiO2, ZnO etc. There is no information that microplastic can enter living algal cells(Li *et al.*, 2015; Zhang *et al.*, 2018b; Hu *et al.*, 2015).New research is needed to prove that microplastics can enter microalgae cells and causes intracellular mechanical damage (Dawson *et al.*, 2018, Da Costa *et al.*, 2016, Nichols *et al.*, 2021; Llorca, *et al.*, 2021).*N. norvegicus'* species might be thought of as sensitive to the possible physical toxicity of microplastics since shape may play a factor in the toxicity of ingested microplastics- long, rod-shaped nanoparticles are thought to be more poisonous than spherules (Wright *et al.*, 2013b).

## Conclusion

Microplastics negatively affect phytoplankton at low ppm concentrations by preventing growth, lowering chlorophyll and photosynthesis, altering morphology, and encouraging the formation of hetero-aggregates. Phytoplankton, on the other hand, appears to recover from these changes through adaptative responses, and current environmental concentrations areunlikely to be harmful. Microplastics, however, have the potential to disrupt phytoplanktonpopulations by lowering the amount of nutrients available, by impeding primary consumers, or by serving as a substrate. All of these modifications depend on particular microplasticcharacteristics, such as polymer type, size, and surface charge. As a result, the delivery ofecosystem services like carbon sequestration and nutrient cycling as well as the ecologicalfunctions of these groups may be compromised by the presence of microplastics in aquaticenvironments. So, further investigations are urgently needed, as this theme is of pivotalimportance for aquatic environments and their ecosystem services.

## Bibliography

Agha, R., Gross, A., Gerphagnon, M., Rohrlack, T., and Wolinska, J. (2018). Fitness and Eco physiological response of a chytrid fungal parasite infecting planktonic cyanobacteria to thermal and host genotype variation. *Parasitology, 145*, 1279-1286.

Agha, R., Saebelfeld, M., Manthey, C., Rohrlack, T., and Wolinska, J. (2016). Chytrid parasitism facilitates trophic transfer between bloom-forming cyanobacteria and zooplankton (Daphnia). *Scientific Reports*, *6*(1), 1-9.

Aki, T., Nara, A., and Uemura, K. (2012). Cytoplasmic vacuolization during exposure to drugs and other substances. *Cell Biology and Toxicology*, *28*(3), 125-131.

Akindele, E.O., Ehlers, S.M., and Koop, J. (2020). Freshwater insects of different feeding guilds ingest microplastics in two Gulf of Guinea tributaries in Nigeria. *Environmental science and pollution research international*, *27*(26), 33373–33379. https://doi.org/10.1007/s11356-020-08763-8

Akindele, E.O, Ehlers, S.M., and Koop, J.H.E. (2019).First empirical study of freshwater microplastics in West Africa using gastropods from Nigeria as bioindicators. *Limnologica*, 78. https://doi.org/10.1016/j.limno.2019.125708

Anbumani, S., and Kakkar, P. (2018). Ecotoxicological effects of microplastics on biota: a review. *Environmental Science and Pollution Research*, *25*(15), 14373-14396.

Avio, C.G., Gorbi, S., and Regoli, F. (2017). Plastics and microplastics in the oceans: from emerging pollutants to emerged threat. *Marine environmental research*, *128*, 2-11.

Bejgarn, S., MacLeod, M., Bogdal, C., and Breitholtz, M. (2015). Toxicity of leachate from weathering plastics: An exploratory screening study with *Nitocraspinipes*. *Chemosphere*, 132, 114–119.

Bellasi, A., Binda, G., Pozzi, A., Galafassi, S., Volta, P., and Bettinetti, R. (2020). Microplastic Contamination in Freshwater Environments: A Review, Focusing on Interactions with

Sediments and Benthic Organisms. *Environments*, *7*(4), 30.https://doi.org/10.3390/environments7040030

Bellingeri, A., Bergami, E., Grassi, G., Faleri, C., Redondo-Hasselerharm, P., Koelmans, A. A., and Corsi, I. (2019). Combined effects of nanoplastics and copper on the freshwater alga *Raphidocelissubcapitata*. *Aquatic Toxicology*, 210, 179-187.

Besseling, E., Foekema, E.M., Van Franeker, J.A., Leopold, M.F., Kühn, S., Rebolledo, E.B., Hebe, E., Mielke, L., Ijzer, J., and Koelmans, A.A. (2015). Microplastic in a macro filter feeder: humpback whale *Megapteranovaeangliae*. *Marine pollution bulletin*, *95*(1), 248-252.

Besseling, E., Wang, B., Lürling, M., and Koelmans, A.A. (2014). Nanoplastic affects growth of *S. obliquus* and reproduction of D. magna. *Environmental science & technology*, *48*(20), 12336-12343.

Bhattacharya, P., Lin, S., Turner, J.P., and Ke, P.C. (2010). Physical adsorption of charged plastic nanoparticles affects algal photosynthesis. *The journal of physical chemistry C*, *114*(39), 16556–16561.https://doi.org/10.1021/jp1054759.

Bhuyan, M.S., Venkatramanan, S., Selvam, S., Szabo, S., Hossain, M.M., Rashed-Un-Nabi, M., and Islam, M.S. (2021). Plastics in marine ecosystem: A review of their sources and pollution conduits. *Regional Studies in Marine Science*, *41*, 101539.

Botterell, Z.L., Beaumont, N., Dorrington, T., Steinke, M., Thompson, R.C., and Lindeque, P.K. (2019). Bioavailability and effects of microplastics on marine zooplankton: A review. *Environmental Pollution*, *245*, 98-110.

Bouwmeester, H., Hollman, P.C., and Peters, R.J. (2015). Potential health impact of environmentally released micro-and nanoplastics in the human food production chain: experiences from nanotoxicology. *Environmental science & technology*, *49*(15), 8932-8947.

Brillant, M.G.S., and MacDonald, B.A. (2000). Post-ingestive selection in the sea scallop, *Placopectenmagellanicus* (Gmelin): the role of particle size and density. *Journal of Experimental Marine Biology and Ecology*, *253*(2), 211-227.

Carbery, M., O'Connor, W., and Palanisami, T. (2018). Trophic transfer of microplastics and mixed contaminants in the marine food web and implications for human health. *Environment international*, *115*, 400-409.

Caron, A.G., Thomas, C.R., Berry, K.L., Motti, C.A., Ariel, E., and Brodie, J.E. (2018). Ingestion of microplastic debris by green sea turtles (*Cheloniamydas*) in the Great Barrier Reef: Validation of a sequential extraction protocol. *Marine Pollution Bulletin*, *127*, 743-751.

Casado-Gavalda, M., Macken, A., and Byrne, H.J. (2013). Ecotoxicological assessment of silica and polystyrene nanoparticles assessed by a multitrophic test battery. *Environment International*, *51*, 97–105. https://doi.org/10.1016/j.envint.2012.11.001.

Cedervall, T., Hansson, L.A., Lard, M., Frohm, B., and Linse, S. (2012). Food chain transport of nanoparticles affects behaviour and fat metabolism in fish. *PloS one*, *7*(2), e32254. http://dx.doi.org/10.1371/journal.pone.0032254.

Chen, G., Li, Y., and Wang, J. (2021). Occurrence and ecological impact of microplastics in aquaculture ecosystems. *Chemosphere*, *274*, 129989.

Chen, P., Powell, B.A, Mortimer, M., and Ke, P.C. (2012). Adaptive interactions between zinc oxide nanoparticles and *Chlorella* sp. *Environmental Science & Technology*, *46*(21), 12178-12185.

Cole, M., Lindeque, P., Fileman, E., Halsband, C., Goodhead, R., Moger, J., and Galloway, T.S. (2013). Microplastic ingestion by zooplankton. *Environmental science & technology*, *47*(12), 6646-6655.

Costa, E., Piazza, V., Lavorano, S., Faimali, M., Garaventa, F., and Gambardella, C. (2020). Trophic transfer of microplastics from copepods to jellyfish in the marine environment. *Frontiers in Environmental Science*, *8*, 571732.

Da Costa, J.P., Santos, P.S., Duarte, A.C., and Rocha-Santos, T. (2016). (Nano) plastics in the environment–sources, fates and effects. *Science of the total environment*, *566*, 15-26.

Davarpanah, E., and Guilhermino, L. (2015). Single and combined effects of microplastics and copper on the population growth of the marine microalgae *Tetraselmischuii*. *Estuarine, Coastal and Shelf Science*, 167, 269–275. https://doi.org/10.1016/j.ecss.2015.07.023.

Dawson, A.L., Kawaguchi, S., King, C.K., Townsend, K.A., King, R., Huston, W.M., and Bengtson Nash, S.M. (2018). Turning microplastics into nanoplastics through digestive fragmentation by Antarctic krill. *Nature communications*, *9*(1), 1-8.

De Bruin, A., Ibelings, B.W., Kagami, M., Mooij, W.M., and Van Donk, E. (2008). Adaptation of the fungal parasite *Zygorhizidiumplanktonicum* during 200 generations of growth on homogeneous and heterogeneous populations of its host, the diatom *Asterionellaformosa*. *Journal of Eukaryotic Microbiology*, *55*(2), 69-74.

Denuncio, P., Bastida, R., Dassis, M., Giardino, G., Gerpe, M., and Rodríguez, D. (2011). Plastic ingestion in Franciscana dolphins, *Pontoporiablainvillei* (Gervais and d'Orbigny, 1844), from Argentina. *Marine pollution bulletin*, *62*(8), 1836-1841.

Desforges, J.P.W., Galbraith, M., and Ross, P.S. (2015). Ingestion of microplastics by zooplankton in the Northeast Pacific Ocean. *Archives of environmental contamination and toxicology*, *69*(3), 320-330.

Espí, E., Salmerón, A., Fontecha, A., García, Y., and Real, A.I. (2006). Plastic films for agricultural applications. *Journal of Plastic Film and Sheeting*, *22*(2), 85–102. doi: 10.1177/8756087906064220.

Farrell, P., and Nelson, K. (2013). Trophic level transfer of microplastic: *Mytilusedulis* (L.) to *Carcinusmaenas* (L.). *Environmental pollution*, *177*, 1-3.

Foekema, E.M., De Gruijter, C., Mergia, M.T., van Franeker, J.A., Murk, A.J., and Koelmans, A.A. (2013). Plastic in North Sea fish. *Environmental science & technology*, *47*(15), 8818-8824.

Frenken, T., Alacid, E., Berger, S.A., Bourne, E.C., Gerphagnon, M., Grossart, H.P., and Agha, R. (2017). Integrating chytrid fungal parasites into plankton ecology: research gaps and needs. *Environmental microbiology*, *19*(10), 3802-3822.

Frenken, T., Wolinska, J., Tao, Y., Rohrlack, T., and Agha, R. (2020). Infection of filamentous phytoplankton by fungal parasites enhances herbivory in pelagic food webs. *Limnology and Oceanography*, *65*(11), 2618-2626.

Gallo, F., Fossi, C., Weber, R., Santillo, D., Sousa, J., Ingram, I., Nadal, A.,and Romano, D. (2018). Marine litter plastics and microplastics and their toxic chemicals components: the need for urgent preventive measures. *Environmental Sciences Europe*, *30*(1), 1-14.

Gerphagnon, M., Agha, R., Martin-Creuzburg, D., Bec, A., Perriere, F., Rad-Menéndez, C., and Wolinska, J. (2019). Comparison of sterol and fatty acid profiles of chytrids and their hosts reveals trophic upgrading of nutritionally inadequate phytoplankton by fungal parasites. *Environmental microbiology*, *21*(3), 949-958.

Gewert, B., Plassmann, M.M., and MacLeod, M. (2015). Pathways for degradation of plastic polymers floating in the marine environment. *Environmental science: processes & impacts*, *17*(9), 1513-1521.

González-Fernández, C., Toullec, J., Lambert, C., Le Goïc, N., Seoane, M., Moriceau, B., Huvet, A., Berchel, M., Vincent, D., Courcot, L., and Paul-Pont, I. (2019). Do transparent exopolymeric particles (TEP) affect the toxicity of nanoplastics on *Chaetocerosneogracile*? *Environmental Pollution*, *250*, 873-882.

Gregory, M.R. (2009). Environmental implications of plastic debris in marine settings-entanglement, ingestion, smothering, hangers-on, hitch-hiking and alien invasions. *Philosophical Transactions of the Royal Society B: Biological Sciences*, *364*(1526), 2013-2025.

Grigorakis, S., Mason, S.A., and Drouillard, K.G.(2017). Determination of the gut retention of plastic microbeads and microfibers in goldfish (*Carassiusauratus*). *Chemosphere*, *169*, 233–238.

Gsell, A. S., de SenerpontDomis, L. N., Verhoeven, K. J., Van Donk, E., and Ibelings, B. W. (2013). Chytrid epidemics may increase genetic diversity of a diatom spring-bloom. *The ISME journal*, *7(10)*, 2057-2059.

Hahladakis, J.N., Velis, C.A., Weber, R., Iacovidou, E., and Purnell, P. (2018). An overview of chemical additives present in plastics: Migration, release, fate and environmental impact during their use, disposal and recycling. *Journal of hazardous materials*, *344*, 179–199.

Hermabessiere, L., Dehaut, A., Paul-Pont, I., Lacroix, C., Jezequel, R., Soudant, P., and Duflos, G. (2017). Occurrence and effects of plastic additives on marine environments and organisms: a review. *Chemosphere*, *182*, 781–793.

Hitchcock, J.N. (2022). Microplastics can alter phytoplankton community composition. *Science of The Total Environment*, *819*, 153074.

Hu, X., Ouyang, S., Mu, L., An, J., and Zhou, Q. (2015). Effects of graphene oxide and oxidized carbon nanotubes on the cellular division, microstructure, uptake, oxidative stress, and metabolic profiles. *Environmental science & technology*, *49*(18), 10825-10833.

Hurley, R.R., Woodward, J.C., and Rothwell, J.J. (2017). Ingestion of Microplastics by Freshwater Tubifex Worms. *Environmental science & technology*, *51*(21), 12844–12851. https://doi.org/10.1021/acs.est.7b03567

Johnson, P.T., Dobson, A., Lafferty, K.D., Marcogliese, D.J., Memmott, J., Orlofske, S.A., and Thieltges, D.W. (2010). When parasites become prey: ecological and epidemiological significance of eating parasites. *Trends in ecology & evolution*, *25*(6), 362-371.

Kagami, M., de Bruin, A., Ibelings, B. W., and Van Donk, E. (2007). Parasitic chytrids: their effects on phytoplankton communities and food-web dynamics. *Hydrobiologia*, *578*(1), 113-129.

Lagarde, F., Oliver, O., Zanella, M., Daniel, P., Hiard, S., and Caruso, A. (2016). Microplastic interactions with freshwater microalgae: hetero-aggregation and changes in plastic density appear strongly dependent on polymer type. *Environmental Pollution*, 215, 331–339. https://doi.org/10.1016/j.envpol.2016.05.006.

Lang, X., Ni, J., and He, Z. (2022). Effects of polystyrene microplastic on the growth and volatile halocarbons release of microalgae *Phaeodactylumtricornutum*. *Marine Pollution Bulletin*, *174*, 113197.

Lazar, B., and Gračan, R. (2011). Ingestion of marine debris by loggerhead sea turtles, *Carettacaretta*, in the Adriatic Sea. *Marine pollution bulletin*, *62*(1), 43-47.

Li, F., Liang, Z., Zheng, X., Zhao, W., Wu, M., and Wang, Z. (2015). Toxicity of nano-TiO2 on algae and the site of reactive oxygen species production. *Aquatic Toxicology*, *158*, 1-13.

Li, J., Yang, D., Li, L., Jabeen, K., and Shi, H. (2015a). Microplastics in commercial bivalves from China. *Environmental pollution*, *207*, 190-195.

Li, Z., Yi, X., Zhou, H., Chi, T., Li, W., and Yang, K. (2020). Combined effect of polystyrene microplastics and dibutyl phthalate on the microalgae *Chlorella pyrenoidosa*. *Environmental Pollution*, *257*, 113604.

Lin, V.S. (2016). Research highlights: impacts of microplastics on plankton. *Environmental Science: Processes & Impacts*, *18*(2), 160-163.

Lithner, D., Larsson, A., and Dave, G. (2011). Environmental and health hazard ranking and assessment of plastic polymers based on chemical composition. *Science of the Total Environment*, *409*(18), 3309–3324. doi: 10.1016/j.scitotenv.2011.04.038.

Liu, Z., Yu, P., Cai, M., Wu, D., Zhang, M., Huang, Y., and Zhao, Y. (2019) Polystyrene nanoplastic exposure induces immobilization, reproduction, and stress defense in the freshwater cladoceran*Daphniapulex*. *Chemosphere*, *215*, 74–81.

Llorca, M., Vega-Herrera, A., Schirinzi, G., Savva, K., Abad, E., and Farré, M. (2021). Screening of suspected micro (nano) plastics in the Ebro Delta (Mediterranean Sea). *Journal of hazardous materials*, *404*, 124022.

Long, M., Paul-Pont, I., Hegaret, H., Moriceau, B., Lambert, C., Huvet, A., and Soudant, P. (2017). Interactions between polystyrene microplastics and marine phytoplankton lead to species-specific hetero-aggregation. *Environmental Pollution*, *228*, 454-463.

Lv, M., Jiang, B., Xing, Y., Ya, H., Zhang, T., and Wang, X. (2022). Recent advances in the breakdown of microplastics: strategies and future perspectives. *Environmental Science and Pollution Research*, 1-17.

Lyu, W., Chen, Q., Cheng, L., and Zhou, W. (2020). Microplastics in aquaculture systems and their transfer in the food chain. *Microplastics in Terrestrial Environments*, 357-382.

Macali, A., Semenov, A., Venuti, V., Crupi, V., D'Amico, F., Rossi, B., Corsi, I., and Bergami, E. (2018). Episodic records of jellyfish ingestion of plastic items reveal a novel pathway for trophic transference of marine litter. *Scientific reports*, *8(1)*, 6105. https://doi.org/10.1038/s41598-018-24427-7

Mao, Y., Ai, H., Chen, Y., Zhang, Z., Zeng, P., Kang, L., and Li, H. (2018). Phytoplankton response to polystyrene microplastics: perspective from an entire growth period. *Chemosphere*, *208*, 59-68.

Nava, V., and Leoni, B. (2020). A critical review of interactions between microplastics, microalgae and aquatic ecosystem function. *Water Research*, 116476.

Nel, H.A., Dalu, T., and Wasserman, R.J. (2018). Sinks and sources: Assessing microplastic abundance in river sediment and deposit feeders in an Austral temperate urban river system. *Science of the Total Environment*, *612*, 950-956.

Nichols, E.C., Lavers, J.L., Archer-Rand, S., and Bond, A.L. (2021). Assessing plastic size distribution and quantity on a remote island in the South Pacific. *Marine Pollution Bulletin*, *167*, 112366.

Nolte, T.M., Hartmann, N.B., Kleijn, J.M., Garnæs, J., Van De Meent, D., Hendriks, A.J., and Baun, A. (2017). The toxicity of plastic nanoparticles to green algae as influenced by surface modification, medium hardness and cellular adsorption. *Aquatic toxicology*, *183*, 11-20.

Oliviero, M., Tato, T., Schiavo, S., Fernández, V., Manzo, S., and Beiras, R. (2019). Leachates of micronized plastic toys provoke embryotoxic effects upon sea urchin *Paracentrotuslividus*. *Environmental Pollution*, *247*, 706–715.

Parker, B., Andreou, D., Green, I.D., and Britton, J.R. (2021). Microplastics in freshwater fishes: Occurrence, impacts and future perspectives. *Fish and Fisheries*, *22* (3), 467-488.

Quigg, A., Chin, W.C, Chen, C.S, Zhang, S., Jiang, Y., Miao, A.J., and Santschi, P.H. (2013). Direct and indirect toxic effects of engineered nanoparticles on algae: role of natural organic matter. *ACS Sustainable Chemistry & Engineering*, *1*(7), 686-702.

Rezania, S., Park, J., Din, M.F.M., Taib, S.M., Talaiekhozani, A., Yadav, K.K., and Kamyab, H. (2018). Microplastics pollution in different aquatic environments and biota: A review of recent studies. *Marine pollution bulletin*, *133*, 191-208.

Rochman, C.M., Tahir, A., Williams, S.L., Baxa, D.V., Lam, R., Miller, J.T., and Teh, S.J. (2015). Anthropogenic debris in seafood: Plastic debris and fibers from textiles in fish and bivalves sold for human consumption. *Scientific reports*, *5*(1), 1-10.

Romera-Castillo, C., Pinto, M., Langer, T.M., Alvarez-Salgado, X.A., and Herndl, G.J. (2018). Dissolved organic carbon leaching from plastics stimulates microbial activity in the ocean. *Nature communications*, *9* (1), 1-7.

Santana, M.F.M., Moreira, F.T., and Turra, A. (2017). Trophic transference of microplastics under a low exposure scenario: insights on the likelihood of particle cascading along marine food-webs. *Marine pollution bulletin*, *121*(1-2), 154-159.

Selakovic, S., de Ruiter, P.C., and Heesterbeek, H. (2014). Infectious disease agents mediate interaction in food webs and ecosystems. *Proceedings of the Royal Society B: Biological Sciences*, *281*(1777), 20132709.

Setälä, O., Fleming-Lehtinen, V., and Lehtiniemi, M. (2014). Ingestion and transfer of microplastics in the planktonic food web. *Environmental Pollution*, 185, 77–83. http://dx.doi.org/10.1016/j.envpol.2013.10.013

Setälä, O., Lehtiniemi, M., Coppock, R., and Cole, M. (2018). Microplastics in marine food webs. In *Microplastic contamination in aquatic environments* (pp. 339-363). Elsevier.

Shen, M., Ye, S., Zeng, G., Zhang, Y., Xing, L., Tang, W., and Liu, S. (2020). Can microplastics pose a threat to ocean carbon sequestration? *Marine pollution bulletin*, *150*, 110712.

Shubin, A.V., Demidyuk, I.V., Komissarov, A.A., Rafieva, L.M., and Kostrov, S.V. (2016). Cytoplasmic vacuolization in cell death and survival. *Oncotarget*, *7*(34), 55863.

Silva, A.B., Bastos, A.S., Justino, C.I., da Costa, J.P., Duarte, A.C., and Rocha-Santos, T.A. (2018). Microplastics in the environment: Challenges in analytical chemistry-A review. *Analyticachimicaacta*, *1017*, 1-19.

Sjollema, S.B., Redondo-Hasselerharm, P., Leslie, H.A., Kraak, M.H.S., and Vethaak, A.D. (2016). Do plastic particles affect microalgal photosynthesis and growth? *Aquatic toxicology*, 170, 259–261.https://doi.org/10.1016/j.aquatox.2015.12.002.

Song, C., Liu, Z., Wang, C., Li, S., and Kitamura, Y. (2020). Different interaction performance between microplastics and microalgae: The bio-elimination potential of *Chlorella* sp. L38 and *Phaeodactylumtricornutum* MASCC-0025. *Science of The Total Environment*, *723*, 138146.

Su, L., Cai, H., Kolandhasamy, P., Wu, C., Rochman, C.M., and Shi, H. (2018). Using the Asian clam as an indicator of microplastic pollution in freshwater ecosystems. *Environmental Pollution*, 234, 347–355. https://doi.org/10.1016/j.envpol.2017.11.075

Sun, X., Li, Q., Zhu, M., Liang, J., Zheng, S., and Zhao, Y. (2017). Ingestion of microplastics by natural zooplankton groups in the northern South China Sea. *Marine pollution bulletin*, *115*(1-2), 217-224.

Teuten, E.L., Saquing, J.M., Knappe, D.R., Barlaz, M.A., Jonsson, S., Björn, A., and Takada, H. (2009). Transport and release of chemicals from plastics to the environment and to wildlife. *Philosophical transactions of the royal society B: biological sciences*, *364*(1526), 2027-2045.

Underwood, G.J., Boulcott, M., Raines, C.A., and Waldron, K. (2004). Environmental effects on exopolymer production by marine benthic diatoms: Dynamics, changes in composition, and pathways of production 1. *Journal of Phycology*, *40*(2), 293-304.

Van Cauwenberghe, L., and Janssen, C.R. (2014). Microplastics in bivalves cultured for human consumption. *Environmental pollution*, *193*, 65-70.

Wang, Q., Wangjin, X., Zhang, Y., Wang, N., Wang, Y., Meng, G., and Chen, Y. (2020b). The toxicity of virgin and UV-aged PVC microplastics on the growth of freshwater algae *Chlamydomonas reinhardtii*. *Science of the Total Environment*, *749*, 141603–141609.

Wang, S., Liu, M., Wang, J., Huang, J., and Wang, J. (2020a). Polystyrene nanoplastics cause growth inhibition, morphological damage and physiological disturbance in the marine microalga *Platymonas helgolandica*. *Marine Pollution Bulletin*, 158, 111403.

Wang, S., Wang, Y., Liang, Y., Cao, W., Sun, C., Ju, P., and Zheng, L. (2020c). The interactions between microplastic polyvinyl chloride and marine diatoms: Physiological, morphological, and growth effects. *Ecotoxicology and Environmental Safety*, *203*, 111000–111007.

Ward, J.E., and Kach, D.J. (2009). Marine aggregates facilitate ingestion of nanoparticles by suspension-feeding bivalves. *Marine environmental research*, *68*(3), 137-142.

Watts, A. J., Lewis, C., Goodhead, R. M., Beckett, S. J., Moger, J., Tyler, C. R., and Galloway, T. S. (2014). Uptake and retention of microplastics by the shore crab *Carcinus maenas*. *Environmental science & technology*, *48*(15), 8823-8830.

Windsor, F.M., Tilley, R.M., Tyler, C.R., and Ormerod, S.J. (2019). Microplastic ingestion by riverine macroinvertebrates. *Science of the total environment*, *646*, 68-74.

Witman, S. (2017). World's biggest oxygen producers living in swirling ocean waters. *Journal of Geophysical Research: Oceans (Sept. 13, 2017)*.

Wright, S. L., Rowe, D., Thompson, R. C., and Galloway, T. S. (2013a). Microplastic ingestion decreases energy reserves in marine worms. *Current Biology*, *23*(23), R1031-R1033.

Wright, S.L., Thompson, R.C., and Galloway, T.S. (2013b). The physical impacts of microplastics on marine organisms: A review. *Environmental Pollution*, 178, 483–492.

Xiang, Y., Jiang, L., Zhou, Y., Luo, Z., Zhi, D., Yang, J., and Lam, S.S. (2022). Microplastics and environmental pollutants: key interaction and toxicology in aquatic and soil environments. *Journal of Hazardous Materials*, *422*, 126843.

Yamashita, R., Takada, H., Fukuwaka, M.A., and Watanuki, Y. (2011). Physical and chemical effects of ingested plastic debris on short-tailed shearwaters, *Puffinustenuirostris*, in the North Pacific Ocean. *Marine pollution bulletin*, *62*(12), 2845-2849.

Yang, M., Zhang, B., Xin, X., Lee, K., and Chen, B. (2022). Microplastic and oil pollution in oceans: Interactions and environmental impacts. *Science of The Total Environment*, 156142.

Yang, W., Gao, X., Wu, Y., Wan, L., Tan, L., Yuan, S., Ding, H., and Zhang, W. (2020). The combined toxicity influence of microplastics and nonylphenol on microalgae *Chlorella pyrenoidosa. Ecotoxicology and environmental safety*, *195*, 110484–110493.

Yi, X., Chi, T., Li, Z., Wang, J., Yu, M., Wu, M., and Zhou, H. (2019). Combined effect of polystyrene plastics and triphenyltin chloride on the green algae *Chlorella pyrenoidosa. Environmental Science and Pollution Research*, *26* (15), 15011-15018.

Yu, P., Liu, Z., Wu, D., Chen, M., Lv, W., and Zhao, Y. (2018). Accumulation of polystyrene microplastics in juvenile *Eriocheirsinensis* and oxidative stress effects in the liver. *Aquatic toxicology*, *200*, 28-36.

Zhang, C., Chen, X., Wang, J., and Tan, L. (2017). Toxic effects of microplastic on marine microalgae *Skeletonemacostatum*: interactions between microplastic and algae. *Environmental pollution*, *220*, 1282-1288.

Zhang, C., Chen, X., Wang, J., and Tan, L. (2018b). Toxicity of zinc oxide nanoparticles on marine microalgae possessing different shapes and surface structures. *Environmental Engineering Science*, *35*(8), 785-790.

Zhang, K., Shi, H., Peng, J., Wang, Y., Xiong, X., Wu, C., and Lam, P.K. (2018a). Microplastic pollution in China's inland water systems: a review of findings, methods, characteristics, effects, and management. *Science of the Total Environment*, *630*, 1641-1653.

Zhao, T., Tan, L., Zhu, X., Huang, W., and Wang, J. (2020). Size-dependent oxidative stress effect of nano/micro-scaled polystyrene on *Kareniamikimotoi. Marine pollution bulletin*, *154*, 111074-111084.

Zheng, X., Zhang, W., Yuan, Y., Li, Y., Liu, X., Wang, X., and Fan, Z. (2021). Growth inhibition, toxin production and oxidative stress caused by three microplastics in *Microcystisaeruginosa. Ecotoxicology and Environmental Safety*, *208*, 111575-111583.

Zhu, L., Bai, H., Chen, B., Sun, X., Qu, K., and Xia, B. (2018). Microplastic pollution in North Yellow Sea, China: Observations on occurrence, distribution and identification. *Science of the Total Environment*, *636*, 20-29.

Zhu, X., Zhao, W., Chen, X., Zhao, T., Tan, L., and Wang, J. (2020). Growth inhibition of the microalgae *Skeletonema costatum* under copper nanoparticles with microplastic exposure. *Marine Environmental Research*, *158*, 105005-105011.

# Chapter - 12

# Environmental Sustainability Using Microbial Aspects Associated With Clearing Up Waste

**Saima Andleeb[1*], Irfan Ahmad[1], Keezia Khurshid[2], Azra Shah[1] and Durdana Qazi[1]**

[1]Division of Fish Genetics and Biotechnology, Faculty of Fisheries, Sher-e-Kashmir University of Agricultural Sciences and Technology of Kashmir (SKUAST-K), Rangil-190006, Ganderbal, India

[2]Division of Aquatic Animal Health Management, Faculty of Fisheries, Sher-e-Kashmir University of Agricultural Sciences and Technology of Kashmir (SKUAST-K), Rangil-190006, Ganderbal, India

Corresponding author: saimaandleeb165@gmail.com

## Abstract

Environmental sustainability is a pressing global challenge that necessitates innovative and eco-friendly waste management solutions. The conventional waste disposal methods have proven to be inadequate in effectively addressing the escalating waste generation and its detrimental impacts on the environment. In this context, harnessing the potential of microbial aspects for waste remediation presents a promising avenue towards achieving environmental sustainability.Microorganisms, such as bacteria, fungi, and algae play a vital role in natural ecosystems, and their abilities can be harnessed to facilitate the clearing up of waste. Microbial-driven processes, such as biodegradation, bioremediation, composting, and bioconversion, offer environmentally sound approaches for managing various types of waste. Biodegradation, conducted by diverse microbial communities, is a fundamental process in breaking down complex organic compounds, reducing waste volumes, and minimizing pollution. Bioremediation employs specific microorganisms to target and degrade hazardous contaminants, transforming them into harmless byproducts. Composting, a natural decomposition process facilitated by microorganisms, converts organic waste into nutrient-rich compost, enhancing soil fertility and supporting sustainable agriculture. This chapter highlights the potential of microbial aspects in mitigating waste-related environmental issues, emphasizing their role in promoting resource efficiency, pollution reduction, and sustainable practices. A holistic approach, involving collaboration between scientists, policymakers, industries, and the public, is necessary to realize the full potential of microbial-driven waste management and pave the way for a greener and more sustainable future. By tapping into the immense capabilities of microorganisms, society can move towards a circular and regenerative waste management paradigm, safeguarding ecosystems and fostering long-term environmental well-being.

**Keywords:** Microbial infections, Xenobiotics, Desulphurization, Microbial fuel cells

## Introduction

A substantial portion of the world's population lacks access to clean water, which is also thought of as a fundamental human right. Additionally, the availability of clean water has a significant influence on the development of individuals. By 2030, the world's urban population is expected to have doubled to more than 5 billion people, implying that if water pollution problems are left uncontrolled, they will not only become worsened but will have an inescapable adverse effect on public health (Oh *et al.*, 2010). As a result, it is of the utmost importance that as the economic systems of the badly affected nations develop, a percentage of their revenue must be invested in sanitary infrastructure. The quantity and quality of trash produced and released into aquatic ecosystems have made attention in a number of globally media channels, and the richest countries spend a lot of resources and time aiming to reduce the pollutants of large populations (Oh *et al.*, 2010). In context with the widespread reports of subsurface water being contaminated with heavy metals, it is crucial to have a comprehensive understanding of the wastewater origin, as well as its biological, chemical, and physical characteristics, in order to implement a successful approach to treatment (Ferrera and Sanchez, 2016). Due to the advancement of research facilities across the globe, several water treatment technology approaches have been established and constantly improved over time.

Water treatment methods may also be classified into biological, physical, and chemical treatment approaches and further categorized as in-situ or ex-situ procedures depending on the categorizations of the relevant wastewater source. In-situ processes of remediation refer to treatment implemented at the specific site, whereas ex-situ entails the removal of pollutants at a distant location. This description can help to clarify the uses of in-situ and ex-situ technologies (Ng *et al.*, 2016). Bacteria, protozoa, viruses, fungi, algae, and helminths are the main microbial organisms identified in wastewater treatment systems. The majority of these organisms are found in water, which promotes the spread of diseases (Akpor *et al.*, 2010). The greatest number of aquatic microbes that infect humans and animals come from feces. These may wash into drinking water bodies or drinking water sources because they carry a diverse range of viruses, bacteria, and protozoa (Kris, 2007). Microbial infections are thought to be significant contributors to an array of waterborne diseases. Many infections caused by microbes in wastewater can cause persistent illnesses with long-term repercussions such as progressive heart disease and ulcers in the stomach. Depending on the severity and spread of diseases in the sewer-connected neighborhood, the density and variety of these pollutants may fluctuate. It is usually challenging, costly, and laborious to find, separate, and identify the many types of microbial contaminants in wastewater. In order to prevent this, biological indicators are always employed to assess the proportionate risk of a certain pathogen's potential exist in wastewater (Paillard *et al.*, 2005).

The organic richness of the substrate under the net cages, which often results in a large stimulation of benthic metabolism, is a significant environmental consequence of marine fish farming (Hall *et al.*, 1990; Holmer and Kristensen, 1992). Contrary to the organic matter typically found in marine sediments, this organic matter is unstable has a low C: N ratio, and has a large amount of nutrients (Kristensen, 1990; Holmer and Kristensen, 1994). Animal and terrestrial sources produce waste products (such as food pellets and excrement), whereas recalcitrant plant material typically makes up marine organic matter (Dauwe and Middelburg, 1998). Both the macro- and meiobenthic communities are impacted when waste materials accumulate in farm debris, which often causes severe changes in the sediment's composition (Brown *et al.*, 1987; Karakassis *et al.*, 1999). According to Pearson and Rosenberg (1978), changes in the organizational structure of benthic communities follow an upward trend of organic enrichment. In the area of the cages, a zone dominated by a few voracious species (such as *Capitella capitata*) in great quantities has frequently been seen (Karakassis *et al.*, 1999). There are also circumstances where the fauna may entirely vanish (azoic zone), such as when the farm is situated in a natural deposition region or when the pace of waste product sedimentation is excessive. With a greater distance from the farm, the community gets more diversified. More species are being discovered, including some that are more vulnerable to pollution. In accordance with the Pearson-Rosenberg model, some investigations have demonstrated faunal changes (Brown *et al.*, 1987), whilst others have only recognised a portion of the sequences. The supply and mineralization of organic substances are both significantly influenced by the benthic fauna. Pelagic and benthic habitats are connected by suspension feeders, whereas benthic deposit feeders disperse organic matter that has been deposited on the sediment surface through sediment reworking and oxidize the deposit through ventilation (Aller, 1982). According to Banta *et al.* (1999), mineralization is substantially improved. Particularly, refractory organic matter's decomposition becomes faster (Andersen and Kristensen 1992; Kristensen e*t al.*, 1992), perhaps as a result of the material's deposition in anoxic sediment layers. According to Kristensen *et al.* (1995; Kristensen and Holmer, 2001), disintegration may increase by an order of magnitude when recalcitrant organic matter is subjected to oxygen during bioturbation. The turnover of nutrients in the sediment is also influenced by benthic organisms. Because of the greater movement of dissolved oxygen and N-compounds resulting from ventilation, N-cycling occurs more rapidly and both nitrification and denitrification speed up in contrast to azoic sediments (Pelegri and Blackburn, 1995; Hansen and Kristensen, 1998). Since phosphorus binds to oxidized iron, it is speculated that benthic fauna will generally improve phosphorus retention (Clavero *et al.*, 1994). However, only a few studies have examined the impact of benthic fauna on phosphorus cycling.

## Significance of microbes in eliminating hazardous industrial chemicals from the environment

By acquiring enzymes that enable them to process numerous man-made substances (also known as xenobiotics), microorganisms have broadened the habitat in which they thrive (Okpokwasili, 2007). Utilizing microorganisms or microbial processes to eliminate and break down environmental pollutants is known as bioremediation. Since a long time ago, microorganisms have been utilized for the regular treatment and transformation of waste materials. The metabolic processes of microorganisms that break down wastes before they enter the treatment facility are an essential requirement for fixed-film and activated sludge treatment systems. In order to treat industrial effluents, a number of waste treatment facilities are specialized that employ carefully selected and habituated microbial populations (Okpokwasili, 2007). Heavy metal contamination of the environment continues to be an imminent danger to the organisms in an ecosystem (Hrynkiewicz & Baum, 2014; Okolo *et al.*, 2016; Siddiquee *et al.*, 2015). Parameterization is described by EPA 2010 (Environmental Protection Agency (EPA), 2010) as an impulsive procedure in which the microbiological process is employed to decompose, break down, or modifies hazardous substances into less harmful or nontoxic forms, thereby, repairing or removing and eradicating contaminants from environmental media. By various metabolic activities, microorganisms utilize chemical pollutants as an energy source throughout the microbiological process. On the other hand, high levels of inorganic nutrients in soil restrict microbial growth (Ahirwar *et al.*, 2016). Particularly bacteria and other microorganisms have the capacity to break down, eliminate, and even accumulate toxic organic and inorganic chemicals. Different sources, including natural, agricultural, industrial solid waste, inland effluent, atmospheric sources, and others, contribute to the accumulation of heavy metals in the environment. Large parts of the world have been polluted as a result of activities like mining, electroplating, metallurgical smelting, and the use of fertilizers and pesticides in agriculture (Zhang *et al.*, 2011).

The biosynthesis of metallothioneins and other proteins which hold metals and convert them to volatile compounds are some of the mechanisms by which microorganisms remove toxic heavy metals from the environment (Meissner and Falkinham, 1984; Mullen *et al.*, 1989). Other mechanisms include adsorption to cell surfaces and the complexation of exopolysaccharides. It has also been documented that *Klebsiella aerogenes* can vaporize mercury (Magos *et al.*, 1964), and that *Streptomyces albus* cells bound in polyacrylamide can remove copper and cobalt as well as some radioactive metals (such as uranium). Metals precipitating out of aqueous solutions are yet another method that microbial activities can be regulated. As a result of interactions between certain bacterial or extrinsic products and unbound or absorbed metal cations, insoluble metal precipitates can occur. The primary processes which trigger this precipitation are the generation of hydrogen sulfide and the immobilization of metal cations as metal sulfides. Metal oxalate crystals and other metal immobilizers can be produced by certain fungi that generate oxalic acid (oxalates) (Okpokwasili, 2007). Additionally, a varicty of metal transformations that are beneficial for waste treatment can catalyze the growth of microbes. Oxidation, reduction, and alkylation processes are some of these alterations. Protozoa, algae, fungus, and bacteria are able to deposit fcrrous and manganese ions during oxidation processes. In mining drainage fluids and polluted ground waters, *Geobacter metallireducens* eliminate uranium, a radioactive waste (Okpokwasili, 2007).

## Fossil Fuel Desulfurization via Microbial Oxidation

Microbial sulphur oxidation facilitates the elimination of inorganic sulphur from coal (Okpokwasili, 2007). Thiobacillus sp. of microorganism is utilized in the direct oxidation of inorganic sulfur for a membrane-bound reaction requiring direct interaction with the substrate with the bacterium. As a result, the culture's relationship to coal particles is an imperative prerequisite. To extract organic sulfur from coal and oil, a range of mixed and pure strains of microorganisms (heterotrophic bacteria) can be utilized (Okpokwasili, 2007).

## The Contribution of Microbes in Environmental Toxicity Bioassays

Both biological waste treatment facilities and the surroundings where they eventually get disposed of are seriously threatened by toxic industrial wastes. Bioassays are therefore essential in order to gather the information that can be utilized for forecasting how waste will affect the ecosystem and to control discharges (Okpokwasili, 2007). It is becoming progressively more prevalent to utilize microorganisms (particularly bacteria) as bioassay organisms since they have several benefits over conventional organisms (Williamson and Johnson, 1981). In comparison to other bioassays, bacteria are simple to handle and demand just a little space for culture and/or testing. The Nitrobacter assay, Microtox tests, the Toxic-chromo test, and the Ames/Salmonella test are among the bacterial bioassay methods that are simple and quick to perform (Ames *et al.*, 1975; Williamson and Johnson, 1981; Okpokwasili and Odokuma, 1996a, 1996b). According to Okpokwasili (2007), the Nitrobacter bioassay depends on quantifying Nitrobacter activity, which is done by monitoring the toxicant's impact on the rate of nitrite uptake. The Microtox assay is based on *Photobacterium phosphoreum*, while the toxichromotest relies on the inhibition of beta-galactosidase biosynthesis in E. coli or the biosynthesis of enzymes like tryptophanse and alpha-glucosidase under the supervision of operons other than the Lac operon by environmental pollutants. The mutagenic potential of contaminants is determined using the Ames/Salmonella test. Salmonella TA 1537, Ta 1538, and TA 98 frameshift and base pair mutations are examined for the presence of histidine-negative, ampicillin-sensitive, and ultraviolet (UV) resistant revertants (Ames *et al.*, 1973).

## Oil Spill Mitigation Using Microbes

By using the rDNA method, microorganisms can now be genetically modified for mineral leaching and recovery, oil recovery, and pollution management. Microorganisms can also be manipulated genetically in petroleum-related industries to generate compounds beneficial for better oil recovery (Daini, 2000). In the future, genetically modified bacteria could be used to clean up oil spills (Su, 1998). Microorganisms with improved leaching abilities might be developed for the mining industry. Metals can be concentrated internally by microorganisms and bound to their surfaces (Daini, 2000). Research is currently being conducted to enhance the naturally present bacteria that can "eat or digest oil" for usage after an oil spill. The intricate oil molecules would be converted into harmless carbohydrates by microbes when applied on oil-covered beaches (Su, 1998). It has been proposed to employ genetically modified microorganisms to treat sewage and effluents, clean up oil spills, and develop some strains of bacteria that can break down fuel hydrocarbons. This technology is now being developed. An alternative technique for degrading industrial waste is bioremediation. In order to lower the toxicity of hazardous and non-toxic compounds by natural biological processes, bioremediation uses microbes and plants to break down pollutants in sewage, residential, agricultural, and industrial effluents (ELC, 2008).

## Remediation of Pesticides by Microorganisms

Researchers are finding novel techniques to break down pesticides in the environment despite advancements in the traditional approach. Microorganisms are used in microbial remediation to remove these substances from the polluted source. Pesticide degradation has been observed by some bacterial species and their strains, actinomycetes, fungi, and algae. These compounds are utilized as a source of energy by microorganisms by metabolizing them and directing their intermediates into energy-generating pathways such as the Krebs cycle. These microbial strains have been collected, isolated, and screened from different rivers, wastewater, and soil (Huang *et al.*, 2018). *Acinetobacter*, *Alcaligenes*, *Klebsiella*, *Bacillus*, and *Flavobacterium* are five bacterial genera that could degrade endosulfan, according to research by Kafilzadeh *et al.* (2015) who identified bacterial strains from water samples. *Alcaligenes*, *Flavobacterium*, *Pseudomonas*, *Bacillus*, *Klebsiella*, *Thiobacillus*, *Escherichia coli*, *Bacillus licheniformis*, and *Clostridium* are just some of the bacteria that have been proven to be able to break down pesticides. Diatoms, *Chlamydomomas*, green algae, microalgae, and fungi all have the ability to break down pesticides, as do the *Anthracophyllum*, *Cladosporium*, *Rhizopus*, *Aspergillus fumigatus*, *Aspergillus*, *Penicillium*, *Mucor*, *Fusarium*, *Mortierella* sp., and *Trichoderma* sp. (Huang *et al.*, 2018). Due to their high levels of adaptability and tendency for swift mutation, bacteria are preferred for bioremediation due to the fact they can quickly adapt to their surroundings. Microbes consume pesticides as nutrients and turn them into carbon dioxide and water through enzymatic processes. These chemicals are initially absorbed by microorganisms, after which metabolic enzymes attack them sequentially until they are entirely broken down into less hazardous or harmless substances. According to Prabha *et al.* (2017), these enzyme-mediated reactions include oxidation (N/P/S-oxidation, epoxidation, hydroxylation reactions on aromatic rings, aliphatic chains, or N-hydroxylation), reduction (reductive dehalogenation, and reduction of nitro or quinone) and hydrolysis of ester bonds, condensation, decarboxylation, dehalogenation, dehydrogenation, and other processes).

## Pesticide Remediation by Bacteria

*Bacillus*, *Micrococcus*, *Arthrobacter*, and *Pseudomonas* have been proven to bioremediate organochlorines. The CS5 strain of *Achromobacter xylosoxidans*, isolated from activated sludge, was found to be capable of degrading α-endosulfan and β-endosulfan by more than 0.0248 g/L and 0.0105 g/L, respectively, after eight days of incubation (Li *et al.*, 2009). *Dehalococcoides* sp. has been demonstrated to anaerobically dechlorinate hexachlorobenzene, and it was suggested that the trichlorobenzene reductive dehalogenase enzyme, a byproduct of the cbrA gene, was essential in the process of degradation (Tas *et al.*, 2011). They demonstrated their potential towards biotransformation and remediation of chlordecone by these two isolates (Chaussonnerie *et al.*, 2016) after isolating and sequencing the genomes of two species of *Citrobacter amalonaticus* that are closely related. Five bacterial species - *Pseudomonas aeruginosa*, *Acinetobacter lwoffii*, *Stenotrophomonas maltophilia*, *Citrobacter amalonaticus*, and *Bacillus atrophaeus* - that were isolated from cockroaches living in pesticide-contaminated areas were reported by Ozdal in 2016; endosulfan bioremediation by these isolates was determined to be 88.5, 80.2, 85.5, 56.7, and 64.4%, respectively. Li *et al.*, 2016 investigated the anaerobic mineralization of pentachlorphenol using multiple bacterial strains. They indicated different dechlorinators (*Sulfospirillum dehalobacter*, *Desulfovibrio*, and *Deslfitobacterium* spp.) and phenol degradation agents (*Syntrophus* and *Cryptanerobacter* spp.) and identified the responsible functional genes; cprA (chlorphenol reductive dehalogenase), bamA (benzoyl-CoA reductase), and seven variants of nitrogenase reductase genes.

## Pesticide Remediation using Fungi

Research was carried out in 2011 by Xiao *et al.* to assess the fungal species that could potentially break down heptachlor and its epoxide. After two weeks of incubation, *P. tremellosa*, *P. brevispora*, and *P. acanthocystis*, respectively, showed remediation of heptachlor to be 71, 74, and 90%. The heptachlor was breakdown to heptachlor epoxide, 1-hydroxy-2, 3-epoxy-chlordene, and 1-hydroxychlordene, with *P. lindtneri*, *P. aurea*, *P. brevispora*, and *P. acanthocystis* removing 22, 25, 16, and 16% after 2 weeks, respectively (Xiao *et al.*, 2011). In *Spingobium japonicum*, *Spingobium francense*, and *Spingobium indicum*, the lingene was revealed to be responsible for the absorption and metabolism of hexachlorocyclohexane (Pal *et al.*, 2005). Dai *et al.*, 2010 reported acetamiperid and thiacloprid remediation by Rhodotorula mucilaginosa strain IM-2. *Aspergillus niger* has been shown to manufacture hydrolase for carbamate hydrolysis (Qing *et al.*, 2003).

## Technology for Microbial Electro-Remediation

A significant amount of wastewater that contains metal is produced primarily as a result of human activity and swift industrialization. Because of the hazardous potentialities and cancer-causing effects of metal(loid) polluted water, strict regulations have been adopted for the discharge of different metal ions in wastewaterswaters in order to prevent environmental contamination.According to the US Environmental Protection Agency's list of priority pollutants, naturally occurring arsenic (As) is a highly toxic metalloid that is catastrophic and affects over 250 million people worldwide, especially in developing countries, causing neurological damage and even death.The most beneficial course of action is to accumulate the metals during the recuperation phase rather than eliminating them. Although there are a number of ways to treat or remove arsenic from water and wastewater, a modern study has focused on developing a variety of technical solutions that seem to be economically feasible, have low prices, enhanced efficiency, and are environmentally benign. In order to further develop the unique method and expand its range of applicability to actual scenarios for pollution removal, intensive advancement in research is therefore critically required in this particular sector. This is actually necessary to discover a solution to the traditional methods of metal removal, which are typically regarded as less effective (Guan *et al.*, 2009). These methods include coagulation, precipitation, and adsorption. For the treatment of enormous quantities of industrial wastewater and sewage containing small amounts of heavy metal ions, activated carbon-based adsorption, ion exchange, and membrane technologies are exorbitantly costly and cannot be used on a large industrial scale (Gautam *et al.*, 2014).

As a result, there is currently a lot of research underway on innovative, appropriate, and affordable methods to get rid of metal from water. Microbial fuel cells (MFCs), in particular, are now being utilized as a novel remediation technique that is predominantly used to mobilize pollutants for removal or reduction in the subsurface. Due to the fact that algae, bacteria, fungus, and yeast cancan lower and deposit the metal ions, this approach glances to be promoting wastewater treatment and metal recovery through bio-electrocatalysis. For bioremediation applications, BES has attracted a lot of fascination (Schroder *et al.*, 2015). It's interesting to note that wastewater containing heavy metals can be managed in MFCs' anode and cathode chambers (Mathuriya and Yakhmi, 2014). The basic principle of the process is that the biological processes that take place at the positively charged electrode (anode) are utilised to transfer electrons to the negatively charged electrode (cathode) and promote reduction with precipitation of metal (loids) through using them as terminal electron acceptors. In order to recover energy, water, and metal (loids), BES offers a distinctive structure for integrated wastewater treatment using oxidation and reduction processes. The removal of different metals during normal MFC operation advances MFCs beyond other waste treatment methods. The environmental benefits of MFCs could be significantly enhanced if pollutants act as electron acceptors in the cathode chamber. The energy required for the treatment procedure is definitely eliminated by recovering metals while generating power (Mathuriya and Yakhmi, 2014).

## Bibliography

Ahirwar, N.K., Gupta, G., Singh, R. and Singh, V. 2016. Isolation, identification and characterization of heavy metal resistant bacteria from industrial affected soil in central India. *International Journal of Pure and Applied Bioscience 4*(6), 88-93.

Akpor, O.B. and Muchie, M. 2010. Bioremediation of polluted wastewater influent: Phosphorus and nitrogen removal. *Scientific Research and Essays 5*(21), 3222-3230.

Aller, R.C. 1982. The effects of macrofauna on chemical properties of marine sediment and overlying water. In Animal–Sediment Relations, Ed. by McCall, P.L. and Tevez, M.J.S. Plenum Publishing Company. pp. 53–102

Ames, B.J, McCann, J. and Yamasaki, E. 1975. Methods for detecting carcinogens and mutagens with the Salmonella/mammalian microsome mutagenicity test. *Mutation Research 31*, 347-363.

Andersen, F.Ø. and Kristensen, E. 1992. The importance of benthic macrofauna in decomposition of microalgae in a coastal marine sediment. *Limnology and Oceanography 37*, 1392–1403.

Banta, G.T., Holmer, M., Jensen, M.J. and Kristensen, E. 1999. The effect of two polychaete worms, *Nereis diversicolor* and *Arenicola marina*, on decomposition in an organic-poor and an organic-enriched marine sediment. *Aquatic Microbial Ecology 9*, 189–204.

Brown, J.R., Gowen, R.J. and Mclusky, D.S. 1987. The effect of salmon farming on the benthos of a Scottich sea loch. *Journal of Experimental Marine Biology and Ecology 109*, 39–51.

Chaussonnerie, S., Saaidi, P.-L., Ugarte, E. *et al.* 2016. Microbial degradation of a recalcitrant pesticide: chlordecone. *Frontiers in microbiology* 7, 2025.

Clavero, V., Niell, F.X. and Fernandez, J.A. 1994. A laboratory study to quantify the influence of *Nereis diver*sicolor O. F. Mu¨ller in the exchange of phosphate between sediment and water. *Journal of Experimental Marine Biology and Ecology 176*(2), 257-267.

Dai, Y.J., Ji, W.-W., Chen, T., et al., 2010. Metabolism of the neonicotinoid insecticides acetamiprid and thiacloprid by the yeast *Rhodotorula mucilaginosa* strain IM-2. *Journal of Agricultural and Food Chemistry 58*(4), 2419–2425.

Daini, O.A. 2000. Fundamentals of Genetic Engineering. Samrol Ventures and Printing Co., Ijebu-Igbo, Ogun State, Nigeria.

Dauwe, B. and Middelburg, J.J. 1998. Amino acids and hexoamines as indicators of organic matter degradation state in North Sea sediments. *Limnology and Oceanography 43*, 782–798.

Environmental Literacy Council 2008. Bioremediation. http://www.enviroliteracy.org/625php.

Environmental Protection Agency (EPA), 2010. The assessment and remediation of contaminated sediments (ARCS) program. *Journal of Great Lakes Research 3*(22), 493-494.

Ferrera, I. and Sanchez, O. 2016. Insights into microbial diversity in wastewater treatment systems: How far have we come? *Biotechnology advances 34*, 790-802.

Gautam, R.K., Mudhoo, G.A., Lofrano, and Chattopadhyaya, M.C. 2014. Biomass-derived biosorbents for metal ions sequestration: Adsorbent modification and activation methods and adsorbent regeneration. *Journal of Environmental Chemical Engineering 2*(1), 239-259.

Guan, X., Dong, H., Ma, J. and Jiang, L. 2009. Removal of arsenic from water: Effects of competing anions on As (III) removal in KMnO4–Fe(II) process. *Water Research 43*(15), 3891-3899.

Guidelines for Drinking-water Quality (Electronic Resource): Incorporating 1st and 2nd Addenda, Volume 1, Recommendations, 3rd ed.; *World Health Organization*: Geneva, Switzerland, 2008.

Hall, P.O.J., Anderson, L.G., Holby, O., Kollberg, S. and Samuelsson, M. 1990. Chemical fluxes and mass balances in a marine fish cage farm. I. Carbon. *Marine Ecology Progress Series 61*, 61–73.

Hansen, K. and Kristensen, E. 1998. The impact of the polychaete Nereis diversicolor and enrichment with the macroalgal (Chaetomorpha linum) detritus on benthic metabolism and nutrient dynamics in organic-poor and organic rich sediment. *Journal of Experimental Marine Biology and Ecology 231*, 201–223

Holmer, M. and Kristensen, E. 1992. Impact of marine fish cage farming on metabolism and sulfate reduction of underlying sediments. *Marine Ecology Progress Series 80*, 191– 201.

Holmer, M. and Kristensen, E. 1994. Organic matter mineralization in an organic-rich sediment: experimental stimulation Organic matter mineralization in fish-farm sediments 433 of sulfate reduction by food pellets. FEMS *Microbiology Ecology 14*, 33–44.

Hrynkiewicz, K. and Baum, C. 2014. Application of microorganisms in bioremediation of environment from heavy metals. In: Malik, A., Grohmann, E., Akhtar, R. (eds) Environmental Deterioration and Human Health. Springer, Dordrecht. 215-227.

Huang, L.Y., Xiao, F., Li *et al.*, 2018. Microbial degradation of pesticide residues and an emphasis on the degradation of cypermethrin and 3-phenoxy benzoic acid: a review. *Molecules 23*(9), 2313.

Kafilzadeh, M.F., Ebrahimnezhad, and Y. Tahery, 2015. Isolation and identification of endosulfan-degrading bacteria and evaluation of their bioremediation in Kor River, Iran. *Osong Public Health and Research Perspectives* *6*(1), 39–46.

Karakassis, I., Hatziyanna, E., Tsapakis, M. and Plaitti, W. 1999. Benthic recovery following cessation of fish farming: a series of successes and catastrophes. *Marine Ecology Progress Series 184*, 205–218.

Kris M. 2007. Wastewater pollution in China. www.dbc.uci/

Kristensen, E. 1990. Characterization of biogenic organic matter by stepwise thermogravimetry (STG). *Biogeochemisty 9*, 135–159.

Kristensen, E., Ahmed, S.I. and Devol, A.H. 1995. Aerobic and anaerobic decomposition of organic matter in marine sediment: which is fastest? *Limnology and Oceanography 40*, 1430–1437.

Kristensen, E., and Holmer, M. 2001. Decomposition of plant materials in marine sediment exposed to different electron acceptors ($O_2$, $NO_3$- and $SO_4^{2-}$), with emphasis on substrate origin, degradation stage and the role of bioturbation. *Geochimica Cosmochimica* Acta (in press) *65*(3), 419-433.

Kristensen, E., Andersen, F.Ø., and Blackburn, T.H. 1992. Effects of benthic macrofauna and temperature on degradation of macroalgal detritus: the fate of organic carbon. *Limnology and Oceanography 37*, 1404–1419.

Li, Z.L., Nan, J. Huang C. et al., 2016. Spatial abundance and distribution of potential microbes and functional genes associated with anaerobic mineralization of pentachlorophenol in a cylindrical reactor. *Scientific reports* *6*(1), 19015.

Li, W., Y. Dai, B. Xue *et al.*, 2009. Biodegradation and detoxification of endosulfan in aqueous medium and soil by *Achromobacter xylosoxidans* strain CS5," *Journal of Hazardous Materials 167*(1-3), 209–216.

Magos, I., Tuffery, A.A. and Clarkson, T.W. 1964. Volatilization of mercury by bacteria. *British Journal of Industrial Med*icine *21*, 294-298.

Mathuriya, A.S and Yakhmi, J.V. 2014. Microbial fuel cells to recover heavy metals. *Environmental Chemistry Letters 12*(4), 483-494.

Meissner, P.S. and Falkinham 3rd, J.O. 1984. Plasmid-encoded mercuric reductase in *Mycobacterium scrofulaceum*. *Journal of bacteriology 157*(2), 669-672.

Mullen, M.D., Wolf, D.C., Ferris, F.G., Beveridge, T.J., Flemming, C.A. and Bailey, G. 1989. Bacterialsorption of heavy metals. *Applied and environmental microbiology 55*(12), 3143-3149.

Ng, K.K., Shi, X., Ong, S.L. and Ng, H.Y. 2016. Pyrosequencing reveals microbial community profile in anaerobic bio-entrapped membrane reactor for pharmaceutical wastewater treatment. *Bioresource technology 200*, 1076-1079.

Oh, S.T., Kim, J.R., Premier, G.C., Lee, T.H., Kim, C. and Sloan, W.T. 2010. Sustainable wastewater treatment: how might microbial fuel cells contribute. *Biotechnology Advances 28*, 871-881.

Okolo, N.V., Olowolafe, E.A., Akawu, I. and Okoduwa, S.I.R. 2016. Effects of industrial effluents on soil resources in Challawa industrial area, Kano, Nigeria. *Journal of Global Ecology and Environment* 5(1), 1-10.

Okpokwasili, G.C. and Odokuma, L.O. 1996a. Response of Nitrobacter specie to toxicity of drilling chemicals. *Journal of Petroleum Science and Engineering 18*, 81 87.

Okpokwasili, G.C. and Odokuma, L.O. 1996b. Tolerance of Nitrobacter specie to toxicity of hydrocarbon fuels. *Journal of Petroleum Science and Engineering 18*, 89-93.

Okpokwasili, G.C. 2007. Biotechnology and Clean Environment. Proc. of the 20th Annl. Conf. of the Biotechnology Society of Nigeria (BSN), 14th-17th, November, 2007 at the Ebonyi State University, Abakaliki, Nigeria.

Ozdal, M. 2016. Isolation and characterization of α-endosulfan degrading bacteria from the microflora of cockroaches. *Polish Journal of Microbiology 65*(1), 63–68.

Paillard, D., Dubois V., Thiebaut R., Nathier F., Hogland E., Caumette P. and Quentine C. 2005. Occurrence of Listeria spp. In effluents of French urban wastewater treatment plants. *Applied and Environmental Microbiology 71*(11), 7562-7566.

Pal, R., Bala, M.S., Dadhwal *et al.* 2005. Hexachlorocyclohexanedegrading bacterial strains *Sphingomonas paucimobilis* B90A, UT26 and Sp+, having similar lin genes, represent three distinct species, *Sphingobium indicum* sp. nov., *Sphingobium japonicum* sp. nov. and *Sphingobium francense* sp. nov., and reclassification of *[Sphingomonas] chungbukensis* as *Sphingobium chungbukense* comb. Nov. *International Journal of systematic and evolutionary microbiology 55*(5), 1965–1972.

Pearson, T.H. and Rosenberg, R. 1978. Macrobenthic succession in relation to organic enrichment and pollution of the marine environment. *Oceanographic Marine Biology, Annual Review, 16*, 229–311.

Pelegri, S.P. and Blackburn, T.H. 1995. Effect of bioturbation by Nereis sp. Mya arenaria and Cerastodermata sp. On nitrification and denitrification in estuarine sediments. *Ophelia, 42*, 289–299

Prabha, D.P. Singh, and Verma, M.K. 2017. Microbial interactions and perspectives for bioremediation of pesticides in the soils. In: Plant-Microbe Interactions in Agro-Ecological Perspectives, pp. 649–671, Springer, Singapore,

Qing, Z., Yang, L. and Yu-Huan, L. 2003. Purification and characterization of a novel carbaryl hydrolase from *Aspergillus niger* PY168," FEMS *Microbiology Letters 228*(1), 39– 44.

Schroder, U., Harnisch, F. and Angenent, L.T. 2015. Microbial electrochemistry and technology: terminology and classification. *Energy and Environmental Science 8*(2), 513-519.

Siddiquee, S., Rovina, K., Azad, S.A., Naher, L., Suryani, S.and Chaikaew, P. 2015. Heavy metal contaminants removal from wastewater using the potential filamentous fungi biomass: A review. *Journal of Microbial and Biochemical Technology* 7(6), 384–395.

Su, S. 1998. Genetic Engineering. Evans Brothers Limited, 2a Portman Mansions, Chiltern Street, London W1M 1LE.

Tas, N., Van Eekert, M.H.A., Wagner, A., Schraa, G., de Vos, W.M. and Smidt, H. 2011. Role of "Dehalococcoides" spp. in the Anaerobic Transformation of Hexachlorobenzene in European Rivers. *Applied and environmental microbiology* 77(13), 4437–4445.

US Environmental Protection Agency, Priority pollutants, Appendix A to 40 CFR Part 423, Available: http://water.epa.gov/scitech/methods/cwa/pollutants.cfm.

Williamson, K.J. and Johnson, O.G. 1981. A bacterial bioassay for assessment of wastewater toxicity. *Water Research 15*, 383-390.

Xiao, P., Mori, T., Kamei, I. and Kondo, R. 2011. Metabolism of organochlorine pesticide heptachlor and its metabolite heptachlor epoxide by white rot fungi, belonging to genus Phlebia. FEMS *Microbiology Letters 314*(2), 140–146.

Zhang, W.J., Jiang, F.B. and Ou, J.F. 2011. Global pesticide consumption and pollution: With China as a focus. *Proceedings of the International Academy of Ecology and Environmental Sciences 1*(2), 125–144.

# Chapter - 13

# Radioactive Waste: Sources and Impact on Environment

[1]Keezia Khurshid, [1]Feroz Ahmad Shah and [2]Arsha Surendran*

[1]Division of Aquatic Animal Health Management, Faculty of Fisheries, Sher-e-Kashmir University of Agricultural Sciences and Technology of Kashmir (SKUAST-K), Rangil - 190006, Ganderbal, India
[2]Karpagam Academy of Higher Education (KAHE), Coimbatore, Tamilnadu
Corresponding Author: achupalathingal1996@gmail.com

## Abstract

Radioactive waste disposal proves to be far more expensive than first anticipated. Although there has been some public criticism, the expense is mostly the result of two factors: the growing understanding of the geological processes that might release radioactive materials from subterranean storage and the sophisticated models required to forecast repository behaviour for 10,000 years or more. Most countries prioritise the safe disposal of nuclear waste, and those that have produced nuclear weapons in the past or are now generating nuclear power will continue to incur significant costs. The handling of radioactive waste until it degrades to harmless levels is being suggested, along with treatment, immobilisation, and confinement.

**Keywords:** Hazardous waste, Transuranic elements, Nuclear reactors, radioactive materials.

## Introduction

Hazardous waste comes in the form of radioactive waste when radioactive material is present. It encompasses any substance that is innately radioactive or has been polluted with radiation and is deemed to be unusable. Radioactive wastes are made up of radioactive materials, which are typically leftovers from the production of nuclear energy and other nuclear fission-related technologies utilized in industries like research, medicine, rare-earth and coal mining, manufacturing, building, and nuclear weapons (Zhang *et al.*, 2019). In the recent years, nuclear energy has been increasingly used to generate power. According to the existing state of affairs, there won't be a material decrease in the use of uranium in the upcoming years, which suggests a rise in the creation of radioactive waste (Ma *et al.*, 2019).

## Sources of Radioactive Waste

All radioactive substances that can be found in the environment are considered naturally-occurring radioactive materials (NORM). The phrase is explicitly used, nonetheless, to refer to any naturally occurring radioactive elements when human actions have raised the risk of exposure in comparison to the pre-altered state. Long-lived radioactive elements like uranium, thorium, and potassium, as well as any of their byproducts of decay, including radium and radon, are examples of NORM. These substances are concentrated in some locations, such as uranium ore deposits that may be extracted, and have always been in the crust and atmosphere of the Earth (Vearrier *et al.*, 2009).

Naturally occurring radioactive materials (NORM) and the nuclear fuel cycle as well as the reprocessing of nuclear weapons are some of the sources of radioactive waste. Other sources include medical and industrial wastes. Nuclear fuel cycle and nuclear weapons are the two sources of waste that are most prevalent among all of them. Both the front end and the back end of the nuclear fuel cycle produce radioactive waste. Alpha-emitting waste from uranium extraction often comes from the beginning end of the nuclear fuel cycle. Since spent fuel rods make up the majority of the nuclear fuel cycle's back end. It frequently contains radium and its decay products as well as fission products that emit beta and gamma radiation (Peluzo and Kraka, 2022).

One of the significant non-military uses of atomic energy is in the field of medicine. Hospitals utilize radioactive isotopes that have not been sealed for both therapeutic and diagnostic purposes. Beta and gamma emitters are typically present in medical waste. Among the isotopes used in medicine is Y-90 for the treatment of lymphoma, I-131 for the treatment of thyroid cancer, Sr-89 for the treatment of bone cancer, Ir-192 for brachytherapy, Co-60 for brachytherapy and external radiation, and Cs-137 for brachytherapy and external radiotherapy (Fassbender, 2020).

## Types of Radioactive Waste

Government regulations determine whether specific items, such spent nuclear fuel and plutonium, are classified as waste. Every radionuclide has a half-life, which is the length of time it takes for half of its atoms to decay and, consequently, for it to lose half of its radioactivity. Radionuclides with long half-lives are more likely to release alpha and beta radiation, which facilitates handling, whereas those with short half-lives are more likely to emit gamma rays, which are more invasive. All radioactive waste eventually decomposes into inert substances. An isotope degrades more rapidly the more radioactive it is.

Based upon the level of radioactivity, wastes have been classified as low-level (LLW), intermediate-level (ILW), or high-level (HLW) (IAEA, 2009):-

**Low-level waste:** Low-level waste (LLW) has a radioactive concentration that is less than four GBq/t of alpha activity or twelve GBq/t of beta-gamma activity. When handled and transported, LLW does not need to be shielded, and it may be disposed-off at facilities that are close to the surface. Hospitals, business, the nuclear fuel cycle, and other sources all produce LLW. It is made up of materials like paper, rags, tools, clothes, filters, etc. that typically have radioactivity that lasts just a short time. Prior to disposal, LLW is frequently compressed or burned to minimize its bulk (Chapman and Hooper, 2012). 90% of all radioactive waste is LLW; however, it only contains 1% of the radiation.

**Intermediate-level waste:** Despite the fact that intermediate-level waste (ILW) is more radioactive than low-level waste (LLW) and produces less heat than LLW (2 kW/m3), it cannot be considered when designing or choosing storage and disposal facilities. ILW needs some shielding due to its higher levels of radiation. ILW frequently consists of resins, chemical sludges, metal fuel cladding, and other things. To dispose of smaller things and any non-solids, concrete or bitumen may be used to solidify them. In terms of volume and radioactivity, it accounts for around 7% of all radioactive waste.

**High-level waste:** High-level waste: High-level waste (HLW) is sufficiently radioactive to dramatically raise both its internal temperature and the ambient temperature (decay heat >2kW/m3). Therefore, cooling and shielding are needed for HLW. The 'burning' of uranium fuel in a nuclear reactor produces high level waste (HLW). The fission products and transuranic elements produced in the reactor core are found in HLW. Only 3% of created trash is HLW, yet it contains 95% of its radioactivity. HLW is made up of both long-lived and short-lived radioactive elements, depending on how long it will take for the radioactivity of a given radionuclide to decline to levels that are deemed safe for people and the environment. When it comes to managing and disposing of HLW, the ability to distinguish between long-lived actinides and relatively short-lived fission products is crucial. The management of HLW reflects the fact that it is the subject of considerable attention in nuclear power.

## Impact of Radioactive Wastes

In terms of anthropogenic sources of radioactive pollution, research into nuclear weapons, nuclear disasters, nuclear power plants, and the disposal of radioactive waste are the main causes of contamination, while the use of radioisotopes in manufacturing and research labs is a small source. Radioactive substances are exposed to people through the employment of nuclear weapons, X-rays, MRIs, and other medical devices. The majority of living forms as well as the environment are negatively impacted by radioactive waste, which is controlled by government organizations to safeguard both.

Water contamination results from the dumping of these radioactive wastes in surface waterbodies. Due to its prolonged exposure to radioactive elements present in the rocks, groundwater is more frequently contaminated with radioactive substances than surface water. Groundwater contamination results from naturally occurring radioactive materials (NORM) percolating through the soil sediments and into the aquifer (Berglund *et al.*, 2013). Radioactive gases are occasionally released into the atmosphere by magma. Furthermore, radioactive pollution is brought on by the deposition of these radioactive gases in waterbodies. Radioactive nuclei are also added to the surface water through atmospheric deposition (dry and wet) of cosmogenic radionuclides. Human-induced radionuclide discharge is mostly caused by nuclear reactors and nuclear weapon tests. Inhaling, injecting, or being exposed to radioactive materials can have an adverse effect on the environment and represent a risk to human health. Polluted water and food can expose human tissues to radiation, which poses major health hazards. Acute radiation syndrome or cutaneous radiation damage can result from high radiation exposures. Radiation exposure may lead to a number of physiological diseases in people, such as chromosomal disruption, cancer, leukemia, genetic mutations, osteonecrosis, cataracts, and cancer.

Nuclear reactors also produce a wide range of unwanted radioisotopic wastes, such as strontium-90, caesium-137, and others. Nuclear power facilities located along the shore release radioactive waste into the ocean, increasing the amount of radiological pollutants. These powerhouses utilize water as coolants as well, which causes contamination. The most often occurring radioactive elements in the sludge produced by sewage treatment facilities are 40K and 7Be. Additionally, surface and groundwater are contaminated by the mining of radioactive materials like uranium and thorium. The natural repository of naturally occurring uranium is oceans and seas where it is present as the ion uranyl carbonate. The higher salinity of the sea water should have a large proportion of uranium. Nuclear submarines contaminate the maritime environment with radioactive material. There have been reports of radioactive contamination caused by submarine mishaps and sinkings. Examples of such nuclear mishaps include the Chernobyl nuclear disaster, the Fukushima nuclear disaster, and the Rocky Flats facility in Colorado.

Nuclear radiation has negative effects on:

- Immediate, reversible effects that include genital and skin irritation as well as hair loss.
- Long-lasting, irreversible effects, including pneumonia, cholera, dysentery, bone marrow death, cataract onset, cancer stimulation, and a variety of illnesses, including cholera, dysentery, TB, and dysentery. These results can sometimes be lethal as well.
- Genetic influences Ionising radiation causes mutations in germ cells (male sperm and female egg cells) or germ cells, resulting in structural changes in germ cell DNA that are passed on to progeny. Hereditary conditions can cause early mortality and serious mental disease.

Today, accurate radioactive pollution analysis and monitoring are also necessary for a safe water supply. The human-caused causes of radioactive contamination in water resources can be checked using prevention and safety measures.

## Treatment and Disposal of Radioactive Waste

Radioactive waste management is a required step to deal with it since the exploitation of radioactive materials was carried out on a massive scale in the recent decades, leading to the generation of an enormous volume of radioactive waste. Irradiation from radioactive waste poses major risks to both people and the environment if it is not properly disposed-off. Implementing current regulatory regulations and procedures in hospitals for the handling of radioactive materials used in patient care requires a suitable model in nations where the expansion of nuclear medicine services is anticipated.

## Future of the Topic

In addition, national and international research and development programmes are required to be developed through international forums in order to ensure the highest safety and security measures in nuclear power plants and waste management. Future topics will centre on issues related to modelling on waste management and estimating potency. A definite trend emerges from the literature review to analyze the design, construction, and operation of simulated nuclear systems. To achieve the highest safety and security standards in nuclear power plants and waste management, it is also necessary to build national and worldwide research and development programmes through international forums. Future discussions will focus on problems with modeling waste management and assessing possible risks exposure. In the realm of radioactive waste management, more research has to be done on the effects of radionuclides on air, water, and soil.

## Bibliography


Berglund, S., Bosson, E., Selroos, J-O. and Sassner, M. 2013. Identification and Characterization of Potential Discharge Areas for Radionuclide Transport by Groundwater from a Nuclear Waste Repository in Sweden. *Ambio* 42: 435-446.

Chapman, N. and Hooper, A. 2012. The disposal of radioactive wastes underground. *Proceedings of the Geologists' Association* 123: 46-63.

Classification of radioactive waste. International Atomic Energy Agency. Vienna: IAEA, 2009. GSG-1

Fassbender, M.E. 2020. Guest Edited Collection: Radioisotopes and radiochemistry in health science. *Scientific Reports* 10:340.

Ma, Z., Gamage, R.P., Rathnaweera, T. and Kong, L. 2019. Review of application of molecular dynamic simulations in geological high-level radioactive waste disposal. 168: 436-449.

Peluzo, B.M.T.C. and Kraka, E. 2022. Uranium: The Nuclear Fuel Cycle and Beyond. *International Journal of Molecular Sciences* 23(9): 4655.

Vearrier, D., Curtis, J.A. and Greenberg, M.I. 2009. Technologically enhanced naturally occurring radioactive materials. *Clinical Toxicology* 47(5): 393-406.

Zhang, X., Gu, P. and Liu, Y. 2019. Decontamination of radioactive wastewater: State of the art and challenges forward. 215: 543-553.

# Chapter - 14

## Nitrates and Phosphates: Boon or Bane for Waterbodies

**[1]Abdul Hakkim.H., [2]Kirthiga, S.S., [2]Sona G Jain, [2]Divya Alex M., [2]Rajeev Raghavan and [2*]Anu Gopinath**

[1]Faculty of Ocean Science and Technology, Kerala University of Fisheries and Ocean Studies, Kochi - 682506, Kerala
[2]Department of Aquatic Environment Management, Kerala University of Fisheries and Ocean Studies, Kochi - 682506, Kerala
*Correspondingemail: dranugopinath@gmail.com

### Abstract

Water bodies are universal resources of high importance that are extremely essential for the survival of all living beings. The health of the waterbody can be easily inferred by the quantity of major nutrients present in it. Two nutrients of paramount importance include nitrate and phosphate which have a major role in the biogeochemical cycles of the ecosystem. Nutrient availability in a waterbody is considered to be a crucial factor in determining the functioning of the aquatic ecosystem as it promotes the growth of phytoplankton and thus regulates the food chain as well as the food web. Besides, the nutrients are also highly tied up with the biogeochemistry of the waterbodies. Despite, nitrates and phosphates being a boon to the water bodies when found in excess concentrations will become a bane as they will lead to harmful negative impacts suchas algal bloom and eutrophication that ultimately result in detrimental effects. By and large, the concentration of Nitrates and Phosphates in water bodies were increasing nowadays mainly because of irresponsible human actions. Frequent monitoring and sustainable use of waterbodies will aid the nutrients to be within the safe limit. Humans are the one and only rationale for maintaining the nutrients as a boon and converting them into bane by his activities and the chapter extensively explains it.

**Keywords:** Nitrate; Phosphate; Nitrogen; Phosphorus; Nutrients, Water bodies; Aquatic environment; Anthropogenic; Biogeochemical cycle; Eutrophication.

## Introduction

The productivity and quality of a water body is mainly determined by the presence of nutrients in it. Among this, nitrogen and phosphorus are the key compounds that is considered to be the limiting nutrients for which there is no substitute at all (RashidiNodeh *et al.*, 2017). Not only in waterbodies but also in fertilizer industries, these pertains to be limiting nutrients which resulted in the tremendous increase of annual demand for phosphorus and nitrogen that is to be increased with the world's population growth. According to predictions by FAO, the long-term global nutrient intake will amount to 199.3 million tons (72.2 percent nitrogen and 27.8 percent phosphorus) in 2030. This excessive use of nutrients by the humans will lead to the increased concentration of nitrogen and phosphorus in the waterbodies as a result of runoff and wastewater discharge in the aquatic environments. Even then, if the concentration of nutrients exceeds the advised limit, are regarded as contaminants. Hence, the nitrogen and phosphate has several key roles as well as some negative effects which will be discussed in this chapter in detail.

## Sources of Nitrates and Phosphates

The major source of nitrates pertains mainly to anthropogenic activities including regular chemical fertiliser use, sewage, landfills, and also household waste that contain high nitrogenous compounds (Shrimali and Singh, 2001). In addition to this, metal and plastics industries, industrial treatment, textile, plywood, home cleaning, and pharmaceuticals also implies the incorporation of nitrogeneous compounds which will further increase the level of contamination in the waterbodies especially by effluents. Human actions that involve the handling, use, and disposal ofmanures and liquid solid waste are the other major anthropogenic sources. The surface drains and wastewater waste water discharge into the waterbodies and also will percolate through the soil cover, and then taints the water in the ground. The agricultural chemicals (both synthetic and natural fertilizers) will also lead to the enrichment of nutrients in the waterbodies.

Phosphorus is never found in its elemental form in nature; instead, it always appears as phosphates, which are made up of one phosphorus atom bound to four oxygen atoms. The two other ways that this can exist is negatively charged phosphate ion ($PO_4^{3-}$), which is how it appears in rocks, or as organophosphates, which have organic molecules bonded to one, two, or three of the oxygen atoms. In soil, phosphorus can take on a variety of shapes. The three main kinds of phosphorus that these sources fall under are (i) organic phosphorus, (ii) soluble phosphorus, and (iii) adsorbed phosphorus. In soil, phosphorus is least mobile in the forms of hydrogen phosphate [$HPO_4^{2-}$] and dihydrogen phosphate [$H_2PO_4^-$]. The routes via which phosphorus enters water and soil are weathering from the rocks by rain, phosphorus mining from rock, and other abiotic methods including agriculture, industrial processes, municipal effluent, etc. Plants and microorganisms can obtain phosphorus to absorb for growth after it reaches the water and soil. After these plants and microbes decompose, the phosphorus that was absorbed will leak back into the water and soil (Meena *et al.*, 2018). Although it led to phosphorus depletion, phosphorus mining in the past was able to meet the demands of the expanding fertiliser industry. Since the aforementioned natural inorganic phosphorus cycle takes millions of years and is not a sustainable source, the depletion of phosphorus might be viewed as a next-generation equity issue (Fan *et al.* 2019). Approximately 27 million tonnes of phosphorus are given to agricultural land each year in the form of fertiliser, yet only 3 million tonnes are absorbed by people each year through food, with the remainder being regarded as lost (Chen *et al.*, 2020).

**Nutrient Retention in Waterbodies**

The relative retention of nutrients depends strongly on the duration of the hydraulic retention time in the lake (Sondergaard, 2007). The longer the retention time, the higher will be the loss percentage of the added nutrients cycling through the water body. Phosphorus (P) is the major limiting nutrient in freshwaters, with Nitrogen (N) as the co-limiting nutrient in some cases. Moreover, high N deposition patterns increase N: P ratios and further amplify P limitation (Elser *et al.*, 2010). Nitrogen recycles rapidly than phosphorus because nitrogen primarily occurs bound in sedimental organic matter, which first has to be decomposed, and not as phosphorus bound and accumulated in inorganic compounds.

In productive waters, the sedimentation of phosphorus may rise in connection with the depositing calcium carbonate. In oligotrophic waters, primary production is limited by nutrients like P and Fe. P-limited algae are deficient in their ability to assimilate nitrate. One of the most likely consequences of these oligotrophic conditions is low denitrification due to limited labile organic carbon (OC) availability in the sediments. This does not cause a large oxygen demand at the sediment-water interface and the surficial sediment (8 to 24 mm) remains oxic and promotes nitrification increasing nitrate concentration in the surficial sediment layer than overlying water and consequent nitrate efflux from sediment to the water column.

This leads to less efficient nitrate sink and consequent nitrate build up. In alkaline lakes, pelagic calcite precipitation is one of the mechanisms stabilising the clear water state (Sondergaard, 2007).

## Crucial Role of Nitrates and Phosphates in Aquatic Environment

Nitrates and phosphates are both beneficial to algal development. These factors, however, impact algal development independently of one another, and there is no interaction between the two. This suggests that nitrates and phosphates are both efficient limiting nutrients that may be adjusted to regulate algal growth.For example, if nitrogen is abundant but phosphorus is not, algae will proliferate until all of the phosphorus is exhausted, even if nitrogen is still available.As a result, phosphorus is the limiting factor on algal development in an aquatic system with high nitrogen levels but low phosphorus levels. So, when both nutrients are present at high concentrations, neither nitrate nor phosphate impedes algal growth, nor an algal bloom form (Fried *et al.*, 2003).

The effects of nitrate on the release of phosphorus from lake sediments are also reported. High nitrate ($NO_3^-$) input may have a dual effect on sediment P release, lowering P release by enhancing sediment oxidation or boosting P release by stimulating phytoplankton growth and increasing decomposition rates and oxygen consumption at the sediment-water interface (Ma *et al.*, 2021). Aside from production management, nitrogen's impact in defining plant species richness is another component of its significance in freshwater (Barker *et al.*, 2008). Nitrogen availability and species richness in plant communities have long been linked by terrestrial ecologists and agronomists. In comparison to controls receiving ambient N deposition, they discovered that chronic low-level nitrogen addition reduced plant species populations by 17% or otherwise, the richness of species decreases as a linear function of the rate of inorganic nitrogen deposition (Stevens *et al.*, 2004; Clark and Tilman, 2008).

## Phosphorus in the Mitigation of Heavy metal Toxicity

Considering phosphorus is a nutrient that limits the biological productivity of the terrestrial environment, it frequently contributes significantly to net carbon uptake in terrestrial ecosystems. Although soil's inorganic phosphorus contributes to eutrophication, which harms the environment, it also has some positive impacts. Zhang *et al.* (2010) reported that the phosphorus in soils aids in reducing the toxicity of heavy metals to plants. Numerous contaminants have been added to the soil as a result of rising population, fast urbanisation, and industrial area expansion, usage of fertilisers, pesticides, and manures, as well as atmospheric decomposition. Even at very low quantities, heavy metals are poisonous and have negative effects. They provide a significant environmental concern since they are persistent and bioaccumulative in nature and do not degrade through microbial and chemical processes and several technologies were emerged to degrade heavy metals (Masindi *et al.*, 2018; Ali *et al.*, 2019). One among that is the phytoremediation process. The advantage of phytoremediation is that it is a natural, inexpensive solution to an environmental issue (Peuke and Rennenberg, 2005). Metal hyperaccumulators include several plants. If the hyperaccumulator plants are edible, the roots will absorb the heavy metals from the soil, which will cause bioaccumulation of heavy metals throughout the entire plant. This will eventually enter the food chain and pose health risks to higher creatures. Because phosphorus will reduce the uptake and accumulation of heavy metals in food crops growing in soil contaminated with heavy metals like arsenic and chromium, it is ideal to add phosphorus to the soil in order to prevent the entry of heavy metals into the food chain.

## Role of Nitrogen Transformations in Aquatic Environments

The form and abundance of N in aquatic systems is partially controlled by two microbially mediated N transformations, nitrification and denitrification. Denitrification rates are mostly controlled by $NO_3^-$ and oxygen availability in sediments while nitrification are controlled by both temperature and pore water $NH4^+$ availability (Bruesewitz, 2012). The shallow edge sediments (<2m deep) of the water bodies are hot spots of N transformation. Shallow edge habitat can be considered as hot spot and profundal sediments as coolspots for denitrification activity. Very shallow edge sediments are subject to windinduced wave action, which likely increases the transport of $NH_4^+$, $NO_3^-$, and oxygen within shallow edge sediment pore waters, and generally creates a more diverse sediment redox profile thereby promoting coupled nitrification-denitrification by bringing oxic and anoxic zones into closer and more frequent contact. N transformation rates are controlled by a complex set of variables that interact and change over space and time. The suite of biological, physical, and chemical variables that change when moving from shallow edge to littoral to profundal sediments include sediment O2 penetration (sediment porosity), organic matter lability, microstratification of the water column, and the presence of periphyton or rooted macrophytes. These variables all act in concert to influence sediment redox conditions, organic carbon availability, and inorganic N availability to the sediments (Bruesewitz, 2012). Sediment organic matter and sediment % N increases with increasing depth.

## Nutrient Uptake Dynamics

One of the most important terms in the N cycle is uptake of inorganic N (i.e., nitrate and ammonium). Knowledge of uptake rates and their environmental controls can provide insights into the controls of N dynamics. Multiple factors including light, and temperature, substrate availability as well as algal biomass could be important in regulating N uptake dynamics. Nitrate uptake is inhibited by ammonium in marine (Eppley *et al.*, 1969) as well as fresh water (Prochazkova *et al.*, 1970) systems. Both nitrate and ammonium uptake rates show light inhibition in freshwater (Dodds and Priscu, 1989) as well as in marine environment therefore the subsurface ammonium uptake maximum. In most aquatic systems under similar conditions, the energy requirements for ammonium assimilation are much less than for nitrate assimilation, because ammonium is already in reduced form and is recycled rapidly.

## Nutrient Status of Waterbodies

Understanding the complex relationship among nutrient loading, lake processes and lake-watershed characteristics, and acid deposition is important in understanding nutrient status and acid-base characteristics and these waterbodies responses to global change and recovery from acidification. In remote lakes under acidification, low ANC (acid neutralizing capacity) leads to P immobiliation by aluminophosphate compounds and limited biodiversity (Dixit *et al.*, 2002). Increases in primary production and retention times improve recovery from acidification as a result of increased in-lake production of ANC from decreased $SO_4^{2-}$ and $NO_3^-$ inputs as well as algal uptake of $NO_3^-$. It could also simultaneously increase the internal supply of TP released (P mobility) from lake sediments which in turn decrease phosphorus limitation of acid-impacted lakes (Gerson *et* al., 2016).

## Aquatic Organisms and Nutrient Recycling

The size and species composition of animal species may have substantial effects on the nutrient dynamics. Different aquatic species sequester and thus recycle nutrients at different N:P ratios according to their metabolic requirements. For example, cladoceran, especially Daphnia, a large herbivorous species, which dominates when zooplanktivorous fish are scarce, has a lower tissue N: P ratio (percentage of dry mass) compared to species such as copepods and rotifers, which often dominate when zooplanktivores are abundant. To maintain nutrient homeostasis (i.e., low body N:P), *Daphnia* would excrete nutrients at a higher N:P ratio than copepods. Fish may therefore regulate stoichiometric processes by controlling zooplankton species composition (Vanni, 1997). Food web structure affect how nutrient recycling done by animals in lakes.

For example, when piscivores are rare and zooplanktivorous fish are abundant, small zooplankton species predominate, which recycle nutrients at a higher rate per unit biomass than larger species. Many "zooplanktivorous" fish increase nutrient transport rates from littoral/benthic habitats to pelagic habitats by feeding on littoral and benthic prey and excreting dissolved inorganic nutrients into the pelagic (open water) zone. Fish biomass gradient can exert major influences on the dynamics, distribution, and ratios of limiting nutrients and their partitioning among ecosystem compartments. Fish increase water column TP relatively more than TN, and thus the TN: TP ratio decrease significantly with increasing fish biomass (Dai *et al.*, 2022).

## Macrophytes and Nutrients

Macrophytes has important role on nutrient budgets. Submerged macrophytes link the sediment with overlying water. This linkage has important implications for nutrient cycling particularly with rapidly growing species such as *Myriophyllum spicatum* (Barko, 1998). For the nutrients Phosphorus and nitrogen, sediment is the primary source for uptake. They mobilize nutrients from sediments by causing marked fluctuations in pH and oxygen through metabolic activities. In particular, high pH values (about 9-10) associated with macrophyte photosynthesis can result in ligand exchange with P adsorbed to iron oxide-hydroxides on sediment particles, thus enhancing the rate of P release from sediments. During night, they promote P release from sediments by causing anoxic conditions, results in substantial accumulation of P in the littoral water column (Barko, 1998). Critical nutrient limiting values for aquatic macrophytes are 0.75 % for nitrogen and 0.07% for (Gerloff, 1975). Moeller *et al.* (1988) have suggested that sedimentation of algae and periphyton from macrophyte leaf surfaces may provide an important link for transfer of nutrients absorbed from the water (by algae) to the sediment.

## Negative Impacts of Nitrogen and Phosphorus in Waterbodies

The conversion of oxygen-carrying pigments to forms incapable of transporting oxygen is the primary harmful action of nitrate. With rising nitrate concentrations and exposure durations, aquatic creatures become more harmful to nitrate. However, when body size, water salinity, and environmental adaption increase, nitrate toxicity may decrease. Nitrate appears to be more toxic to freshwater creatures than to marine animals. Fresh water invertebrates, amphibians, and fish can be harmed by a nitrate content of 10 mg $NO_3$-N/1 l (the statutory limit threshold for drinking water in the United States). To protect delicate freshwater creatures from nitrate contamination, safe limits below this nitrate content are advised. Furthermore, safeguarding the most vulnerable freshwater species would need a maximum concentration of 2 mg $NO_3^-$ - N/l.Some marine invertebrates, which are well adapted to low nitrate concentrations, may, nevertheless, be as vulnerable to nitrate as delicate freshwater invertebrates in their early larval phases (Camargo *et al.*, 2006).

Three primary environmental issues caused by inorganic nitrogen contamination in aquatic habitats are as follows: (1) it can raise hydrogen ion concentrations in freshwater habitats without considerable acid-neutralizing ability, resulting in system acidification; (2) it can stimulate or enhance the development, maintenance and proliferation of primary producers, resulting in eutrophication of aquatic ecosystems; (3) It can accumulate to dangerous amounts, making aquatic species unable to survive, thrive, or reproduce. The researchers found that total nitrogen levels of less than 0.5–1.0 mg TN/L might avoid acidification and eutrophication in aquatic environments (excluding those with naturally high N levels), at least due to inorganic nitrogen pollution.Because surface waters typically contain relatively high concentrations of dissolved oxygen in the absence of eutrophication, those relatively low TN levels may also protect aquatic animals from the toxicity of inorganic nitrogenous compounds, with most inorganic reactive nitrogen being in the form of nitrate. Furthermore, inorganic nitrogen pollution's negative impacts on human health and the economy would be avoided (Camargo *et al.*, 2005).

### Eutrophication

Eutrophication is one of the most serious water quality challenges facing lakes worldwide. The ageing of a lake is indicated by eutrophication of a water body. The accumulation of nutrients, sediments, silt, and organic matter in the lake from the surrounding watershed causes it. Eutrophication refers to changes in aquatic systems caused by nutrient enrichment, with the end result being an increase in primary production to nuisance levels (Marsden *et al.*, 1989). The primary reason is excessive phosphate and nitrogen loading into the system, which results in high algal biomass, cyanobacterial dominance, and macrophyte loss.

## Conclusion

Both the nitrogen and phosphorus as well plays a crucial role in the aquatic environment. Naturally, serves as the limiting nutrients and thus promoting the productivity of the lake. Also, it is fair to state that nitrates and phosphates are probably the key nutrients in controlling aquatic plant growth. The cycling of these nutrients is essential for aquatic as well as terrestrial systems to maintain the ecological nature. Despite being important and a boon to waterbodies, these nutrients can also be a bane to them, if their level exceeds beyond a certain limit. The level of nutrients gets increased mainly because of the irresponsible anthropogenic activities. The United States Environmental Protection Agency (USEPA) advises that phosphate levels be kept below 0.1mg/l and nitrate levels be kept below 10mg/l in order to maintain a healthy water system and limit algal blooms. The responsibility to keep these nutrients within the desired limit and to monitor the status of these nutrients in the waterbodies frequently and consistently, prevailed to everyone of us and also it is indispensable to realize that making the nutrients boon or a bane to is ultimately in the hands of humans.

## Bibliography

Ali, H., Khan, E. and Ilahi, I. 2019. Environmental chemistry and ecotoxicology of hazardous heavy metals: Environmental persistence, toxicity, and bioaccumulation. *Journal of Chemistry* 6730305.

Barker, T., Hatton, K., O'Connor, M., Connor, L. and Moss, B. 2008. Effects of nitrate load on submerged plant biomass and species richness: results of a mesocosm experiment. *Fundamental and Applied Limnology* 173(2), 89.

Barko, J.W. and James, W.F. 1998. Effects of submerged aquatic macrophytes on nutrient dynamics, sedimentation, and resuspension. In *The structuring role of submerged macrophytes in lakes* (pp. 197-214). Springer, New York, NY.

Bruesewitz, D.A., Tank, J.L. and Hamilton, S.K., 2012. Incorporating spatial variation of nitrification and denitrification rates into whole-lake nitrogen dynamics. *Journal of Geophysical Research: Biogeosciences 117*(G3).

Camargo, J. A., Alonso, A. and Salamanca, A. 2005. Nitrate toxicity to aquatic animals: a review with new data for freshwater invertebrates. *Chemosphere* 58(9), 1255-1267.

Camargo, J.A. and Alonso, Á. 2006. Ecological and toxicological effects of inorganic nitrogen pollution in aquatic ecosystems: a global assessment. *Environment International* 32(6), 831-849.

Chen, J., van Groenigen, K.J., Hungate, B.A., Terrer, C., van Groenigen, J.W., Maestre, F.T., Ying, S.C., Luo, Y., Jørgensen, U., Sinsabaugh, R.L. and Olesen, J.E. 2020. Long-term nitrogen loading alleviates phosphorus limitation in terrestrial ecosystems. *Global Change Biology* 26(9), 5077-5086.

Clark, C. M. and Tilman, D. 2008. Loss of plant species after chronic low-level nitrogen deposition to prairie grasslands. *Nature* 451(7179), 712-715.

Dai, M., Xiao, Y., Wang, T., Xu, J. and Wang, Y. 2022. Influence of N:P Ratio of Water on Ecological Stoichiometry of *Vallisneria natans* and *Hydrilla verticillata*. *Water* 14(8), 1263.

Dixit, S., Gupta, S. K. and Tiwari, S. 2005. Nutrient overloading of fresh water lake of Bhopal, India. *Electronic Green Journal* 1.

Dodds, W.K. and Priscu, J.C. 1989. Ammonium, nitrate, phosphate and inorganic carbon uptake in an oligotrophic lake: seasonal variations among light response variables. *Journal of Phycology* 25, 699–705.

Elser, J.J., Fagan, W.F., Kerkhoff, A.J., Swenson, N.G. and Enquist, B.J. 2010. Biological stoichiometry of plant production: metabolism, scaling and ecological response to global change. *New Phytologist* 186(3), 593-608.

Eppley, R.W., Rogers, J.N. and Carthy, Mc. 1969. Half-Saturation Constants for Uptake of Nitrate and Ammonium by Marine Phytoplankton. *Limnology and Oceanography* 14(6), 912-920.

Fan, Y., Zhong, X., Lin, F., Liu, C., Yang, L., Wang, M., Chen, G., Chen, Y. and Yang, Y., 2019. Responses of soil phosphorus fractions after nitrogen addition in a subtropical forest ecosystem: Insights from decreased Fe and Al oxides and increased plant roots. *Geoderma* 337, 246-255.

Fried, S., Mackie, B. and Nothwehr, E. 2003. Nitrate and phosphate levels positively affect the growth of algae species found in Perry Pond. *Tillers* 4, 21-24.

Gerloff, G.C. 1975. Nutritional ecology of nuisance aquatic plants. National Environmental Research Center, Office of Research and Development, *US Environmental Protection Agency.*

Gerson, J.R., Driscoll, C.T. and Roy, K.M. 2016. Patterns of Nutrient Dynamics in Adirondack Lakes Recovering From acid Deposition. *Ecological Applications* 26(6), 1758-1770.

Ma, S. N., Wang, H. J., Wang, H. Z., Zhang, M., Li, Y., Bian, S. J., and Jeppesen, E. (2021). Effects of nitrate on phosphorus release from lake sediments. *Water Research* 194, 116894.

Marsden, M.W. 1989. Lake restoration by reducing external phosphorus loading: the influence of sediment phosphorus release. *Freshwater biology* 21(2), 139-162.

Masindi, V. and Muedi, K.L. In: Saleh, H.E-D.M. and Aglan, R.F. 2018. Environmental Contamination by Heavy Metals, Heavy Metals. *IntechOpen.*

Meena, R.S., Vijayakumar, V., Yadav, G.S. and Mitran, T. 2018. Response and interaction of Bradyrhizobium japonicum and arbuscular mycorrhizal fungi in the soybean rhizosphere. *Plant Growth Regulation* 84, 207–223.

Moeller, R.E., Burkholder, J.M. and Wetzel, R.G. 1988. Significance of sedimentary phosphorus to a rooted submersed macrophyte (Najasflexilis (Wild.) Rostk. and Schmidt) and its algal epiphytes. *Aquatic Botany* 32(3), 261-281.

Peuke, A.D. and Rennenberg, H. 2005. Phytoremediation. *EMBO Reports* 6(6), 497-501.

Prochazkova, L., Blazka, P. and Kraeva, M. 1970. Chemical changes involving N metabolism in water and particulate matter during primary production experiments. *Limnology and Oceanography* 15, 797-807.

RashidiNodeh, H., Sereshti, H., Afsharian, E. Z. and Nouri, N. 2017. Enhanced removal of phosphate and nitrate ions from aqueous media using nanosized lanthanum hydrous doped on magnetic graphene nanocomposite. *Journal of Environmental Management* 197, 265–274.

Shrimali, M. and Singh, K.P. 2001. New Methods of Nitrate Removal from Water. *Environmental Pollution* 112(3), 351-359.

Sondergaard, M. 2007. Nutrition dynamics in lakes- with emphasis on phosphorus, sediment and lake restoration. https://www2.dmu.dk/pub/dsc_ms_uk.pdf

Stevens, C. J., Dise, N. B., Mountford, J. O. and Gowing, D. J. 2004. Impact of nitrogen deposition on the species richness of grasslands. *Science* 303(5665), 1876-1879.

Zhang, M.K., Liu, Z.Y. and Wang, H. 2010. Use of single extraction methods to predict bioavailability of heavy metals in polluted soils to rice. *Communications in Soil Science and Plant Analysis* 41(7), 820-831.

# Chapter - 15

# Factors Modifying Toxicity of Pesticides in Lakes and Methods of Decontamination

**[1]Imtiyaz Qayoom*, [1]Adnan Abubakr, [1]Sameena Khan, [1]Taranum Mushtaq, [1]Mehjabeena Akhtar, [2]Farheen Saba and [3]Uzma Siddique**

[1]Division of Aquatic Environmental Management, Faculty of Fisheries, SKUAST-K, Rangil-190006, Ganderbal, J&K
[2]Department of Zoology, Moulana Azad National Urdu University, Hyderabad - 500032, Telangana
[3]Department of Zoology, DSB Campus, Kumaun University, Nainital-263001
*Corresponding email: drsheikhimtiyaz@gmail.com

## Abstract

Pesticides are washed away from application sites into waterbodies, wherein they end up in accumulating in the compartments of aquatic ecosystems. More oftenly, they get move through food chains and affect vital organisms like fishes and subsequently reach humans to intoxicate them. However, there are various factors that govern the environmental fate of pesticides in the water body. Those factors determine the movement, kinetics and toxicity end points of pesticides in lake ecosystems which differ from chemical to chemical and compartment to compartment. This chapter describes toxicity caused by pesticides, their fate and factors responsible for governing their toxicity in aquatic ecosystems and environmental fate up to their toxic end points.

**Keywords:** Pesticides, Environmental Fate, Toxicity, Aquatic Ecosystem, $LC_{50}$

## Introduction

Chemical agents used to eliminate or control pests is known as a pesticide. The Latin term "cida," which means to kill, serves as the root of the word. The general phrase "pesticides" can refer to a variety of compounds, including insecticides, rodenticides, herbicides, fungicides, biocides, and similar substances.1,1,1-trichloro-2,2-bis[4-chlorophenyl]ethane commonly known as DDT is the first synthesised organic substance to be applied as a pesticide. The preparation of this chemical as an experiment began in 1874, and its insecticidal qualities were identified in 1939 by Muller and colleagues. But the development of new chemical analysis techniques at the start of the 1960s brought to light the bioaccumulation of DDT in the environment and its involvement in upsetting the natural equilibria of insects, birds, and fish in the environment. Widespread environmental pollution by persistent pesticides, including the accumulation of chemical residues in wildlife, drinking water, and people, is an issue that is more pervasive ecologically.

## Environmental toxicity caused by pesticides

Accidental spills, industrial effluent, surface runoff and transport from pesticide-treated soils, washing of spray equipment after spray operations, drift into ponds, lakes, streams, and river water and aerial spray to control water-inhibiting pests are all ways pesticides might end up in water (Singh and Mandal, 2013). These paths can be categorized as either unintentional or deliberate. Pesticides can reach surface and groundwater bodies through both diffuse and point-source pollution. Low pesticide concentrations are detected widely due to pesticide diffuse contamination, whereas high pesticide concentrations are detected by point sources. Pesticide-contaminated water is extremely dangerous to aquatic life as it has the potential to harm aquatic vegetation, reduce dissolved oxygen levels in the water, and cause physiological and behavioural abnormalities in fish populations (Mahmood *et al.*, 2016). Pesticides that are water soluble dissolve in water and enter ground water, streams, rivers, and lakes, harming untargeted species. Fat-soluble pesticides, on the other hand, enter the bodies of animals through a process known as "bio-amplification." They are absorbed by animal's fatty tissues, leading to pesticide persistence in food chains for long periods of time. Pesticide exposure in fish and other aquatic species is determined by biological availability, bio-concentration, biomagnification, and environmental persistence. There are three main ways that fish and aquatic animals are exposed to pesticides as shown in figure 1. First is the direct skin-to-skin contact with pesticides while swimming in contaminated waters, secondly, when the pesticides get consumed orally through ingestion of pesticide-contaminated food or water or direct uptake through the gills during respiration (Helfrich *et al.*, 2009). Fish exposure can also result through eating another species that has previously been poisoned by a chemical, among other secondary reasons.

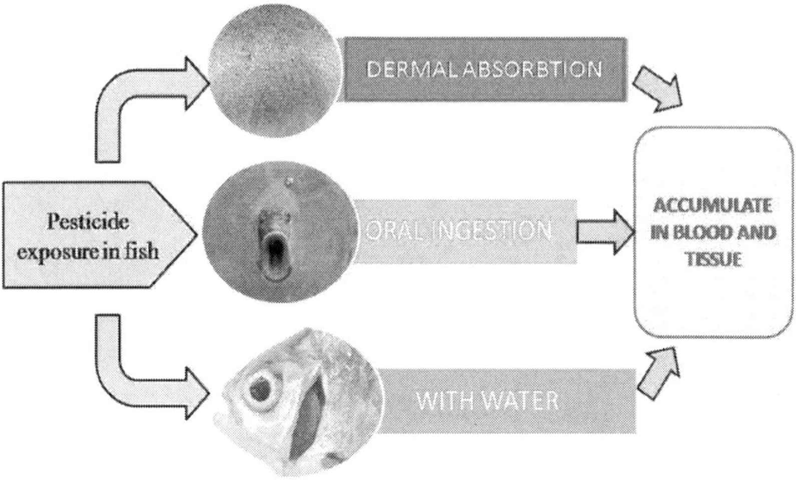


**Fig. 1: Pesticide exposure routes in fish**

Acetylcholinesterase(ACHE) enzyme inhibition has been used to show fish exposure and effects from pesticides. Fish raised in heavily contaminated areas with heavy metals and pesticides showed inhibition of both cerebral ACHE and muscle (Lionetto*et al.*, 2003). ACHEinhibition results in CNS consequences such as rapid twitching of voluntary muscles, which finally leads to paralysis. Fish exposed to organophosphate for an extended period of time reveal decreased swimming ability and peroxidative damage to the brain and gills (Cattaneo*et al.*,2008). Pesticide-exposed fish show signs of reduced swimming performance, peroxidative damage, immune response disruption, excess coagulated mucus on gills, hyperactivity, zigzag movement, loss of buoyancy, higher cough, loss of schooling behaviour, and swimming toward the upper surface. Pesticides affect mitochondrial electron transport, impede the generation of ATP, and decrease oxidative phosphorylation in postsynaptic cells via disrupting ACH binding in nicotinic ACH receptors (Sabra and Mehana, 2015). Pesticide exposure results in chromosomal abnormalities in fish, such as centromeric gaps, chromatid gaps, chromatid breaks, sub-chromatid breaks, attenuation, extrafragements, pyknosis, and stubbed arms, among others (Rishi and Grewal, 1995). For instance, Dichlorvos toxicity causes changes in DNA replication, which result in local mutations and cellular hyperproliferation (Benford*et al.*, 1994). In addition, fish exposed to pesticides had noticeably lower protein levels in their liver, muscles, intestines, gills, and blood (Tiwari and Singh, 2004). Immune system in fish can be hampered by pesticide exposure, even at low doses. Low amounts of pesticides have the potential to imitate or block sex hormones, leading to atypical sex development, male feminization, abnormal sex ratios, and strange mating behaviour.

Fish can be indirectly affected by pesticides by having their food source disrupted or their habits changed (Satyavardhan, 2013).

The term "toxicity" describes how dangerous or poisonous a material might be. A substance's propensity to produce negative consequences that manifest quickly, or within a day or a few hours of exposure, is known as acute toxicity and in case of aquatic organism it is usually expressed in terms of $LC_{50}$. The $LC_{50}$ is typically evaluated for a certain exposure period and is the concentration of a pesticide that kills 50% of a population of test animals (e.g., inhalation for 4 hours). Shorter exposure times typically require larger pesticide concentrations to have harmful effects hence the length of exposure is crucial. The $LC_{50}$ values for pesticides in water are expressed as parts per million (ppm) or parts per billion (ppb) ratios of the pesticide to the water. Johnson and Finley, 1980 evaluated the $LC_{50}$ value of different pesticides in Rainbow trout and largemouth bass as depicted in Table 1.

**Table 1: 96 Hr LC 50 of pesticides in rainbow trout and Largemouth bass**

| Fish species | Insecticide | 96 Hr LC 50 |
|---|---|---|
| **Rainbow trout** | Azodrin | 4.9-50 ppm |
| | Aldrin | 2.5-53 ppm |
| | Carbaryl | 0.9-39 ppm |
| | Chlordane | 0.72-11.9 ppm |
| | DDT | 1.5-21.5ppb |
| | Dieldrin | 1.2-19ppb |
| | Diflubenzuron | 25-240ppm |
| **Largemouth bass** | Dioxanthion | 22-110ppb |
| | Disulfoton | 0-4700ppb |
| | Fenthion | 1.1-3.4ppm |
| | Trichlofon | 1.1-3.4ppm |

(Johnson and Finley, 1980)

**Fate of pesticides in water bodies**

Both pointand non-point sources allow pesticide to enter water bodies. Chemical runoff during inappropriate storage, loading, and disposal as well as the incorrect application of pesticides to water bodies are examples of point sources that come from a fixed site. Urban insecticide use is regarded as a point pesticide source in surface waters. The transport of pesticides from wide areas across the watersheds and their eventual arrival in the water bodies over time constitute the non-point source. Pesticides are a non-point source because they gradually seep into the ground and surface water from agricultural fields as a result of runoff and erosion occurrences (Aydinalp and Porca, 2004). Pesticides are deposited mostly in soil and sediments, which has an impact on pesticide dispersion to other environmental compartments. The spatial and temporal distribution of pesticide concentrations in surface water is determined by physical, chemical, and biological processes in the water sediment environment. The pH of a pesticide has a considerable impact on its solubility. Most natural waters have an alkaline pH, and some pesticide chemicals, such as organophosphates, are produced in an alkaline medium and are particularly vulnerable to hydrolysis, resulting in their high solubility in water and seepage into surface waters. Organochlorine insecticides, on the other hand, are synthesised in slightly acidic pH media and only rarely hydrolyze in natural water (Obidike*et al.*, 2020). Pesticides can move into sediments, remain floating in the water column or bio-concentrate in organisms once they enter the aquatic system. The persistence of a pesticide in an aquatic system, which can be defined by its Log $K_{ow}$ value and persistent nature, is its fate in the environment. It is sometimes stated in terms of "half-life" which is defined as the amount of time taken by a pesticide to decay to one-half of its initial concentration. The longer the period of persistence of a pesticide in water body, higher the risk of exposure. The ability of these pesticides to persist and degrade is governed by a number of biotic and abiotic factors. Several microorganisms, such as fungus, bacteria, and actinomycetes, cause structural changes in pesticides, resulting in their total degradation into obsolete or non-toxic metabolic products (Nayak *et al.*, 2018). The degradation of pesticide by microorganisms is governed by the metabolic capacity of the microbe, the complexity of the pesticide molecule, and several factors such as pH of the water, temperature, and the concentrations and surface area of microorganisms contained in water, as well as oxygen demand (Isiuku*et al.*, 2019; Yichen*et al.*, 2018). Sunlight is one of the most important abiotic factors in pesticide degradation in surface waters. When they receive light energy/ photons from the sun, they become exceedingly unstable and easily break down (Huntscha*et al.*, 2008).

**Factors modifying toxicity of pesticides in waterbodies**

There are various factors that govern the toxic effects of pesticides in aquatic body. Beside physical factors such as pH, temperature etc, several other parameters like half life, accumulation, and synergy influence their toxicity.

a. **Abiotic factors:** The toxicity of pesticides is believed to be considerably affected by water temperature, pH, oxygen etc (Figure 2). Depending on the type, species, response, and specific process, temperature changes in one direction may enhance, reduce, or have no effect on chemical toxicity. Several pesticides have been found to be more harmful at higher temperatures, whereas others have been found to be more dangerous at lower temperatures. Chemical toxicity to aquatic species increases when the dissolved oxygen level of water decreases. The toxicity of pesticides that ionize under the influence of pH may be affected more by pH. Chemicals that have not been separated are usually more hazardous to organisms since they can easily pass-through cell membranes.

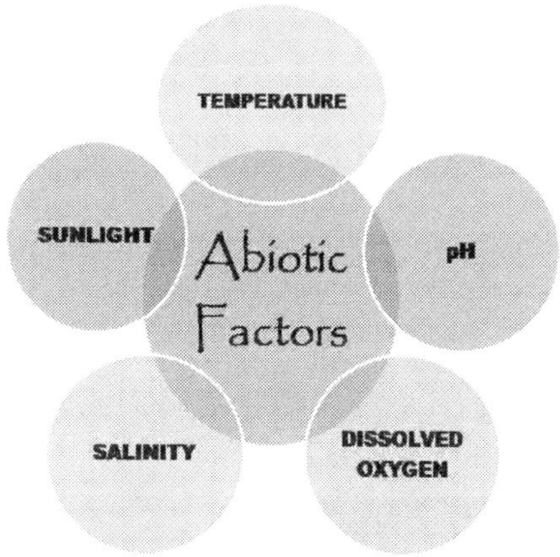


**Fig. 2: Abiotic factors governing pesticide toxicity**

b. **Half -life and persistence**: The half-life of a pesticide is the amount of time it takes for a given amount of the pesticide to be decreased by half. This happens due to their dissipation or break down in water body. After a single half-life, a pesticide will typically break down to 50% of its original level. The half-life of pesticide is affected by a number of factors, including sunshine, temperature, the availability of oxygen, and microbial activity. Pesticides having shorter half-life are less likely to accumulate since they last for small time period in the environment. Pesticides with longer half-lives, on the other hand, are more prone to accumulate after repeated applications. The residence time of a chemical species in a specifically defined compartment of the environment is defined as persistence. The persistence of a compound's should be determined after taking into account a variety of elements that influence its behavior. The inherent qualities of the chemical and its surroundings are the most important considerations. These include the physical and chemical qualities of pesticide, as well as meteorological circumstances, composition, and features. Partitioning and transformation in the tissues of aquatic biota have an impact on pesticide chemical transport and persistence in the hydrologic system. The bioconcentration factor has been the most popular method for describing the distribution of pesticides and other organic chemicals between water and biological tissues. Ferrando*et al.*, 1992 evaluated half-life of some most common pesticides in water that are given in table 2.

**Table 2: Half-life of pesticides in water**

| Pesticide | Half life (hours) in Natural waters |
|---|---|
| Lindane | 64.98(63.99-65.97) |
| Endosulfan | 50.31(47.32-53.30) |
| Methylparathion | 11.10(9.90-12.30) |
| Diazinon | 70.54(70.25-70.82) |
| Thiobencarb | 74.27(73.22-75.32) |

(Ferrando*et al.*, 1992)

c. **Extracellular polymeric substances (EPS):** Extracellular polymeric substances are polymers produced by a variety of bacteria as shown in figure 3. The creation of these slimes, which are mostly composed of polysaccharides, proteins, and DNA, is predominantly triggered by environmental signals (Costa *et al.*, 2018). The presence of charged moieties or hydrophobic moieties in extracellular polymeric substances allows them to bind to pesticides. As a result, pesticides are efficiently transferred tropically to consumer species that devour EPS. Animals eat the sorbed pesticides when they ingest the matrix, which happens due to their grazing on sediments, cells, or flocs. For sorbed pesticides, EPS has been found to be an effective trophic-transfer carrier (Widenfalk*et al.*, 2008).

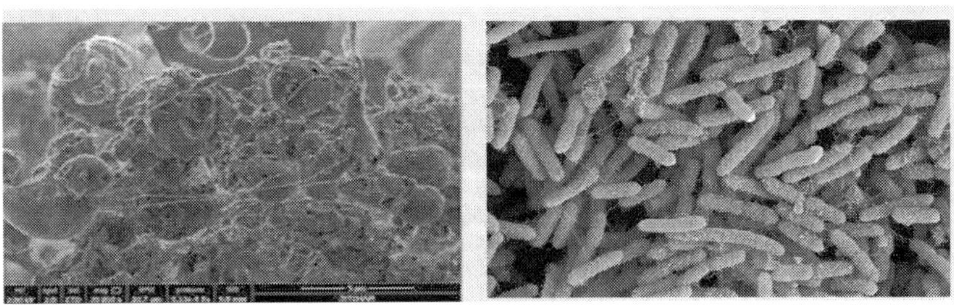

**Fig. 3: Extracellular polymeric substances**

(Solmaz*et al.*, 2018; Quantamagazine)

d. **Bioconcentration factor and bioaccumulation factor:** Bioconcentration factor (BCF) and bioaccumulation factor (BAF) can be used to determine the concentration of pollutants in aquatic organisms. The amount of pesticide residue in the tissue of an organism in relation to the concentration in the organism's environment is referred to as bio-concentration (USEPA, 2008). Pesticide tissue concentrations are compared to environmental pesticide concentrations to calculate bio-concentration factors (BCFs). BCF values larger than one indicates that the pesticide concentration in the organism is higher than the pesticide concentration in the medium (water). The net uptake of a pesticide from the environment via all potential channels (e.g., respiration, diet, cutaneous) from any source (e.g., water, sediment, and other species) is known as bioaccumulation (Spacie*et al.*, 1995). The bioaccumulation factors (BAF) are derived by comparing pesticide tissue concentrations to pesticide concentrations in the environment. The process of bioaccumulation, which results from the intake and retention of components in organisms, is difficult to explain. Environmental factors predominantly affect element intake, whereas biological characteristics of the organisms have a greater impact on element retention(Karlsson*et al.*, 2002). Both these factors determine the toxicity of pesticide in aquatic ecosystem.

Tomza-Marciniak and Witczak, 2009 examined the bioaccumulation of DDT in fish. DDT and its metabolites were found to be more accumulated in the liver and gonads, while in the gills and muscles their content was several-fold lower. The level of DDT found out in different organs of *Rutilus rutilus*as given by Tomza-Marciniak and Witczak, 2009 are given in table 3.

**Table 3: DDT in different organs of *Rutilus rutilus***

| SAMPLE | DDT (μg/kg w.w.) |
|---|---|
| Gill | 6.549 ± 1.005 |
| Gonad | 19.815 ± 4.854 |
| Muscle | 9.073 ± 3.248 |
| Liver | 45.823 ± 9.845 |

(Tomza-Marciniak and Witczak, 2009)

e. **Photolysis:** In the aquatic environment, photo-degradation is one of the most important abiotic transformations for pesticides (Katagi, 2018), and the rate of photolysis of a pesticide is determined by its chemical structure, water properties, and bioavailability. The rate of photolysis in water determines the toxicological effects of a pesticide and its principal degradates on aquatic species.When a pesticide has a UV-visible absorption at >290 nm, its threshold wavelength is sunlight at the earth's surface, and it is possible for a photoproduct with a structure very different from the original pesticide to be formed. Direct photolysis of the pesticide proceeds via excited state(s) in the environment (Burrows *et al.*, 2002). In contrast, regardless of the water absorption profile, a pesticide reacts with a photo-generated reactive oxygen species (ROS) in indirect photolysis especially in natural water containing dissolved organic matter (DOM) and/ or $NO_3$ (Katagi, 2004).

f. **Synergy of pesticides:**Synergy is a term used to describe when two or more pesticides have a higher effect than expected individual results.Let's take an example: A population exposed to neither "Pesticide A" nor "Pesticide B" has a 2% background level of a certain health consequence. The effect is evident at 5% (3 percent + the 2% background) in a group exposed just to "Pesticide A." The effect is seen at 8% (6 percent + the 2% background) in a group exposed exclusively to "Pesticide B." If the two pesticides are just additive and not synergistic, the effect should be noticed at a rate of 11% (3 percent + 6 percent + the 2% background). The combination is synergistic if the observed benefit is more than 11%.

These chemical interactions can alter bioavailability, absorption, internal transit, metabolization, binding at the target site, and excretion, all of which are significant for a pesticide's toxicity to an organism (Cedergreen, 2014).

g. **Octanol/Water Partition Coefficient ($K_{ow}$):** The n-Octanol/Water Partition Coefficient ($K_{ow}$) is the ratio of a chemical concentration in n-octanol to its concentration in water at equilibrium at a certain temperature.

$$Kow = \frac{\text{Concentration in Octanol}}{\text{Concentration in water}}$$

$K_{ow}$ values are unitless and are commonly given as $\log K_{ow}$, which is a relative measure of an organic compound's potential to adsorb to soil and living organisms. $\log K_{ow}$ are inversely proportional to water solubility and directly proportional to the molecular weight of a substance. The $K_{ow}$ value of pesticide determines the component in which it will remain in the water body. Pesticides with large $K_{OW}$ values (> 5), are hydrophobic and tend to be associated with organic materials especially sediments. On the other hand, pesticides with large $K_{OW}$ values (<5), are hydrophilic and tend to be remain in water column. For example, Chlorpyrifos has $K_{OW}$ value of 4.70

h. **Chemical structure of pesticide:** The chemical structure of a substance—what it is comprised of, what atoms and molecules it contains, and how they are arranged—is the most essential determinant in toxicity. Substances having similar structures are frequently linked to the same health issues. Slight variations in chemical structure, on the other hand, can result in significant variances in the type of health effect caused. The chemical structure of pesticide doesn't only impact its health hazard but it also governs its rate of degradation in the environment. Pesticides can be exceedingly unstable or incredibly stable, and their underlying structure has a major impact on how long they last. No matter how the environment changes a pesticide that is readily metabolized degrades quickly, and a chemical that is stable will persist for a long time even under the most unfavorable environmental condition.

**Case studies**

The water quality of Huangpu River (China) was significantly impacted by agricultural non-point source pollution, as well as industrial and urban waste, resulting in a drop in water quality to Class III–IV, compromising its basic ecological functions and services. 18 pesticides were found in every sample taken from the Huangpu River, out of a total of 29 pesticides tested. Target pesticide concentrations in water samples ranged from less than limit of quantitation (buprofezin) to 607.30 ng/L (carbendazim). Because carbendazim is widely employed as a broad-spectrum fungicide in the cultivation of crops such as rice, wheat, and cotton, its concentration was found to be high (Xu *et al.*, 2020).

Similarly, organochlorine pesticides identified in the Selangor River in Malaysia included lindane, heptachlor, endosulfan, dieldrin, endosulfan sulphate, DDT, and DDE, while organophosphate pesticides included chlorpyrifos and diazinon. During the years 2002–2003, the concentration range of identified residual pesticides in the Selangor River, Malaysia, was 10.1 ng/L for lindane and 1848.7 ng/L for endosulfan. Agriculture, urban, and industrial activity in Selangor, combined with significant population growth, had resulted in deterioration of river water quality, according to the study. Pesticides discovered in raw river water intake were shown to be unaffected by ordinary water treatment (Leong *et al.*, 2007).

In India, Yamuna has become one of the world's most polluted river as a result of high-density population expansion and rapid industrialization. Hexachlorocyclohexane and DDT concentrations in the river varied between 12.76 - 593.49 ng/L and 66.17

- 722.94 ng/L, respectively. Pesticide residues in the Gomti River water varied from 2.16 to 567.49 ng/L, while those in the bed sediments ranged from 0.92 to 813.59 ng/L (Syafrudin*et al.*, 2021).

**Precaution and decontamination of lake from pesticide toxicity**

Pesticides should only be used as much as is absolutely essential to protect plants. In sensitive locations like water-saturated and flooded terrain, as well as in vegetation strips established along surface waters to limit inputs of substances by surface runoff, the use of pesticides, all types of fertilisers, and organic waste should be regulated. Setting the impurity threshold values for pesticides should be done with extreme caution. Any approach to prevent and lessen the harmful effects of pesticides on the environment should include safe handling, storage, and disposal of pesticides, as well as the implementation of a disposal system for pesticide residues and washings. Pesticides should not be used outside of the treated area, especially near water. Strict measures should be used. To determine the best way to apply pesticides in order to lessen the risk they pose to surface waters and groundwaters, information on the physico-chemical properties of a pesticide should be gathered along with other pertinent information, such as: human health considerations; a pesticide's behaviour in soil and water, and its likely effects on birds, fish, earthworms, etc; usage data; time of application; soil and climate conditions. This suggests that environmental concentration should be predicted using destiny modelling, and the outcomes should be compared to environmental standards and objectives.

Due to their extensive usage in modern agriculture, pesticide levels in water have skyrocketed. The type of pesticide and the effectiveness of the treatment process determine the best water treatment method for removing pesticides (Saleh *et al.*, 2020). There are numerous physical, biological, and chemical treatment options available for the remediation of pesticide-contaminated waters, and the most significant ones are covered below:

1. **Activated sludge:** The most popular biological wastewater treatment method is an activated sludge system. Aerobic, mesophilic, and suspended cultures provide the system's foundation. It contains a wide variety of microorganisms. The type of microorganism in activated sludge also varies depending on the type, concentration, and other environmental elements (pH, temperature, oxygen concentration, poisonous material, etc). The principal microorganisms that are frequently discovered in activated sludge are species like bacteria, fungi, protozoa, and occasionally nematodes.

Activated sludge is a sewage and industrial wastewater treatment method. In the activated sludge process, organic matter-containing wastewater is aerated in an aeration basin, where microorganisms digest the suspended and soluble organic matter. To generate energy, a portion of organic matter is synthesised into new cells and a portion is oxidised to $CO_2$ and water. In activated sludge systems, the new cells generated during the reaction are removed from the liquid stream and deposited in settling tank as flocculent sludge. A portion of the settled biomass is returned to the aeration tank as activated sludge, while the residue is discarded as waste or excess sludge. When compared to chlorination, it is a very efficient treatment process that does not require a large amount of land, is less expensive, and is more environmentally friendly.

2. **Membrane bioreactor:** Membrane bioreactor (MBR) technology has surpassed the activated sludge technique as the preferred wastewater treatment method. Membrane bioreactors combine a biological wastewater treatment method with a membrane technology such as microfiltration or ultrafiltration. There are two main types of MBR configurations: side stream membrane bioreactor and immersed membrane bioreactor (Goswami *et al.*, 2018). The filtering element is put in either the primary bioreactor vessel or a different tank in the immersed membrane bioreactor arrangement. The filtering modules are located outside the aerobic tank in side stream membrane bioreactor technology, therefore the name side-stream arrangement. For a specific MBR unit, the structure and microbial community's composition varies over time and from one MBR plant to another. The significant characteristics of the microorganisms in environmental engineering systems, especially MBR plants, are the main cause of this difference. Diverse microorganisms are organised into a variety of communities under the influence of atmospheric wastewater that is fed into the bioreactor (Nguyen *et al.*, 2020). However, a particular type of microorganism can be enriched in the bioreactor by altering operative conditions and reactor design. The fouling propensity of a membrane is influenced by the material it is comprised of. Ceramic membranes are very hydrophilic, making them less prone to fouling (Mutamim*et al.*, 2013).

3. **Bio-augmented activated sludge:** Bioaugmentation is the technique of putting cultivated microorganisms into the subsurface with the purpose of biodegrading certain soil and groundwater pollutants. In general, wastewater discharge contains a wide variety of naturally occurring bacterial species. Assuming a steady-state, they gradually gain the ability to handle the degradation of pollutants over time. Some pollutants can withstand biological breakdown, nevertheless, because of characteristics including high toxicity, low water solubility, and complicated

chemical structures. They must therefore be biodegraded by a variety of bacteria. Here, bio-augmentation is useful. Bio-augmented activated sludge enhances sludge breakdown and reduces sludge-related concerns. The microbial ecology in the activated sludge process has been shown to be negatively affected by low temperatures (Martin *et al.*, 2005). Bioaugmentation proved successful in stabilising the performances of three biological municipal wastewater treatment processes at low temperatures and accelerating the start-up of activated sludge at low temperatures (Jingbo*et al.*, 2010).

4. **Advanced oxidation process:** Advanced oxidation methods are recognised as environmentally friendly solutions for the remediation of pesticide-contaminated water (Figure 4). The main concept of advanced oxidation processes in water treatment is based on the in-situ generation of highly reactive hydroxyl radicals that indiscriminately oxidise a wide range of recalcitrant organic pollutants for complete mineralization of organic contaminants to carbon dioxide, water, and mineral salts, as well as the transformation of pesticide compounds into more biodegradable species (Ribeiro *et al.*, 2015). It has recently been used as a water purification technique due to its thermodynamic viability and wide range of applications (López-Loveira*et al.*, 2017).

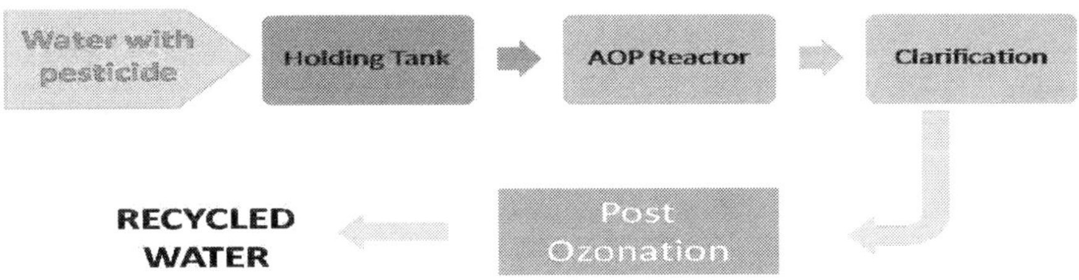


**Fig.4: Process for advanced oxidation method**

## Conclusion

Pesticide addition to water bodies has an effect on aquatic life as well as human health due to the biomagnification and bioaccumulation processes. Pesticides present a danger to human health when they are present in water, and the severity of this danger relies on the solubility, half-life, adsorption capability, and biodegradability of the pesticide components. To reduce the spread of pesticides into aquatic systems, appropriate preventative measures should be taken. If necessary, potable water should go through advanced water treatment procedures, and the water bodies where pesticide chemicals have been found should be monitored constantly.

## References


Aydinalp, C., &Porca, M. M. (2004). The effects of pesticides in water resources. *J. Cent. Eur. Agric.*, 5, 5-12.

Benford, D. J., Price, S. C., Lawrence, J. W., & Bremmer, J. N. (1994). Investigations of the genotoxicity and cell proliferation activity of dichlorvos in mouse forestomach. *Toxicology*, 92, 203-215.

Burrows, H. D., Canle, M., Santaballa, J. A., and Steenken, S. (2002). Reaction pathways and mechanisms of photodegradation of pesticides. *J. Photochem. Photobiol.*, 67B, 71–108.

Cattaneo, R., Loro, V. L., Spanevello, R. S., Moraes, B. S., &Clasen, B. (2008). Metabolic and histological parameters of silver catfish (*Rhamidquelen*) exposed to commercial formulation of 2, 4-dichlorophenoxi-acetic acid (2, 4-D) herbicide. *Pesticide Biochemistry and Physiology*, 92, 133-137.

Cedergreen, N. (2014). Quantifying synergy: a systematic review of mixture toxicity studies within environmental toxicology. *PloS one*, 9(5), e96580.

Costa, O. Y., Raaijmakers, J. M., &Kuramae, E. E. (2018). Microbial extracellular polymeric substances: ecological function and impact on soil aggregation. *Frontiers in microbiology*, 9, 1636.

Ferrando, M. D., Alarcon, V., Fernandez-Casalderrey, A., Gamon, M., & Andreu-Moliner, E. (1992). Persistence of some pesticides in the aquatic environment. *Bulletin of environmental contamination and toxicology*, *48*(5), 747-755.

Goswami, L., Vinoth, K., Borah, R., Narayan, S., Manikandan, A., Kannan, P., Pugazhenthi, G. (2018). Membrane bioreactor and integrated membrane bioreactor systems for micropollutant removal from wastewater: A review. *Journal of Water Process Engineering*. 26, 314–328.

Helfrich, L. A., Weigmann, D. L., Hipkins, P.,& Stinson, E. R. (2009). Pesticides and aquatic animals: a guide to reducing impacts on aquatic systems. In: Virginia Polytechnic Institute and State University.

Huntscha, S., Singer, H., Canonical, S., Schwarzenbach, R.P. and Fenner, K. (2008). Pesticides *environ. Sci. Technol.* 42, 5507-5513.

Isiuku, B. O.,&Enyoh, C. E. (2019). Water pollution by heavy metal and organic pollutants: brief review of sources, effects and progress on remediation with aquatic plants. *Anal Methods Environ Chem J*, 2(3), 5–38.

Jingbo, G., Wang J., Cui, D., Wang, L., Ma, F., Chein-Chi, C. and Jixian, Y. (2010). Application of bioaugmentation in the rapid start-up and stable operation of biological processes for municipal wastewater treatment at low temperatures. Bioresource technology. *Elsevier*, 6622–6629

Johnson, W.W., &Finley, M. T. (1980). Handbook of acute toxicity of chemicals to fish and aquatic invertebrates: Resources publication 137, USDI, "Fish and Wildlife Service", Washington, DC, 89P, 1980.

Karlsson, S., Meili, M., &Bergström, U. (2002). Bioaccumulation factors in aquatic ecosystems. A critical review.

Katagi, T. (2004). Photodegradation of Pesticides on Plant and Soil Surfaces. In: Ware, G.W. (eds) Reviews of Environmental Contamination and Toxicology. Continuation of Residue Reviews, vol 182. Springer, New York, NY. https://doi.org/10.1007/978-1-4419-9098-3_1

Katagi, T. (2018). Direct photolysis mechanism of pesticides in water. *Journal of pesticide science*, 43(2), 57-72.

Leong, K. H., Benjamin Tan, L. L.,& Mustafa, A.M. (2007). Contamination levels of selected organochlorine and organophosphate pesticides in the Selangor River, Malaysia between 2002 and 2003. *Chemosphere*, 66, 1153–1159.

Lionetto, M. G. R., Caricato, M. E., Giordano, M. F., & Schettino, T. (2003). Integrated use of biomarkers (acetylcholineestrase and antioxidant enzymes activities) in *Mytilus galloprovincialis*and*Mullus barbatus* in an Italian coastal marine area. *Mar. Pollut. Bull.*, 46, 324-330.

López-Loveira E., Ariganello F., Medina M.S., Centrón D., Candal R., &Curutchet G. (2017). Degradation alternatives for a commercial fungicide in water: Biological, photo-Fenton, and coupled biological photo-Fenton processes. *Environ. Sci. Pollut. Res.*, 24, 25634–25644.

Mahmood, I., Imadi, S. R., Shazadi, K., Gul, A., & Hakeem, K. R. (2016). Effects of pesticides on environment. In *Plant, soil and microbes* (pp. 253-269). Springer, Cham.

Martin Jr., R.W., Baillod, C.R., and Mihelcic, J.R. (2005). Low-Temperature inhibition of the activated sludge process by an industrial discharge containing the azo dye acid black. *Water research*, 39(1), 17–28.

Mutamim, N.S.A., Noor, Z.Z., Hassan, M.A.A., Yuniarto, A.,& Olsson, G. (2013). Membrane bioreactor: Applications and limitations in treating high strength industrial wastewater. *Chem. Eng. J.*,225, 109–119.

Nayak, S.K., Dash, B., &Baliyarsingh, B. (2018). Microbial Remediation of Persistent Agro-chemicals by Soil Bacteria: An Overview. *Microb. Biotechnol.*, 275–301. doi: 10.1007/978-981-10-7140-9_13.

Nguyen, L. N., Commault, A. S., Kahlke, T., Ralph, P. J., Semblante, G. U., Johir, M. A. H., & Nghiem, L. D. (2020). Genome sequencing as a new window into the microbial community of membrane bioreactors–A critical review. *Science of the Total Environment*, 704, 135279.

Obidike, B. M., Verla, A. W., Enyoh, C. E., Verla, N., & Onyekachi, O. N. (2020). Fate and Distribution of Pesticides Residues in Water Sources in Nigeria: A Review. *International Journal of AgroChemistry*, 6(2), 29-48.

Ribeiro A.R., Nunes O.C., Pereira M.F., &Silva A.M. (2015). An overview on the advanced oxidation processes applied for the treatment of water pollutants defined in the recently launched Directive 2013/39/EU. *Environ. Int.*,75, 33–51. doi: 10.1016/j.envint.2014.10.027.

Rishi, K., & Grewal, S. (1995). Chromosomal aberration test for the insecticide, dichlorvos, on fish chromosomes. *Mutation research*, 344, 1-4.

Sabra, F. S., &Mehana, E. S. E. D. (2015). Pesticides toxicity in fish with particular reference to insecticides. *Asian Journal of Agriculture and Food Sciences*, *3*(1).

Saleh, I. A., Zouari, N., & Al-Ghouti, M. A. (2020). Removal of pesticides from water and wastewater: Chemical, physical and biological treatment approaches. *Environmental Technology & Innovation*, 19, 101026.

Satyavardhan, K. (2013). A comparative toxicity evaluation and behavioral observations of freshwater fishes to fenvalerate.*Middle-east Journal of Scientific Research*, 13(2), 133-136.

Singh, B.,& Mandal, K. (2013). Environmental impact of pesticides belonging to newer chemistry. In: Dhawan, A. K., Singh, B., Brar-Bhullar, M.,& Arora, R. (eds.). Integrated pest management. Scientific Publishers, Jodhpur, India, p152-190.

Solmaz, K. B., Ozcan, Y., Mercan Dogan, N., Bozkaya, O., & Ide, S. (2018). Characterization and production of extracellular polysaccharides (EPS) by Bacillus pseudomycoides U10. *Environments*, 5(6), 63.

Spacie, A., McCarty, L. S., & Rand, G. M. (1995). Bioaccumulation and bioavailability in multiphase systems. In: Fundamentals of Aquatic Toxicology (Rand, G. M., ed). Washington, DC:Taylor& Francis, 493–521.

Syafrudin, M., Kristanti, R. A., Yuniarto, A., Hadibarata, T., Rhee, J., Al-Onazi, W. A., ... & Al-Mohaimeed, A. M. (2021). Pesticides in drinking water—a review. *International Journal of Environmental Research and Public Health*, 18(2), 468.

Tiwari, S.,& Singh, A. (2004). Toxic and sublethal effects of oleandrin on biochemical parameters of freshwater air breathing murrel, *Channa punctatus* (Bloch).*IndianJ of Experimenta Biology*, 42, 413-418.

Tomza-Marciniak, A., &Witczak, A. (2009). Bioaccumulation of DDT and Its Metabolites in the Międzyodrze Ecosystem, Poland. *Polish Journal of Environmental Studies*, 18(3).

Widenfalk, A., Lundqvist, A., and Goedkoop, W. (2008). Sediment microbes and biofilms increase the bioavailability of chlorpyrifos in *Chironomus riparius* (Chironomidae, Diptera). *Ecotox. Environ. Saf.*, 71, 490–497. doi: 10.1016/j.ecoenv.2007.10.028.

Xu, L., Granger, C., Dong, H., Mao, Y., Duan, S., Li, J., &Qiang, Z. (2020). Occurrences of 29 pesticides in the Huangpu River, China: Highest ecological risk identified in Shanghai metropolitan area. *Chemosphere*, 251, 126411.

Yichen, H., Lijuan, X., Feiyu, Li., Mengshi, X., Derong, L., Xiaomei, L., and Zhijun, W. (2018). Microbial Degradation of Pesticide Residues and an Emphasis on the Degradation of Cypermethrin and 3-phenoxy Benzoic Acid: A Review. *Molecules*, 23(9), 2313. doi: 10.3390/molecules23092313.

Made in the USA
Las Vegas, NV
19 September 2026

20f62422-da5f-4afb-ae3a-62a9e175539fR01